Springer Series on Bio- and Neurosystems

Volume 14

The Springer Series on Bio- and Neurosystems publishes fundamental principles and state-of-the-art research at the intersection of biology, neuroscience, information processing and the engineering sciences. The series covers general informatics methods and techniques, together with their use to answer biological or medical questions. Of interest are both basics and new developments on traditional methods such as machine learning, artificial neural networks, statistical methods, nonlinear dynamics, information processing methods, and image and signal processing. New findings in biology and neuroscience obtained through informatics and engineering methods, topics in systems biology, medicine, neuroscience and ecology, as well as engineering applications such as robotic rehabilitation, health information technologies, and many more, are also examined. The main target group includes informaticians and engineers interested in biology, neuroscience and medicine, as well as biologists and neuroscientists using computational and engineering tools. Volumes published in the series include monographs, edited volumes, and selected conference proceedings. Books purposely devoted to supporting education at the graduate and post-graduate levels in bio- and neuroinformatics, computational biology and neuroscience, systems biology, systems neuroscience and other related areas are of particular interest.

All books published in the series are submitted for consideration in Web of Science.

More information about this series at https://link.springer.com/bookseries/15821

Bernard Widrow

Cybernetics 2.0

A General Theory of Adaptivity and Homeostasis in the Brain and in the Body

Springer

Bernard Widrow
Stanford, CA, USA

ISSN 2520-8535 ISSN 2520-8543 (electronic)
Springer Series on Bio- and Neurosystems
ISBN 978-3-030-98139-6 ISBN 978-3-030-98140-2 (eBook)
https://doi.org/10.1007/978-3-030-98140-2

This Springer imprint is published by the registered company Springer Nature Switzerland AG
The registered company address is: Gewerbestrasse 11, 6330 Cham, Switzerland

This book is dedicated to the memory of Norbert Wiener and to the memory of Donald Hebb. They are long gone but their works live on.

Foreword by Nikola Kasabov

Homeostasis and *adaptivity* are major features of living organisms that make them survive in changing environments, from single cells, organs and individual organisms to entire species and populations. According to Oxford Dictionary, homeostasis is about "physiological processes by which the internal systems of the body (e.g. blood pressure, body temperature, acid-base balance) are maintained at equilibrium, despite variations in the external conditions". Adaptivity is the ability of an organism to change in order to continue existing. Learning is a way of adaptation. Our brains learn all the time from the data provided by the surrounding environment and the society we live in and maintain homeostasis. So do the plants.

These unique biological properties of a living organism are themselves evolving, adapting to make the organism adapt and survive. Their evolution/adaptation at an individual level and at a population/species level is still to be better understood not only as part of the progress in biology and neuroscience but also for the sake of the creation of more intelligent, adaptive engineering systems.

And here comes this monograph book by the pioneer of artificial neural networks and engineering learning systems Bernie Widrow, to cover various aspects of homeostasis and adaptivity in both living and engineering systems in a systematic way.

Widrow is the originator (with Hoff) of the LMS (Least Mean Square) learning algorithm for artificial neural networks (1960) implemented in electronic machines called ADALINE and MADALINE. Widrow introduced also the Hebbian-LMS (with Kim and Park, 2015) as a continuation of the Hebbian learning rule introduced by Donald Hebb in 1949 as a synaptic neuronal learning rule. It is shown that the Hebbian-LMS algorithm, which realizes clustering adaptability, initially for modeling synaptic plasticity, can be used to model homeostasis and adaptability at different levels and organs of the human brain and body, also presumed generally applicable to every living organism.

The monograph book by Widrow is a creative and original continuation of the previous seminal works by the great pioneers Norbert Wiener and Donald Hebb. And the title of the book (Cybernetics 2.0) clearly points to this continuation after the book by Wiener, titled Cybernetics, published in 1948.

The book covers a wide spectrum of material, from biological neurons, synapses and neuronal learning algorithms, to applications across neuroscience and psychology (cognition), psychiatry (addiction, anxiety, bi-polar, depression, Alzheimer, Parkinson), physiology (body homeostasis, autonomic nervous system, blood glucose, immune system, viruses, cancer), plants (plants growing) and engineering (adaptive filters), all explained in a clear way with simple diagrams, as only an engineer can do that. The style of presentation is easy to read, unique and interesting. It is a perfect science communication book that can be understood by both undergraduate students and practitioners in areas of neuroscience, biology, health, engineering. Many of its chapters end with questions like in tutorial books.

So, does the Hebbian-LMS learning rule represent a general and unifying theory that can be used to model homeostasis and adaptation of living organisms at different levels of functionality? Yes, it seems so, at least to certain degree of granularity. Chapter 26 is more autobiographical, where the author describes with a lot of respect and also with a sense of humor some encounters with famous scientists working at the same time in the USA. Stories, such as the absent-mindedness of Wiener, or Marvin Minsky's "pregnant" dog, or entering by mistake a "strip" club while discussing fuzzy logic with the father of fuzzy logic Lotfi Zadeh, or Richard Feynman being "full of the devil". Furthermore, old photos of ADALINE and MADALINE, and personal family photos, all are highly entertaining and informative. We cannot move to the future without knowing the past, indeed!

The book makes an intriguing reading, showing how starting from a single cell learning rule one can create a general theory and this is what Widrow has done. The encyclopedic scope of the book reminds us of the work by Aristotle 24 centuries ago, where he collected thousands of facts from all areas of science to come up with simple and unifying theory of inductive and deductive reasoning. But Widrow did it differently. He showed the applicability of a simple learning rule of a single cell to model and explain homeostasis and adaptation in living organisms.

December 2021

Nikola Kasabov
Series Editor, Life Fellow IEEE, Fellow
RSNZ, Fellow INNS College of Fellows

Nikola Kasabov is a Professor of neural networks and knowledge engineering and Founding Director of the KEDRI Institute at Auckland University of Technology (AUT), New Zealand and also George Moore Chair Professor at Ulster University UK. He is a Life Fellow of IEEE, Fellow of the Royal Society (Academy) of New Zealand, Fellow of the INNS College of Fellows, Honorary Professor at the Teesside University UK and the University of Auckland NZ, Doctor Honoris Causa from Obuda University, Hungary. Kasabov holds a Master's degree in Engineering and Ph.D. in Mathematics from TU Sofia, Bulgaria. He has published more than 700 works in the areas of neural networks, computational intelligence, evolving connectionist systems, neuro-informatics and bioinformatics. He is a Past President of the INNS and APNNS (Asia-Pacific Neural Network Society). He is the Editor-in-Chief of the Springer Series on Bio- and Neurosystems.

Foreword by Henry Yin

In 1948, Norbert Wiener published Cybernetics, Or Control and Communication in the Animal and the Machine, a book that has exerted a tremendous influence on science as well as popular culture. The title of Professor Widrow's new book, Cybernetics 2.0: A general Theory of Adaptivity and Homeostasis in the brain and in the body, pays homage to Wiener. It is not praising the book too highly to say that it is a worthy successor to Wiener's work.

Wiener's original Cybernetics attempted to apply ideas from engineering control theory and information theory to the study of life. Cybernetics 2.0 addresses similar topics, but with a fresh perspective and using a different set of tools. The two authors could not be more different. Wiener, a great mathematician with a philosophical bent, was often obscure and diffuse, and for that reason his book is more famous than actually read. Widrow, on the other hand, is a classic example of the pragmatic, level-headed engineer. This contrast is made dramatically clear in the delightful story of their encounter in a parking lot, as Widrow assisted Wiener in starting his car. Incidentally, driving a car is a famous example used by Wiener to illustrate the concept of feedback control, though he himself may have struggled with this task.

We are fortunate that, compared to his distinguished predecessor, Widrow is not only more capable at operating a vehicle but also more skilled in the art of exposition. He provides a masterful introduction to the basic concepts in neural networks. He also proposes a new hypothesis for the rules governing synaptic plasticity based on the least mean square (LMS) algorithms widely used in adaptive filters. Wiener was the pioneer to introduce least squares thinking into engineering, and Hoff and Widrow, standing on Wiener's shoulder, discovered the LMS algorithm that is now used everywhere. Widrow argues that the famous Hebbian postulate on synaptic plasticity, with a few key modifications, can be incorporated into the LMS framework. The result is what he calls the % Hebbian-LMS, which is hypothesized to be nature's basic learning algorithm. This idea promises to reconcile observations on spike-timing-dependent plasticity and homeostatic synaptic scaling. I believe it will have a major impact on future thinking on synaptic plasticity.

Having introduced the % Hebbian-LMS algorithm, Widrow then attempts to apply it to a wide range of subjects, ranging from neurological disorders to body temperature control, viral infection and even cancer. As a pioneer who has made fundamental contributions to the study of signal processing and neural networks, Widrow is unafraid of speculation or controversy, venturing into areas that the experts are reluctant to tread. A unifying theme in these more speculative chapters is that the same adaptive algorithm, Hebbian-LMS, which is hypothesized to control synaptic plasticity, might also play a comparable role in regulating receptors all over the body, including those for hormones. Such a universal adaptive mechanism may link neuroscience, endocrinology and other fields.

By far, the most personal section of the book tells stories of various famous men Widrow has known, ranging from his heroes like Wiener, Shannon and Feynman, to his rivals like Rosenblatt and Minsky. These stories offer unique perspectives on some of the major shapers of the twentieth century. This is followed by a brief history of how the LMS algorithm was discovered, and of the first generation of neural network models, Adaline and Madaline, which form the foundation for modern work on artificial intelligence. As a major contributor to this field, Widrow's account is especially valuable.

This is a unique book that combines a rigorous presentation of new scientific hypotheses with autobiographical sketches and a personal account of the development of a new scientific field. Its scope and eclecticism indeed rival Wiener's original book, and show a side of Widrow not seen in his previous work. Astonishingly, at the time of completing this book, Widrow is 91 years old, which makes his achievement all the more impressive. I do not know of any other example of significant intellectual contribution made at this age. While the occasional reminiscent tone betrays the age of the author, the originality of the perspective, the insatiable curiosity and the rigor of the analysis all reveal a still youthful spirit, always venturing into the unknown and always striving to understand.

December 2021

Henry Yin
Professor of Psychology and Neuroscience,
Professor of Neurobiology
Duke University
Durham, USA

Foreword by Fredric B. Kraemer

You might rightly wonder what a physician-scientist who is a clinical endocrinologist is doing writing a foreword for a book on cybernetics authored by an electrical engineer, Prof. Bernard Widrow, who arrived at Stanford University in 1959. The answer might be surprising. Even though I have been at Stanford University since 1978 and a member of the faculty in the School of Medicine since 1983 and the Chief of the Division of Endocrinology, Gerontology and Metabolism since 2001, I did not meet Bernie until 2018 when our wives, Linda Kraemer and Ronna Widrow, who met through the Stanford Women's Club when they were serving as co-presidents of the Club, arranged for us all to get together for drinks and nibbles at the Stanford Faculty Club. It is certainly not unusual for faculty located within different schools of the University and who have very different academic interests not to have interactions. Moreover, Bernie had moved to emeritus status many years prior to our meeting, further curtailing possible academic encounters. Bernie was very engaging and humorous at our initial meeting and suggested that he and I should get together one-on-one at another time to discuss his personal and scientific insights into human biology, specifically neural function. I must admit that I was not quite sure what to expect and was somewhat hesitant, but intrigued, for this to occur; however, it did eventually happen and resulted in my reading an earlier draft of this book.

Since I have, at best, only a very rudimentary knowledge of, or interest in, electrical engineering, I was pleasantly surprised that I thoroughly enjoyed reading the draft and learned a great deal from it. Bernie has taken a very personal approach in the book to explain to readers the basis of the LMS (least mean square) learning algorithm, which he invented in 1959 and which is used in a wide variety of electronics that involve signal processing, control systems, communication systems and artificial neural networks. In reflecting on the utility of the supervised LMS algorithm combined with the unsupervised Hebbian algorithm, Bernie provides arguments how the Hebbian-LMS learning algorithm underlies the functional control of a variety of homeostatic mechanisms in biology, thus suggesting that concepts developed by electrical engineers are, in fact, intrinsic to living multicellular organisms. The book is organized into very clear, precise chapters, each of which focuses on a particular biological function, with each chapter similarly organized with an introduction, explanations, simplified graphical displays, a summary and even questions for students to address. If I extrapolate from the chapters covering aspects of endocrinology and metabolism, obviously areas where I have in depth knowledge and understanding, Bernie has made some simplifications of the basic biology, but has, for the most part, sufficiently captured the general concepts, enabling him to explain how the Hebbian-LMS learning algorithm applies and is utilized in the individual homeostatic function being discussed. For instance, the number of hormone receptors that are expressed by a cell is critical for determining how that cell will respond to a specific hormone. Bernie proposes that the Hebbian-LMS algorithm is the way that nature controls the number of hormone receptors in cells by upregulating or downregulating the number of receptors to achieve optimal function. Thus, the book provides a unique perspective for understanding facets of human biology through the eyes and formulations of electrical engineering.

As much as I found the core of the book illuminating and provocative, I must admit that I most enjoyed Bernie's personal stories and the descriptions of his interactions with mentors, mentees and 'famous' people. These anecdotes provide not only a glimpse into Bernie's professional experiences, as well as his personality and humorous side, but also a fascinating depiction of the people with whom Bernie had interactions. I trust that other readers will also find the book equally enlightening.

December 2021

Fredric B. Kraemer, M.D.
Professor of Medicine, Chief of Division of Endocrinology
Gerontology and Metabolism
Stanford University School of Medicine
Stanford, USA

Preface

In writing this book, I have been strongly influenced by the teachings of Donald Hebb and Norbert Wiener. Hebbian learning is a fundamental precept of neuroscience. Wiener's introduction of least square methodology to the world of engineering and filter design was a profound development. These concepts are explained and absorbed herein.

Hebb was one of the first to suggest that learning is dependent on synaptic change. The basic idea was suggested earlier by Cajal and Sherrington. Hebb presented the idea as a learning rule. This was an amazing supposition. It has been accepted since he published his book in 1949 until the present. Deeper introspection reveals shortcomings to Hebb's learning rule. It needs to be modified or extended in order to apply to both excitatory and inhibitory synapses and to allow the phenomenon of homeostasis to exist.

Wiener's book entitled Cybernetics was published in 1948. He was one of the first to suggest that self-regulating mechanisms are present all over the body creating homeostasis. Wiener learned about homeostasis from Rosenbleuth, a Mexican physiologist who was his friend and collaborator. Negative feedback is the key. Wiener's book is philosophical, not detailed. An important element in this is adaptivity, not present in his writing. Adaptive algorithms did not exist when Wiener wrote his book. The least square methods that he introduced are fundamental to the development of adaptive and learning algorithms, however. Extending Hebb's work and Wiener's work is the subject of this book.

I wrote a paper entitled "The Hebbian-LMS learning algorithm" which was published in the IEEE Computational Intelligence Magazine of November, 2015. This algorithm is an extension of the Hebbian learning rule. Hebbian learning was often simplified to the sing-song "neurons that fire together wire together". What this means is that if a neuron signals another neuron by way of a synaptic coupling, the strength of that coupling increases if both neurons are firing at the same time. The rationale is that if the presynaptic neuron's firing contributes to the postsynaptic neuron's firing, then the synaptic coupling strengthens. But what happens if the presynaptic neuron is firing and the postsynaptic neuron is not firing? The extended rule requires that the coupling weakens. What happens when the presynaptic neuron is not firing? The extended rule requires that the coupling remains fixed.

The original Hebbian learning rules, if practiced by nature, would only increase synaptic coupling. To achieve learning, some synapses would need to strengthen, while others would need to weaken. Hebb's rule would eventually cause all synapses to strengthen until saturation and a useless neural network would be the result. Clearly, Hebbian learning needs extension.

The extended learning rules are obtained in a simple logical way. For example, fire together wire together applied only to excitatory synapses. Hebb did not talk about inhibitory synapses. The extended rule for inhibitory synapses is opposite that for excitatory synapses. The result is "fire together unwire together".

The extended Hebbian rules can be perfectly implemented by an unsupervised version of the LMS learning algorithm. The LMS algorithm comes from the fields of artificial neural networks, machine learning and adaptive signal processing. In usual applications, LMS learning is supervised. Here, LMS learning is unsupervised.

Hebbian rules are qualitative. LMS can be expressed mathematically and formulated as an implementable algorithm. Hebbian-LMS is Hebbian learning expressed in a form that can be treated analytically and that can be computer simulated.

In the IEEE paper, it is shown that the information needed to perform the Hebbian-LMS algorithm is available at the synapse, whether excitatory or inhibitory. Nature could implement the Hebbian-LMS algorithm, and an intrinsic property of Hebbian-LMS is homeostasis. Hebbian-LMS is homeostasis, and a lot more. Wherever there is homeostasis, there is Hebbian-LMS. The IEEE paper suggests that Hebbian-LMS may be nature's learning algorithm. Although nothing is for sure, it seems to fit observed phenomena. "If it walks like a duck, looks like a duck, and quacks like a duck, who knows, maybe it is a duck".

Further work on the subject motivated the writing of another paper. The subject became too extensive to fit into a single paper. So, it became two companion papers. Further work suggested yet another paper and at that point, it became clear that a book is needed to express the ideas. It started with my interest in trying to understand nature's learning algorithm for control of the synapses of the brain. There are approximately a thousand trillion of them. My interest stems from many years of work on artificial neural networks and engineering many different types of learning algorithms and their practical applications.

One day I was reading an article in the New York Times about opioid addiction. The article mentioned something about neurotransmitters and neuroreceptors. My antenna went up. I became interested in the subject. From that subject, my interest wandered to other mental disorders such as anxiety, depression and bipolar disorder. The Hebbian-LMS algorithm when applied to brain systems gave insight into the problems, made them easier to understand. Homeostasis plays a key role with the disorders, and the Hebbian-LMS algorithm incorporates homeostasis.

With further study, I began to realize that all the organs of the body are controlled with neurotransmitters and neuroreceptors and hormones and hormone receptors. Adaptivity and homeostasis are crucial elements for the body's control processes. Assuming that the Hebbian-LMS algorithm not only controls the synapses of the brain but that the same algorithm controls the body's organs, the kidneys, the heart, the thermoregulation system, etc., the behavior of this algorithm fits the behavior of the body's systems. Countless systems all over the body are controlled with adaptivity and homeostasis. It is likely that nature utilizes Hebbian-LMS all over the body, wherever there is homeostasis and receptors.

My objective in writing this book has been to take the notions of adaptivity and learning from the realm of engineering into the realm of biology and natural processes. I have done this by using the Hebbian-LMS algorithm as a model for adaptivity and homeostasis in living systems. The algorithm and its natural applications involve many scientific disciplines, some of which are: electrical engineering, control engineering, computer science, artificial intelligence, neurobiology, cell biology, psychology, psychiatry, endocrinology, cardiology, nephrology, internal medicine, etc. Also, we must not forget the fundamentals: mathematics, physics and chemistry. This book is written for scientists and practitioners in these fields.

Finding a name for a book with such a range of subject matter was not easy, until I thought of my hero Prof. Norbert Wiener. When I was an MIT undergraduate, he was the most famous professor. His book "Cybernetics" discusses feedback control and homeostasis in man and machine. This represents a wide range of subject matter. The subject matter of this book overlaps much of that of Wiener's book. I decided to call this book Cybernetics 2.0. Since Wiener's book was written 73 years ago, biology has advanced tremendously. Some of the new knowledge has had direct application to the examples of this book. The main advance of this work over the original Cybernetics is the introduction of adaptivity and the description of the mechanism of homeostasis. Wiener's book is more philosophical. This book gets into the "nitty gritty" and tells how things work.

I am a 91-year-old Stanford Professor of Electrical Engineering, Emeritus. I have had a long career in electrical engineering. I was researching artificial neural networks and learning algorithms when there were only a handful of people doing this all over the world. I believe I have had a successful career. Some of my inventions affect everyone everywhere who use computers, cell phones and the internet. Engineers at Apple have told me that every iPhone, since the iPhone 5, uses the LMS algorithm all over the device. They could not tell me where because, if they did, they would have to shoot me. Apple keeps secrets.

I wrote this book as a believer in the Hebbian-LMS algorithm. I was lulled into this state of mind by the number of places where Hebbian-LMS fits the data and where I have been able to predict functioning and behavior as I looked deeper into the biology. Hebbian-LMS is suggested as the universal natural algorithm of adaptivity in the body (this sometimes feels like the hammer that looks everywhere for nails to hammer down). Could some other algorithm or algorithms account for the observed phenomena? I do not know, but Hebbian-LMS seems to be a good bet as nature's adaptive algorithm.

This work is my first incursion into biology. All my life I have worked on engineering problems finding solutions that are analytic and provable. Studying questions in biology is not as clear cut. Getting provable answers is often much more difficult. The complexity of biology overshadows the complexity of engineering. Design of the innards of a cell phone is no "walk in the park". But compared to biology, cell phone design is a "piece of cake". Engineers and physical scientists can learn a lot from study of biology. Nature is the greatest engineer of all time.

Stanford, CA, USA
2021

Bernard Widrow

Acknowledgements

When developing the subject of this book and preparing its manuscript, I was helped with advice, council, suggestions, debates, encouragement from a group of friends whose contributions are noted here.

Dramatis Personae

Abhipray Sahoo
I met Abhi during my spring quarter 2019 class EE373b, adaptive signal processing. Abhi was one of the students who after class would stay with me for further discussion and questions. Toward the end of the quarter, I introduced the Hebbian-LMS algorithm and expressed my belief that it is likely to be Nature's learning algorithm. Abhi was working for a neuroscience company. He seemed to be captivated by the idea of Hebbian-LMS. For the last two years, he has worked closely with me as I wrote the chapters. He was the one that I needed to convince with every stage of development. He took my hand-written text and computerized it. Without his help, I could not have written this book.

Dr. Henry Yin
Henry is a professor of neuroscience at Duke University. He contacted me by email with compliments on a paper that I wrote with two of my Ph.D. students, Youngsik Kim and Dookun Park. The paper was published in the IEEE Computational Intelligence magazine of November, 2015. He said that this paper was one of the best pieces of work in neuroscience that he has seen over a number of years. We exchanged emails back and forth. I told him I was writing a book on the subject and he volunteered to read it and provide feedback. He has been helping and advising me ever since. He read the manuscript from "cover to cover" and provided detailed comments, suggestions, and corrections. He has made great contributions to this work and I thank him profoundly.

Dr. Fredric Kraemer
Rick is a professor of medicine at Stanford University. He is chief of Endocrinology. His specialty is highly relevant to the subject of this book. He too read the book from "cover to cover" and provided detailed feedback in the area of his specialty.

His suggestions have enhanced major portions of the manuscript. His help is greatly appreciated.

Dr. Ferid Murad, Dr. Jeff Kuret, Thomas Kuret
Ferid Murad, 1998 Nobel laureate in physiology and medicine, paid me a visit with his son-in-law Jeff Kuret and his grandson Thomas Kuret. Jeff is a professor of neuroscience at Ohio State University and Thomas is a graduate student in neuroscience at the Salk Institute and UC San Diego. The subject of our discussion was synaptic plasticity and Hebbian learning. This was a very useful session and very helpful to me in clarifying my own thinking about Hebbian-LMS. Since then I have had emails back and forth with Ferid discussing many of the subjects of this book. I greatly appreciate his interest and help.

Dr. Thomas Kailath
Tom is professor emeritus of electrical engineering at Stanford. He read the manuscript and provided useful feedback. I am very grateful.

Dr. Stephen Boyd
Stephen is professor of electrical engineering and department chair. He too read the manuscript and gave useful advice and help. I thank him very much.

Dr. Srabanti Chowdhury
Srabanti is a professor of electrical engineering at Stanford. She too read the work from "cover to cover" and provided very useful feedback. She and Abhi will join with me in teaching a course at Stanford on the subject of this book, once the COVID-19 pandemic subsides.

Dr. Edward Katz
I first met Ed Katz when both of us worked on local IEEE business. We have been good friends ever since. Ed has read a number of drafts of this book and provided detailed feedback on how to improve clarity of presentation. He has a long history on fuzzy logic, but he has stepped away from his comfort zone to work in the field of neural networks. His comments, suggestions, and corrections are greatly appreciated.

Helen Gordan
Helen is a third year undergraduate student in electrical engineering at Stanford. She has a strong engineering background and has also studied biology and life sciences. She became very interested in the subject of this book. When studying the manuscript, she was able to take my hand drawing diagrams and make them into computer drawings. Most of the figures in this book were drawn by Helen. I am most grateful for her fine work.

Audrey Bloom
Audrey is a third-year undergraduate student in human biology at Stanford. Her family are my next door neighbors. She introduced me to her friend Helen Gordan. With her background in biology, she was to read the manuscript and provide useful feedback on the biological side. She also suggested ways to make the engineering clearer to biologists. My many discussions with her were very helpful. Thanks to Audrey for the comments and feedback.

Siddharth Sharma

Siddharth is a high school student who lives in the San Jose area. He has just been accepted as a Stanford undergraduate and will begin in September, 2021. I had received an email from him asking me if I could suggest a research project that he could do for his high school class. He sent me a CV and a link to a book manuscript that he wrote on AI. I was amazed that a high school student could do such work. I asked him to join with Abhi and me on our Hebbian-LMS project. Explaining the subject to him helped with my thinking. Thank you, Siddharth.

Life Support

During the pandemic, my wife and I stayed home. We were dependent on younger people to help us with grocery shopping. I would like to especially thank Linda Kramer, and Ali Sharafat. Linda is the wife of Rick Kraemer. Ali is a Ph.D. student in electrical engineering. His father was one of my Ph.D. students years ago. Thank you Linda and Ali. You kept the virus away.

Google

During the pandemic, it was not possible to visit the Stanford libraries. Google has much of humanity's knowledge available at the keyboard. Thanks to Google, I was always able to find what I needed to progress with this book.

Ronna Lee Widrow

Thanks to Ronna Lee, my wife, who kept me fed during the pandemic. She was pleased that I was writing this book. This kept me out of her kitchen.

Contents

Part III Regulation and Control of Physiological Variables and Body Organs

Part VI Growth of Plants

Part VII Norbert Wiener

Part VIII Conclusion

About the Author

Bernard Widrow is a Professor Emeritus in the Electrical Engineering Department at Stanford University. His research focuses on adaptive signal processing, adaptive control systems, adaptive neural networks, human memory, cybernetics, and human-like memory for computers. Applications include signal processing, prediction, noise cancelling, adaptive arrays, control systems, and pattern recognition. He received the Doctor of Science Degree from MIT in 1956 and was appointed professor from the same university. He has been active in the field of artificial neural networks since 1957, when there were only a half-dozen researchers working on this all over the world. In 1959, he moved to Stanford University. In the same year, together with his student Ted Hoff; he invented the Least Mean Square (LMS) algorithm, which has been the world's most widely used learning algorithm to date. Since 2010, he has expanded his interest to living neural networks and biological adaptivity. A Life fellow of the Institute of Electrical and Electronic Engineering (IEEE), he was awarded with the IEEE Alexander Graham Bell Medal in 1986 and with the Benjamin Franklin Medal for Electrical Engineering in 2001. He has been inducted into both the US National Academy of Engineering and the Silicon Valley Engineering Hall of Fame, in 1995 and 1999, respectively.

When I was a young teenager, my father called me aside to have a serious discussion. He said, what do you want to be when you grow up? I said that I wanted to be an electrician. He said no, you want to be an electrical engineer. I asked, what's that? He said, I don't know but that is what you will be, and you will go to MIT. I asked, What's that? He said that it is a college in Boston where you learn to be an electrical engineer.

Somehow this all came to pass. I received the S.B. degree in 1951, the S.M. degree in 1953, and the Sc.D. degree in 1956, all in electrical engineering at MIT.

Upon completion of the doctorate, I joined the MIT EE faculty with a three year appointment. This is where I began research on learning systems. The idea of learning systems was not well regarded at MIT at that time since, as everyone knew, only humans learn. In 1959, I joined the EE faculty at Stanford. When I left MIT, there were 60 faculty in EE. At that time, in 1959, Stanford had only 15 EE faculty members. It felt like going from a pressure cooker to something like a vacuum.

In the 1960's, Stanford was undergoing a rapid buildup in all departments, under the leadership of Provost Frederick Terman. Some new faculty were home grown, but most were stolen from the great universities of the East. They came to Stanford, rolled up their sleeves, and got to work. My ideas about learning systems were met with great enthusiasm and expectation. The Stanford EE Department was looking for new bold ideas.

With such a small EE Department, every man counted (the EE faculty was all men). Stanford EE had a larger Ph.D. program than MIT, so there were many Ph.D. students that needed research advisors. As an Assistant Professor, I was able to have Ph.D. students. My first Ph.D. student was Marcian E, Hoff, Jr., known as "Ted". We began work. It was only a few weeks after my arrival at Stanford. At our first meeting, I was explaining some things about learning, quadratic mean square error surfaces and stochastic gradients, and the LMS algorithm popped out of the blackboard. It was a Friday afternoon in the Autumn of 1959.

More students joined me and we formed into a lab. Ted completed the Ph.D. and accepted a job at a new start-up named Intel. He was employee number 12. Soon there, he proposed to Bob Noyce and Gordon Moore the development of a programmable computer on a chip. They agreed and gave him two colleagues to form a development group. They created the first microprocessor. For this work, they received the Kyoto Prize from the Emperor of Japan and the US National Medal of Technology and Innovation from President Obama. Ted has had a brilliant career. And he was my first Ph.D. His work at Intel did not involve learning systems but the lab experience probably helped him.

I was on the Stanford faculty for 50 years before retirement. I was principal research advisor of 89 Ph.D.'s. It was a privilege to work with these talented students. Over the years, I learned a great deal. We developed adaptive digital filters and adaptive neural networks, all based on the LMS algorithm. With this work, I authored or co-authored numerous research papers and patents. With co-authors, I wrote three books: "Adaptive Signal Processing", "Adaptive Inverse Control", and "Quantization Noise". A collection of my papers can be found on my Stanford web site. Click on publications, then click on the individual reference and this will obtain the entire paper.

I have been honored by the IEEE (Institute of Electrical and Electronic Engineers), of which I am a Life Fellow. I received the 1986 IEEE Alexander Graham Bell Medal. The citation was "For Fundamental Contributions to Adaptive Filtering, Noise and Echo Cancellation, and Adaptive Antennas." In 1984, I received The IEEE Centennial Medal. I received the 1991 IEEE Neural Networks Pioneer Medal for "Adaptive Networks". I received the 1998 Society Award of the IEEE Signal Processing Society with the citation "For His Pioneering and Lasting Contributions to the Fields Of Adaptive Signal Processing, Adaptive Control, and Artificial Neural Networks."

From the Franklin Institute of Philadelphia, I received the 2001 Benjamin Franklin Medal for Engineering, with the citation "For Pioneering Work in Adaptive Signal Processing as Exemplified by the LMS Algorithm, Adaptive Filters, Adaptive Control, Adaptive Antennas, Noise Cancellers, Artificial Neural Networks, and Directional Hearing Aids." At the gala awards ceremony at the Franklin Institute, ten other scientists and engineers received awards for various fields of science, medicine, and technology. Marvin Minsky received the Benjamin Franklin for computer science. They called us "laureates."

From the International Neural Networks Society, I received the 2009 Donald Hebb Award. The citation read "In Recognition of His Outstanding Achievements In Biological Learning."

In 1995, I was inducted into the National Academy of Engineering. This was a very great honor.

I have been honored with many other awards over the years.

Although I am now working on several new projects, I must admit that at age 92, I have reached the twilight of my career. I have had a pretty good run. Thanks go to my students and colleagues. Thanks, thanks.

Part I
Synaptic Learning, Homeostasis, Synaptic Plasticity

Introduction to Part I

The human brain has about 100 billion neurons. Each neuron has about 10 thousand synapses. The total number of synapses is in the trillions. A synapse is the link that connects one neuron to another. The strength of the synaptic coupling can vary and, according to Hebb, synaptic change is associated with learning.

What is the mechanism or algorithm that nature uses to control the strength of coupling of trillions of synapses, and how is this control related to learning? These are issues to be considered in Part I. Some answers are given. With each answer, further questions are raised. This is like peeling an onion. The deeper you go, the more of the unknown is exposed. There are no absolute answers.

The % Hebbian-LMS algorithm is proposed as nature's synaptic learning algorithm. It is a realization of the Hebbian learning rules by means of the LMS algorithm from the field of adaptive signal processing.

Neurons whose synapses are controlled by % Hebbian-LMS can be connected in a network. Synaptic learning takes place independently with each neuron. Network inputs could come from sensors supplying visual, tactile, auditory, olfactory, etc. patterns. Connected in a network, Hebbian-LMS neurons perform clustering, automatically making connections and associations between objects observed in input patterns. Clustering seems to be a fundamental element of human thinking and learning.

The purpose of Part I is to develop a theory of synaptic plasticity, control of the coupling strength or the synaptic weighting of the brain's synapses. The change and the rate of change of the synaptic weights is at issue. What signals, what information is needed at the synapses in order to effect learning? This is a fundamental question. Answers are proffered revolving around the Hebbian-LMS algorithm. Biological evidence supported by experiment is presented.

The effectiveness of Hebbian-LMS learning might be questioned. Demonstration of the Hebbian-LMS algorithm showing a learning curve, showing convergence, and showing clustering is presented by computer simulation.

Part I explains the Hebbian-LMS algorithm. The observed behavior of this algorithm corresponds with observed phenomena in living neural systems. This algorithm provides an understanding of the phenomenon of homeostasis and makes clear the connection between synaptic scaling and homeostasis. This connection has been elusive in neurobiology.

The % Hebbian-LMS algorithm introduces adaptivity to the modeling and analysis of living neural systems. Adaptivity is what controls the synaptic weights.

The synaptic weight is proportional to the number of its neuroreceptors. Adaptivity controls the number of receptors, increasing or decreasing their population. This is called receptor upregulation or downregulation.

Part I introduces a way of thinking that will be of interest to neuroscientists. In part II the theory of the % Hebbian-LMS algorithm will explain phenomena related to mood disorders such as anxiety, depression, bipolar disorder, and opioid addiction. This will be of interest to psychiatrists. The same algorithm will be shown in Part III to be involved with control of body organs, control of the kidneys, the heart, and body thermoregulation. This will be of interest to biologists and medical scientists. Feedback control systems are examined in Part III having Hebbian-LMS neurons within their feedback loops. This will be of interest to control engineers. This wide range of subjects falls under the general umbrella of cybernetics as defined by Wiener.

Advice to Life Scientists and Engineers

Given your background and experience, you will probably find material in this book that is outside your sphere of knowledge. Read these parts to get an overview, then continue your reading into parts that are more familiar. Every chapter is self-contained. There is a common thread of adaptivity and homeostasis throughout. Deep study of all parts is not absolutely necessary to understand the more familiar parts. Further background could be acquired by googling the various subjects. Blending engineering with the life sciences will surely lead to creativity in both areas.

Chapter 1
Introduction

Abstract The purpose of this chapter is to describe neurons, and synapses, and how they connect and work together. A model of the neuron and its synapses, suitable for computer simulation, is presented.

1.1 The Beginning

In 1949, at McGill University, D.O. Hebb introduced the idea of synaptic learning which today is called synaptic plasticity. Hebbian learning is widely accepted, fundamental in the fields of psychology, neurology, and neurobiology [1]. Ten years after Hebb's publication, Widrow and Hoff at Stanford university in 1959 invented the LMS (least mean square) learning algorithm [2, 3]. Analogous to synaptic plasticity, this algorithm varies weighting coefficients to perform learning and least squares optimization. It became the world's most widely used adaptive algorithm, fundamental to the fields of signal processing, control systems, communication systems, pattern recognition and above all, artificial neural networks.

These two learning paradigms are very different. Hebbian learning is unsupervised. LMS learning is supervised. In spite of their differences, there is a connection between them. A form of the LMS algorithm can be constructed that implements Hebbian learning. We call this the Hebbian-LMS algorithm [4]. It is a clustering algorithm and it has been used in artificial neural networks. In this book, we propose that Hebbian-LMS may in fact be nature's learning algorithm.

In living networks, signal flow from one neuron to another takes place via nature's coupling device, the synapse. The coupling weight, called "efficiency" by Hebb, is increased by increasing the number of neuroreceptors in the synapse, decreased by reducing the number of neuroreceptors. This is synaptic plasticity. There are many types of receptors, and only certain types of excitatory neurotransmitters that bind to their corresponding receptors would have this effect.

At the time of Hebb's writing, neurotransmitters and neuroreceptors had not yet been discovered. The synapse is a self-contained two terminal device having an input and output. The question is, what kind of information does the synapse require in order to know whether to increase its weight or to decrease its weight and at what

B. Widrow, *Cybernetics 2.0*, Springer Series on Bio- and Neurosystems 14,
https://doi.org/10.1007/978-3-030-98140-2_1

rate? Hebb's learning rule implies that this depends on the firing of the presynaptic neuron and the firing of the post-synaptic neuron. Hebb's rule is often simplified to: "neurons that fire together wire together." i.e., their coupling weight increases when both the presynaptic neuron and the post-synaptic neuron are firing at the same time. In his book, Hebb actually said "*when an axon of Cell A is near enough to excite a cell B and repeatedly or persistently takes part in firing it, some growth process or metabolic change takes place in one or both cells such that A's efficiency, as one of the cells firing B is increased*". This is the original Hebbian learning rule and it applies to excitatory synapses. It needs to be extended for conditions that cause a decrease in efficiency (weight) as well as for conditions that cause increase. It also needs to be extended to apply to inhibitory synapses as well as to exhibitory synapses. In the following sections, Hebb's rule will be extended, LMS neural training and the Hebbian-LMS algorithm will be explained, and Hebbian-LMS learning will be shown to implement extended Hebbian learning. The question will be, is Hebbian-LMS nature's learning algorithm? Evidence will be presented indicating that this is likely.

1.2 The Neuron

The neuron is the basic computing element of the brain. A sketch of a neuron is shown in Fig. 1.1. Shown is its cell body, its dendrites, its axon, and the axon's terminal.

The neuron receives input signals primarily from the dendrites. It delivers an output signal via its axon. The output signal excites other neurons by their connections with the axon terminal. The neuron can have thousands of input signals. It delivers a single output signal by way of its axon.

Input signals are applied through synapses that are attached to the neuron's dendrites and cell body. Most synapses are electrochemical devices. The neuron resem-

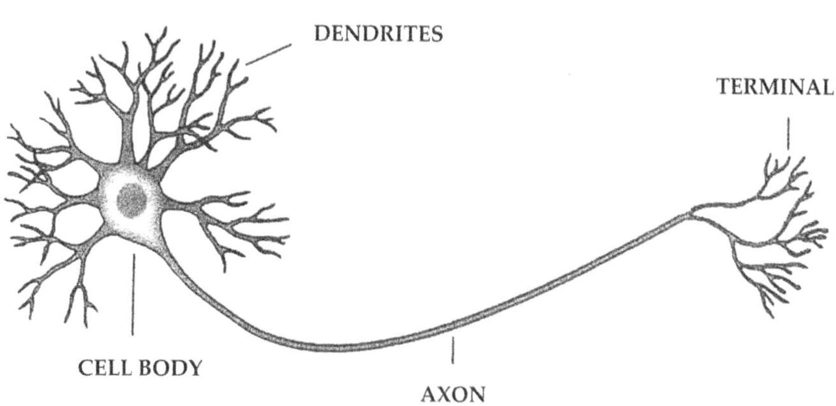

Fig. 1.1 The neuron, its cell body, dendrite, axons, and an axon terminal

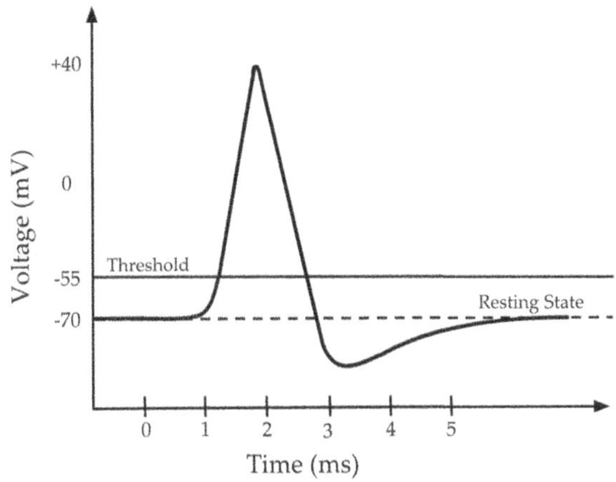

Fig. 1.2 An action potential

bles an octopus with a rounded head, arms, each with many suckers along their lengths. The suckers are analogous to the synapses, the arms to the dendrites, and the head to the cell body.

The neuron is electrically active, firing or not, depending on the voltage of its nucleus relative to that of the surrounding fluid. Inputs to the neuron are excitatory that raise the voltage of the nucleus and increase the probability of firing, or they are inhibitory and lower the voltage of the nucleus thus lowering the probability of firing.

Like every cell in the body, the neuron is enveloped by a membrane that insulates it electrically and chemically from the surrounding fluid, the cerebral spinal fluid. The membrane of the neuron is a thin coating of lipid proteins and other proteins that cover the neuron cell body and all the dendrites. The synapses are embedded in this membrane and are part of it. The interior of the neuron within the membrane is electrically conductive and is generally at the same voltage everywhere, an equipotential.

A voltage develops between inside the membrane and outside the membrane and the voltage difference is called the membrane potential. When the membrane potential is more positive than −55 millivolts (mV), the neuron fires. The firing rate is roughly proportional to the amplitude of the membrane potential above −55 mV. When the membrane potential is lower than the threshold of −55 mV, the neuron does not fire. The firing rate is the neuron's output signal, a signal that could be transmitted to other neurons. A neuron is in its resting state when all synaptic inputs are zero. The resting state membrane potential is −70 mV. In its resting state it can be triggered with an electric impulse if strong enough to make it fire. When the neuron fires, it produces a pulse in the membrane potential that then propagates along the axon. This pulse is called an action potential. It is also referred to as a "spike." A sketch of an action potential's voltage as a function of time is shown in Fig. 1.2.

Assume now that the neuron was in a resting state. The membrane potential was −70 mV. The trigger pulse caused the membrane potential to increase until

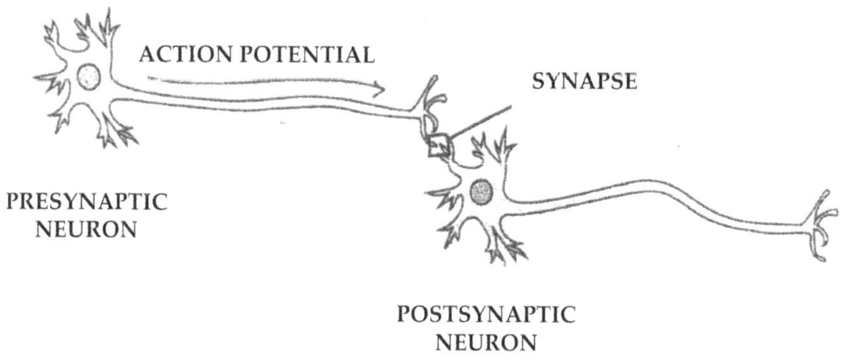

Fig. 1.3 Neuron-to-neuron communication via a synaptic connection

the threshold of −55 mV was reached. Then the neuron fired generating an action potential, a spike. The voltage spike peaks at about 40mV, then subsides and finally returns to the resting state. When action potentials are generated in a rapid sequence, under usual operating conditions, the neuron's output signal is the pulse rate.

Neurons deliver their output signals to other neurons via synapses. Figure 1.3 is a sketch showing one neuron, the presynaptic neuron, connecting via a synapse to another neuron, the postsynaptic neuron. When the presynaptic neuron fires, an action potential travels down its axon to the axon terminal to the synapse. The synapse in turn provides stimulation, either excitatory or inhibitory, to the postsynaptic neuron.

1.3 The Synapse

The axon terminal has a tree-like structure. The ends of the branches are rounded nubs called buttons. When a neuron is being "installed" in a circuit, during development or during replacement, these buttons reach out seeking and synapsing to other neurons and their dendrites. Wherever they make close enough proximity, a synapse forms. The result of this is a neural network with neurons randomly placed in three dimensions having what seems to be random inter-connections. It is amazing that with all the randomness, the network adapts and learns, and does useful things.

A synapse is pictured in Fig. 1.4. It has a presynaptic part which is a button of one of the branches of the axonal terminal of a presynaptic neuron. It has a postsynaptic part which is the membrane of a dendrite or of the cell body of the postsynaptic neuron. The button and the membrane are separated by a 0.02 micron gap called the synaptic cleft. The cleft is filled with cerebral spinal fluid containing varying concentrations of neurotransmitter.

The button, being part of the presynaptic neuron, is enveloped by a membrane. Vesicles at the bottom of the button store and release neurotransmitter into the synaptic cleft. The neurotransmitter is generated by the presynaptic neuron and is carried

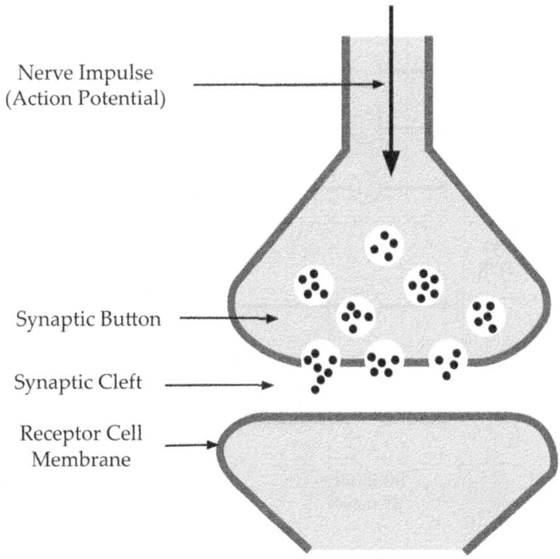

Fig. 1.4 A synapse

down the axon to the vesicles in the button. When an action potential from the presynaptic neuron arrives at the button, the vesicles release neurotransmitter into the cleft. Neuroreceptors embedded in the postsynaptic membrane are capable of bonding to neurotransmitter molecules, thereby carrying signals from the presynaptic neuron to the postsynaptic neuron.

Each time an action potential arrives, a finite amount of neurotransmitter is released. The neurotransmitter does not remain permanently. Some of it will be lost by diffusion away from the synapse. Some of it will be lost by re-uptake back into the vesicles. Some will be lost due to degradation. In any event, the amount of neurotransmitter remaining after the arrival of a single activation pulse will decay exponentially over time.

When the presynaptic neuron is firing periodically sending a signal, the concentration of neurotransmitter in the cleft will go up and down, but the average concentration will be proportional to the firing rate of the presynaptic neuron. Therefore, the average concentration of neurotransmitter in the cleft is proportional to the output signal of the presynaptic neuron. The average concentration of neurotransmitter in the cleft is therefore representative of the synapse's input signal.

The neurotransmitter molecules in the cleft seek out neuroreceptors located on the postsynaptic membrane. The neurotransmitter and neuroreceptors are selective mates, tuned to each other like a lock and key. When neurotransmitter binds to a neuroreceptor, the neuroreceptor "gets excited" and opens a co-located gate through the postsynaptic membrane, allowing ions to flow from the cerebrospinal fluid outside the membrane to inside the membrane. The ions are charged particles. Ionic flow contributes to a raising or lowering of the membrane potential of the postsynaptic neuron. For example, a net inflow of positive ions contributes to raising the

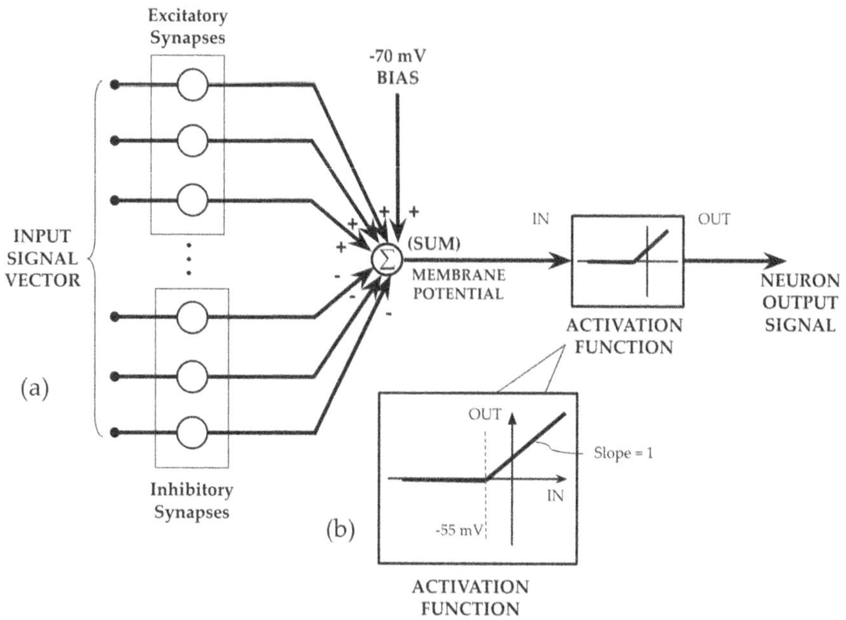

Fig. 1.5 **a** Schematic diagram of a postsynaptic neuron, its excitatory inputs, inhibitory synapses, its soma, and its activation function. **b** Details of the activation function

postsynaptic membrane potential and is excitatory. Ionic flow is proportional to the concentration of neurotransmitter in the cleft multiplied by the number of receptors. Note that a net outflow of positive ions reduces the postsynaptic membrane potential and is inhibitory.

A neuron typically has thousands of attached synapses. The synapses are designed to allow either the inflow of positive ions and is excitatory, or to allow the inflow of negative ions and is inhibitory. The effect of the ionic flow at all the synapses, some raising the voltage, some lowering the voltage, is embodied in a summation. The voltages add by Kirchhoff addition.

Signal flow runs one way, from the presynaptic neuron via the synapse to the postsynaptic neuron. Inside the synapse, the input signal is the neurotransmitter concentration. The signal is carried across the cleft by neurotransmitter. The signal flow path from presynaptic neuron to postsynaptic neuron is weighted by the number of neuroreceptors on the postsynaptic side. Thus, the synaptic weight that Hebb called "the efficiency" is proportional to the number of neuroreceptors in the synapse. A schematic diagram of a postsynaptic neuron and its synapses is shown in Fig. 1.5. This is a model of an actual neuron.

The schematic of Fig. 1.5a shows excitatory and inhibitory inputs. Corresponding excitatory and inhibitory synapses (the synaptic weights) carry the input signals to the soma (a summation). The sum, hereby called (SUM), is the membrane potential. Added to the (SUM) is a bias of −70 mV. This is included so that when all input signals have zero values and the neuron is at rest, the membrane potential will be −70 mV.

A nonlinear activation function is shown in Fig. 1.5b. Its purpose is to maintain the output signal at zero as long as the membrane potential is less than the firing threshold of −55 mV. Above that threshold, the neuron produces an output that increases linearly with the membrane potential. The firing rate of an actual neuron increases approximately linearly with membrane potential above −55 mV. As the membrane potential increases however, the activation function departs from linearity and begins to saturate in accord with a sigmoidal function. This will be discussed in subsequent chapters.

The synaptic weights are adjusted up or down by an adaptive algorithm, a learning algorithm. The learning rules are based on Hebb's work and will be described in the next chapter.

1.4 Summary

The synapse is the connecting link between neurons. Signalling through the synapse involves neurotransmitter that is generated by the presynaptic neuron, and neuroreceptors that are attached to the postsynaptic neuron. Neurotransmitter binds to neuroreceptors like a key in a lock. The strength of the synaptic connection, the synaptic weight, is proportional to the number of neuroreceptors. Learning involves weight change, increase or decrease, in accord with an adaptive algorithm.

1.5 Questions and Experiments

1. Check the literature to verify the description of the neuron as described in this chapter. What references did you find useful?
2. As the voltage of the (SUM) increases, the activation function departs from linearity and begins to roll off and saturates. The function is sigmoidal. Modify the diagram of Fig. 1.5 accordingly.

References

1. Hebb, D.O.: The Organization of Behavior: A Neuropsychological Theory. Psychology Press (2005)
2. Widrow, B., Hoff, M.E.: Adaptive switching circuits. Technical report, Stanford Univ Ca Stanford Electronics Labs (1960)

3. Widrow, B.: Adaptive inverse control. In: Adaptive Systems in Control and Signal Processing 1986, pp. 1–5. Elsevier (1987)
4. Widrow, B., Kim, Y., Park, D.: The Hebbian-LMS learning algorithm. IEEE Comput. Intel. Mag. **10**(4), 37–53 (2015)

Chapter 2
Hebbian Learning

Abstract The learning rule "neurons that fire together wire together" is extended in several ways, to cover both excitatory and inhibitory synapses, to cover synaptic coupling that can increase or decrease, and to determine the rate of change of synaptic coupling. The extended rules apply to synaptic plasticity. Classic in-vitro experiments by the neuroscientists, G Q Bi and M M Poo lend credibility to the extended Hebbian rules, although their data is presented to support a "spike timing" learning rule. Actually their data casts doubt on the spike timing hypothesis. Another more consistent analysis suggests that the extended Hebbian rules are more likely to be nature's learning rules.

2.1 Extended Hebbian Rules

The original Hebbian learning rule cannot adequately describe natural learning phenomena. It only allows weights to increase, and it does not apply to inhibitory synapses. In the spirit of Hebb's teaching, the rule can be extended as follows. These extensions are based on intuitive reasoning.

A. Excitatory synapses

   (1) When the presynaptic neuron is not firing, there will be no weight change
   (2) When the presynaptic neuron is firing and the postsynaptic neuron is firing, the weight increases (Hebb's original rule).
   (3) When the presynaptic neuron is firing and the postsynaptic neuron is not firing, the weight decreases [1, 2].

B. Inhibitory synapses

   (1) When the presynaptic neuron is not firing, there will be no weight change
   (2) When the presynaptic neuron is firing and the postsynaptic neuron is firing, the weight decreases
   (3) When the presynaptic neuron is firing and the postsynaptic neuron is not firing, the weight increases.

B. Widrow, *Cybernetics 2.0*, Springer Series on Bio- and Neurosystems 14,
https://doi.org/10.1007/978-3-030-98140-2_2

The excitatory and inhibitory synapses have similar but opposite purposes. They differ primarily with the chemistry. The neuroreceptor signals that propagate through an excitatory synapse contribute in a positive way to the firing of the postsynaptic neuron. Signals that propagate through an inhibitory synapse contribute in a negative way to the firing of the postsynaptic neuron. The learning rules for excitatory and inhibitory synapses are opposites.

The Hebbian learning rules need to be augmented further to include the issue of rate of learning. The issue involves increase or decrease in the number of synaptic neuroreceptors. How fast does this take place, and what are the factors that determine the rate? These are good questions. The current literature does not provide answers. Herewith are proposed answers.

A. Excitatory synapses

(1) When the weight is increasing, the rate is proportional to the product of the presynaptic firing rate and the postsynaptic firing rate. Translation: The rate of weight increase is proportional to the product of the neurotransmitter concentration in the cleft and the magnitude of the postsynaptic membrane potential above the −55 mV threshold.
(2) When the weight is decreasing, the rate is proportional to the product of the neurotransmitter concentration in the cleft and the magnitude of the postsynaptic membrane potential below −55 mV.

B. Inhibitory synapses

(1) When the weight is decreasing, the rate is proportional to the product of the presynaptic firing rate and the postsynaptic firing rate. Translation: the rate of weight decrease is proportional to the product of the neurotransmitter concentration in the cleft and the magnitude of the postsynaptic membrane potential above −55 mV.
(2) When the weight is increasing, the rate is proportional to the product of the neurotransmitter concentration in the cleft and the magnitude of the postsynaptic membrane potential below −55 mV.

The above learning rules are a natural extension of Hebb's original rules. The extended rules are supported by experimental evidence which will be described below.

2.2 Synaptic Signal Transmission

Signals propagate from the presynaptic neuron to the postsynaptic neuron, while at the same time the synaptic weight can be changing. Key to all this activity is the neurotransmitter, whether excitatory or inhibitory. Without neurotransmitter, there is no signaling and no adaptivity. The input signal to the synapse is the firing rate of the presynaptic neuron. Each firing causes an activation pulse or spike at the presynap-

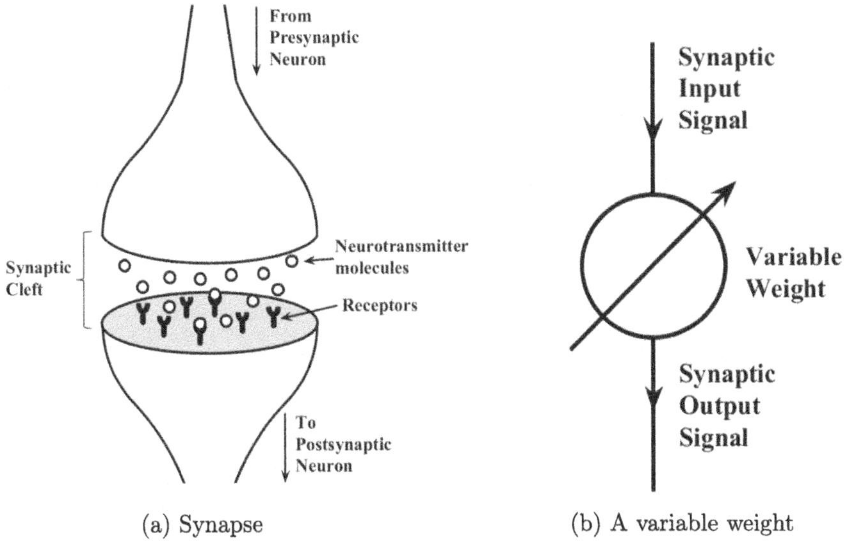

Fig. 2.1 A synapse corresponding to a variable weight. Reproduced from B. Widrow, Y. Kim, and D. Park, The Hebbian-LMS Learning Algorithm, IEEE Computational Intelligence Magazine, vol. 10, no. 4, pp 37–53, Nov. 2015 with permission from IEEE @ IEEE 2021

tic side of the synapse, which in turn causes a finite amount of neurotransmitter to be emitted into the synaptic cleft. The neurotransmitter of the individual activation pulse gradually disappears, exponentially over time. Continual firing of the presynaptic neuron results in an average concentration of neurotransmitter in the cleft which is proportional to the presynaptic firing rate. Thus, the synaptic input signal is represented by the time average of the neurotransmitter concentration in the cleft.

Signaling is caused by neurotransmitter molecules attaching to receptors on the postsynaptic side of the cleft. A symbolic representation of the synapse is shown in Fig. 2.1a. The receptors are attached to the membrane which envelopes the postsynaptic neuron. This membrane acts as a barrier and an insulator separating the nucleus and the soma from the surrounding fluid. The receptors are gates that when activated by attached neurotransmitter molecules, open to allow ions from the fluid to enter into the postsynaptic neuron. For an excitatory synapse, typical ions are of sodium, potassium, and calcium. These are positive ions that, when they enter the neuron, make the voltage of the soma more positive. The output signal of the synapse is this voltage raising effort which is proportional to the average neurotransmitter concentration and the number of receptors attached to the membrane. The number of receptors is the synaptic weight. So, the output signal of the synapse is proportional to the product of the input signal and synaptic weight. Figure 2.1b is a representation of input signal, weight, and output signal.

The postsynaptic neuron generally has thousands of attached synapses, all providing input signals. These signals combine by addition. The voltage raising effects add. Thus, the soma voltage is a weighted sum of the synaptic inputs. If the soma voltage is greater than a threshold voltage of $-55\,\text{mV}$, the postsynaptic neuron fires. The firing rate is the output signal of the neuron and it is roughly proportional to the membrane potential above $-55\,\text{mV}$.

Signal can flow from a presynaptic neuron to a postsynaptic neuron via the connecting synapse that acts as a variable weight. Signals from thousands of presynaptic neurons flow into their respective synapses, then combine and control the firing rate of the postsynaptic neuron. This is represented symbolically by the diagram of Fig. 1.5.

2.3 Synaptic Weight Change—Hebbian Rules

Within a synapse, the neuroreceptors are embedded in the postsynaptic neuron's membrane. In the presence of neurotransmitter, the number of receptors will increase or decrease at a rate determined by the membrane potential and the neurotransmitter concentration. The number of receptors will remain constant in the absence of neurotransmitter with the presynaptic neuron not firing.

A simplified diagram of a neuron with a single synapse attached to the membrane on one of its dendrites is shown in Fig. 2.2. The postsynaptic membrane potential is correlated with the firing or not firing of the postsynaptic neuron. With an excitatory

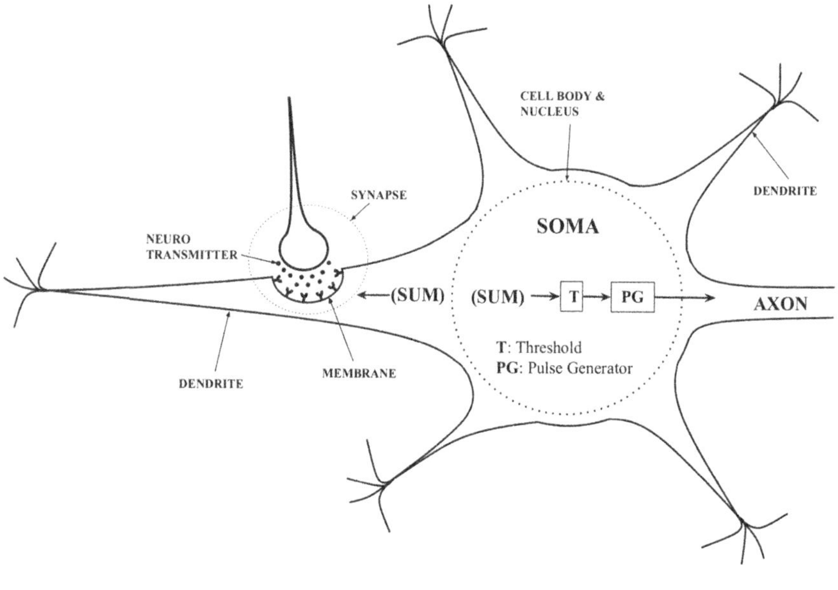

Fig. 2.2 A neuron dendrite, and a synapse. Reproduced from B. Widrow, Y. Kim, and D. Park, The Hebbian-LMS Learning Algorithm, IEEE Computational Intelligence Magazine, vol. 10, no. 4, pp 37–53, Nov. 2015 with permission from IEEE @ IEEE 2021

synapse, if both presynaptic and postsynaptic neurons are firing, the number of ligand binding receptors will increase and do this at a rate proportional to the product of the presynaptic firing rate and the magnitude of the membrane potential above the firing threshold of −55 mV. If the presynaptic neuron is firing and the postsynaptic neuron is not firing, the postsynaptic membrane potential will be lower than −55 mV. The number of ligand binding receptors will decrease at a rate proportional to the product of the presynaptic neuron's firing rate and the magnitude of the membrane potential below −55 mV. The ligand is of course the neurotransmitter.

With an inhibitory synapse, conditions reverse. When the presynaptic neuron is firing and the postsynaptic neuron is not firing, the rate of increase of the number of receptors is proportional to the presynaptic firing rate and the magnitude of the membrane potential below −55 mV. With both neurons firing, the rate of decrease of the number of receptors is proportional to the presynaptic firing rate and the magnitude of the membrane potential above −55 mV.

The information that the synapse needs in order to decide whether to increase or decrease the weight and at what rate is available to it. The receptors sense the membrane potential, being embedded in the membrane, and they sense the amount of neurotransmitter present, being exposed to it. The synaptic weight change mechanism described above is Hebbian, in accord with the extended Hebbian rules. These rules are original and are based on thought experiments (in Einstein's terminology), i.e. intuitive reasoning. Nevertheless, these rules are supported by the physical experiments of Bi and Poo, described next.

2.4 Bi and Poo

Classic experiments were performed at the University of California, Berkeley by Guo-qiang Bi and Mu-ming Poo [3] on the subject of spike timing and plasticity. Their experimental results show that if the presynaptic neuron fires just before the postsynaptic neuron fires, the weight of their connecting synapse increases. If the postsynaptic neuron fires just before the presynaptic neuron fires, the synaptic weight decreases. Their conclusion was that the relative firing times of the presynaptic and the postsynaptic neurons is critical to controlling the increase or decrease of the synaptic weight. Their conclusion is widely accepted in the fields of neurobiology and artificial neural networks.

Bi and Poo presented data for their work, experimenting with cultures from two rat hippocampal neurons. The connecting synapse was excitatory. They triggered firing with an electronic signal generator providing repetitive correlated impulse pairs at a rate 1 Hz, one impulse for the presynaptic neuron and the second for the postsynaptic neuron. The signal generator was a high impedance current source. The time spacing between the pulse pairs was precisely controllable. Their measurements of synaptic weight change is given in Fig. 2.3. The data is from their 1998 paper. Each data point was cumulative, taken with 60 pulses. The horizontal axis gives timing difference Δt in milliseconds. The positive side corresponds to the presynaptic neuron firing

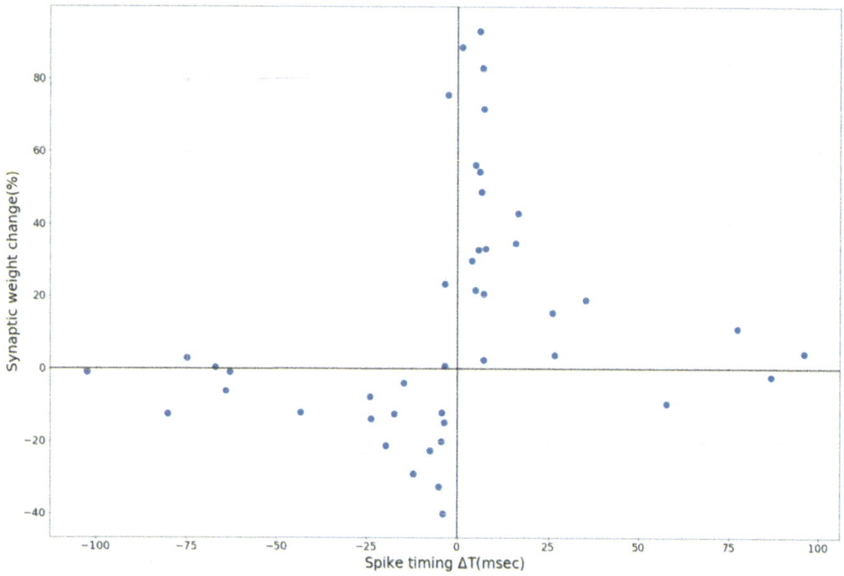

Fig. 2.3 Bi and Poo 1998 data. % weight change after 60 pulses

first, the negative side corresponds to the postsynaptic neuron firing first. Figure 2.4, from their 2001 paper [4], shows the same experimental data but now interpolated. From inspection of these figures, it is easy to be convinced of the importance of spike timing.

2.5 Results of Bi and Poo Can Be Anticipated

The Bi and Poo experiments are useful for understanding synaptic learning. By pulsing at the low frequency 1 Hz, they were able to isolate the synaptic response to an individual pair of trigger pulses, while controlling the time spacing between these pulses, the presynaptic and postsynaptic triggers. Following the extended Hebbian rules and the learning phenomena described in Sect. 2.4 above, it is possible to anticipate the Bi and Poo findings and at the same time to verify the Hebbian rules.

Figure 2.5 shows how this can be done. Assume that the presynaptic neuron is triggered once at time $t = 0$ and it fires. Neurotransmitter is suddenly injected into the synaptic cleft and maximum concentration occurs at the outset. From then on, the neurotransmitter concentration diminishes exponentially over time as a result of diffusion and reabsorption. Figure 2.5a shows a typical plot of neurotransmitter as a function of time. This is an exponential step function. We can designate this as $f_1(t)$. Figure 2.5b shows a plot of membrane potential verses time for the case where the postsynaptic neuron is triggered Δt milliseconds after the presynaptic neuron

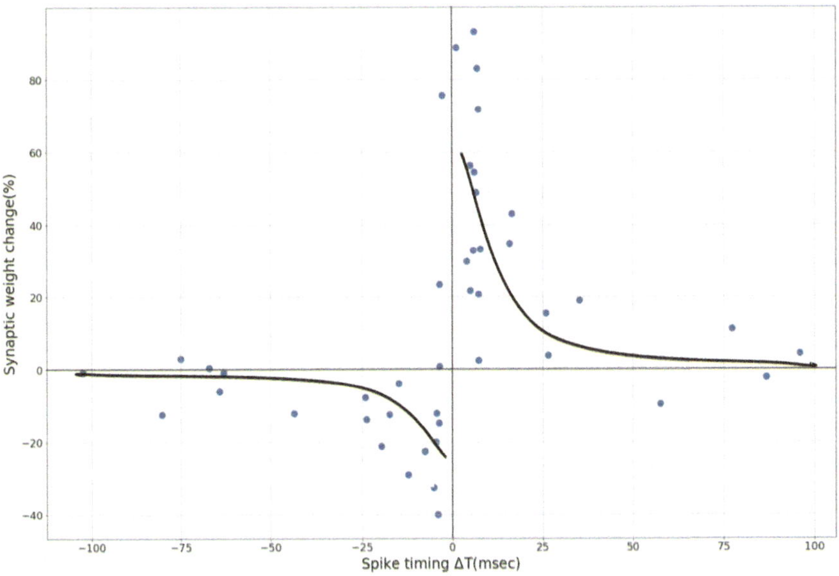

Fig. 2.4 Bi and Poo data interpolated in their 2001 paper. % weight change after 60 pulses

has fired. This is an action potential plot. We can designate this as $f_2(t - \Delta t)$. The postsynaptic neuron was in the resting state at −70 mv when it was triggered at $t = \Delta t$ milliseconds. A horizontal time axis for the action potential plot of Fig. 2.5b was established so that the firing threshold at −55 mV is now designated to be 0 V. This is a 55 mV shift. On the new vertical scale, membrane potentials above 0 correspond to a neuron firing and membrane potentials below 0 correspond to a neuron not firing.

Now assume that the given synapse is excitatory. Assume that the postsynaptic neuron fires after the presynaptic neuron fires. Let Δt be fixed. In accord with Hebbian theory, the rate of increase of the synaptic weight is proportional to the product of the neurotransmitter concentration and the membrane potential. In this case the change in synaptic weight is proportional to the time integral of the product of the concentration curve and the membrane potential curve. The change is equal to the time integral of the rate of change.

If Δt is changed, the action potential curve will accordingly slide along in time and the integral will change. The value of the integral as a function of Δt is a cross-correlation of the two curves. For these curves, the cross-correlation was computed and is plotted in Fig. 2.5c. The cross-correlation function is

$$C(\Delta t) = \int_{-\infty}^{+\infty} f_1(t) f_2(t - \Delta t) dt \tag{2.1}$$

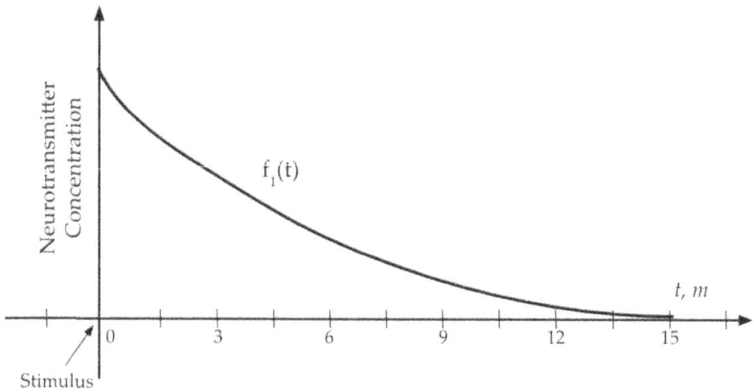

(a) Neurotransmitter concentration in cleft after stimulation of presynaptic neuron.

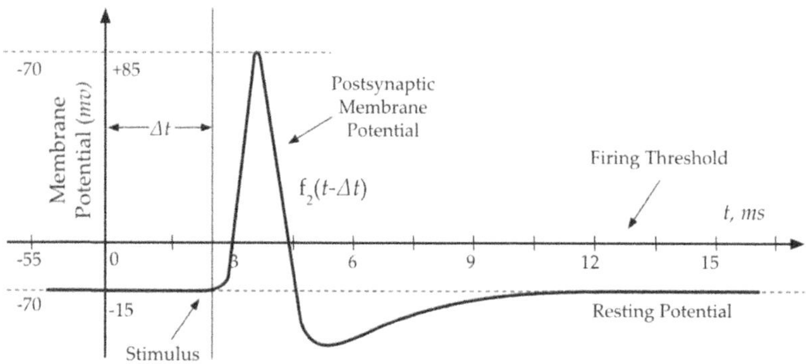

(b) Action potential of postsynaptic neuron after stimulation

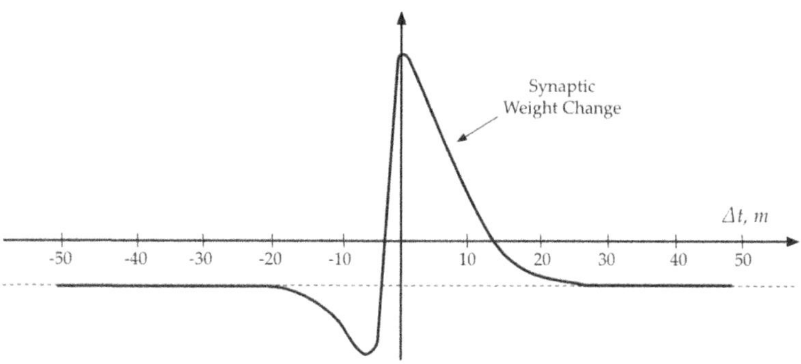

(c) Cross-correlation theoretical anticipation of experimental results of Bi and Poo

Fig. 2.5 Neurotransmitter concentration in cleft after stimulation of presynaptic neuron. Action potential of postsynaptic neuron after stimulation. Cross-correlation theoretical anticipation of experimental results of Bi and Poo

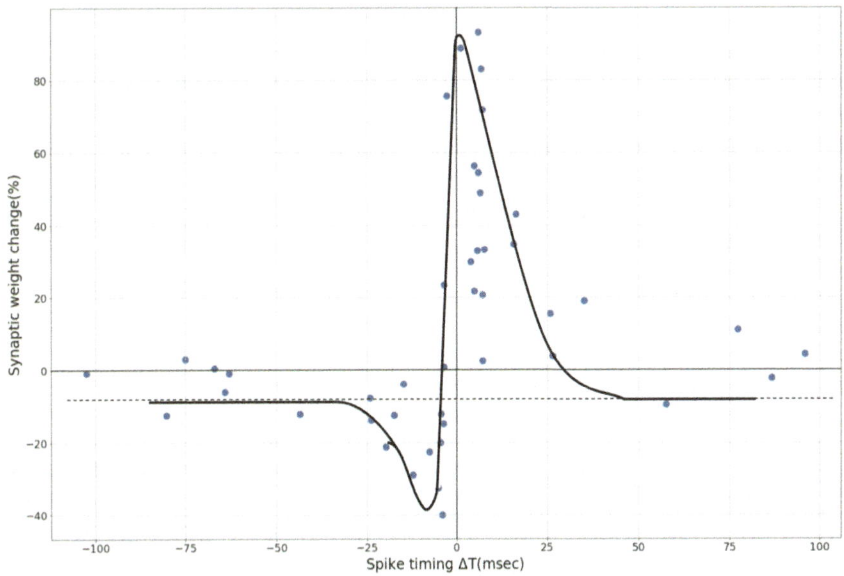

Fig. 2.6 Bi and Poo data with extended Hebbian interpolation, excitatory synapse

The time constant for the exponential step was not known for the synapse of the experimental rat. Also, since everything is proportional, the scale of the plot of Fig. 2.5a was not known. These two parameters were adjusted to the Bi and Poo data.

The resulting cross-correlation plot shows synaptic growth as a function of spike timing, the difference in time between presynaptic and postsynaptic firing, and is a theoretical anticipation of the findings of Bi and Poo [3]. The original Bi and Poo data points are shown in Fig. 2.3. In Fig. 2.4, data points are interpolated as two exponentials. There is no question about their original data points, although they are somewhat sparse, and somewhat noisy as would be expected with natural measurements. Their two-sided exponential interpolation, is arbitrary and questionable.

We propose instead the cross-correlation curve of Fig. 2.5c as an interpolation of the data points. This is shown in Fig. 2.6 with original data points. The Bi and Poo interpolation and the cross-correlation function differ in two main aspects. Bi and Poo show indeterminate behavior in the vicinity of $\Delta t = 0$. The cross-correlation shows what this behavior is and fills in the gap, the unknown. The second difference has to do with asymptotic behavior for large $|\Delta t|$. The Bi and Poo interpolated curves approach zero for large $|\Delta t|$. The cross-correlation function disagrees with this. It shows finite negative asymptotic behavior values for large $|\Delta t|$ which, it turns out, have physical significance. With a large $|\Delta t|$, the presynaptic neuron fires and the postsynaptic neuron fires a long time after or a long time before. In either case, the situation is equivalent to the postsynaptic neuron not firing. The Hebbian rules say that this will cause a weight reduction for an excitatory synapse. The negative asymptotic levels of the cross-correlation function curve reflect the weight reduction.

The morphology of the cross-correlation curve fits the data points. More data points would be needed to make this claim even stronger. More experiments would need to be done to measure weight change when the presynaptic neuron fires and postsynaptic neuron does not fire. By and large, the Bi and Poo data points confirm the extended Hebbian theory for excitatory synapses. Their interpolation of the data points on the other hand is in direct conflict with extended Hebbian learning.

For an inhibitory synapse, there would be a reversal of the y-axis scale of the cross-correlation. The two kinds of synapse are similar in structure and function, only they have opposite effects on firing and wiring.

2.6 The Spike Timing Hypothesis

The Bi and Poo spike timing hypothesis offers one explanation for synaptic plasticity. It is based on a certain interpretation of Hebb's original rule, that if the presynaptic neuron fires first, it contributes to the firing of the postsynaptic neuron and contributes to a positive increase in the synaptic coupling. The Bi and Poo data also show that if the postsynaptic neuron fires first, the later firing of the presynaptic neuron contributes to a decrease in synaptic weight. With enough delay, this is like the presynaptic neuron firing and the postsynaptic neuron not firing, and the weight decreases. All this agrees with the extended Hebbian learning rules.

A closer examination of the Bi and Poo data shows that if the postsynaptic neuron fires first and the presynaptic neuron fires with a small delay, that the synaptic weight increases. This is contradictory to the spike timing hypothesis. When both presynaptic and postsynaptic neurons fire at the same time, the spike timing hypothesis gives no idea of what will happen to the synaptic weight. The Bi and Poo data points show that nothing drastic happens; there is a continuity of synaptic response as Δt goes from small positive to small negative. Their two-sided exponential interpolation avoids this issue.

The extended Hebbian rules are based on a different interpretation of Hebb's original rule. These rules are based on membrane potential and neurotransmitter concentration in the cleft. These are overriding factors, and they could be invoked to support spike timing with an experiment like that of Bi an Poo, but the spike timing per se is not the controlling factor although it may appear to be the key.

The Bi and Poo experiments could be repeated and extended. Firing the presynaptic neuron and not firing the postsynaptic neuron should result in a decrease in synaptic weight. The spike timing hypothesis gives no idea of what would happen in this case.

The presynaptic neuron could be fired periodically at a rate as in a living neural network. The postsynaptic neuron could be fired periodically at the same time at a different rate. With an excitatory synapse, the weight should increase. The opposite should happen with an inhibitory synapse. Spike timing theory would not apply here.

The Bi and Poo experiments could be repeated with many more data points. The objective would be to determine how well the cross-correlation curve would fit the enhanced data. A nice fit would add further credibility to the extended Hebbian rules.

Bi and Poo did excellent work, but they did not go far enough. If they went deeper, they could have made discoveries that would have been troublesome for the spike timing theory. As far as it goes, their work does provide experimental credibility to the extended Hebbian learning rules.

2.7 Addendum

The Hebbian learning rules presented above are not complete. If nature were to practice them as is, there would be instability. The (SUM) of the postsynaptic neuron could in magnitude grow without bound and its output pulse rate would be out of control. Nature does not allow this. When the magnitude of (SUM) exceeds an equilibrium point, the direction of adaptation reverses to drive the (SUM) magnitude back toward the equilibrium point. This is called homeostasis. It stabilizes the firing rate of the postsynaptic neuron.

The extended Hebbian rules have been presented in simplified form, without discussion of homeostasis, in order to keep explanations easy to understand. The simplified rules work well as long as the magnitude of (SUM) is only a fraction of the value of the equilibrium point. Otherwise, for larger magnitudes of (SUM), homeostasis must be a part of the Hebbian rules. This is quite distant from Hebb's original fire together, wire together.

In Chap. 3, the LMS algorithm is described. In Chap. 6, the Hebbian-LMS neuron will be introduced. Homeostasis and equilibrium points will be explained. Hebbian-LMS is conceptually simple and will serve as a model that encompasses the various forms of Hebbian learning. In this sense, it will present a unified theory of synaptic plasticity.

2.8 Summary

Hebbian learning has been extended to apply to both excitatory and inhibitory synapses and to both upregulation and downregulation of synaptic neuroreceptors. These extended rules give direction to synaptic weight changing. Further extension gives rules for rate of change. The extended Hebbian rules are supported by experimental data recorded by Bi and Poo. An in-vitro experiment measured synaptic weight change as a function of the time difference between presynaptic and postsynaptic stimuli. They plotted weight change vs timing difference. Their plot can be anticipated theoretically as the cross-correlation function between an exponential step and an action potential curve. The theoretical basis is the extended Hebbian rules. They provide a more complete and more precise explanation of the Bi and Poo phenomena.

2.9 Questions and Experiments

1. Reproduce the work of Bi and Poo.
2. Reproduce the work of Bi and Poo, but trigger every half second and obtain twice as many points. Check data points with cross-correlation curve. How long can the neurons and synapse remain viable? How many data points could be obtained?
3. Start with the synaptic weight in mid range. Fire the presynaptic neuron without firing the postsynaptic neuron. Demonstrate weight reduction for an excitatory synapse and the opposite for an inhibitory synapse.
4. With an excitatory synapse, trigger the presynaptic rate at 20 Hz and trigger the postsynaptic rate at some other rate and observe synaptic weight increases. Repeat experiment with pre and post firing rates equal. Observe the effect of timing difference or synaptic weight increase. Repeat with an inhibitory synapse and observe opposite effects.
5. Devise your own experiments with synaptic plasticity.

References

1. Sejnowski, T., TJ, S.: Statistical constraints on synaptic plasticity (1977)
2. Sejnowski, T.J.: Storing covariance with nonlinearly interacting neurons. J. Math. Biol. **4**(4), 303–321 (1977)
3. Bi, G.-Q., Poo, M.-M.: Synaptic modifications in cultured hippocampal neurons: dependence on spike timing, synaptic strength, and postsynaptic cell type. J. Neurosci. **18**(24), 10464–10472 (1998)
4. Bi, G.-Q., Poo, M.-M.: Synaptic modification by correlated activity: Hebb's postulate revisited. Ann. Rev. Neurosci. **24**(1), 139–166 (2001)

Chapter 3
The LMS Algorithm

Abstract The adaptive linear combiner is the fundamental building block of all neural networks, and of all adaptive filters in signal processing and control applications. The coefficients or weights of the linear combiner are adjusted automatically by an adaptive algorithm. In most cases, this algorithm is LMS (Least Mean Squares). The linear combiner is trainable. Input vectors or patterns are presented to the combiner, as well as corresponding desired responses. The objective of training is to adjust the parameters to cause the pattern responses to match the desired responses as well as possible in the least square sense. The mean square error is a quadratic function of the parameters. A stochastic gradient is used with the method of steepest descent to seek the optimal solution at the bottom of the quadratic "bowl." This is like a ball rolling down a hill. The resulting LMS algorithm is perhaps the simplest of all adaptive algorithms. It is used widely in engineering applications and appears to be a natural algorithm, part of nature's process for learning and adaptation.

3.1 Introduction

The fields of artificial neural networks and neuroscience have progressed over the years more or less independently. Although the original ideas and motivation for artificial neural networks came from early knowledge of neurobiology, modern neural networks do not work like real ones. Connections between the fields have been proposed. This book follows that tradition. An unsupervised learning algorithm, the Hebbian-LMS algorithm has been devised to have engineering applications but it will very likely be a good model for natural synaptic plasticity. This algorithm fits the Hebbian learning rules. It is an implementation of Hebbian learning with the LMS algorithm. The LMS algorithm is well known in computer science and electrical engineering.

B. Widrow, *Cybernetics 2.0*, Springer Series on Bio- and Neurosystems 14,
https://doi.org/10.1007/978-3-030-98140-2_3

3.2 Contribution of Norbert Wiener

Least squares methods were probably invented by Gauss, or possibly by Leibniz before him. Modern statistical methods of regression are based on this. An example would be fitting a parabolic curve to a set of data points. The parameters of the parabola would be adjusted so that the errors at each data point, would be minimized. A "best fit" would minimize the sum of squares of the errors.

Why minimize mean square error? It makes sense. Positive errors are weighted equivalently to negative errors. Why not minimize the mean of the magnitude of errors, or minimize the mean of the magnitude of the errors cubed or minimize the mean fourth of the error? The answer is that they all make sense but minimizing the mean square error leads to simple elegant analytic solutions to problems.

Norbert Wiener chose the least square criterion and brought it into the world of engineering. He was first to design statistically optimal signal processing filters for signal prediction and noise reduction. Nothing like this had ever been done before. This was revolutionary. In Wiener's day, filters were analog. They were usually designed to be low pass, bandpass or highpass, but nothing statistical.

Following in Wiener's footsteps, Widrow and his students at MIT in 1957 began simulating adaptive filters and adaptive neurons that learned from example. These were all digital, simulated on an IBM mainframe. Learning was based on gradient descent.

In 1959, at Stanford, Widrow and his first Ph.D. student Marcian (Ted) Hoff, Jr. invented or discovered the LMS algorithm. If they invented it, they were doing engineering and applied mathematics. If they discovered it, then nature invented it eons ago, and they were doing physics.

3.3 The Linear Combiner and the LMS Algorithm

In the discussion to follow, it should be noted that signals are not represented by the frequency of pulse trains, but by variable (dc) levels. Learning is not done continuously, but cyclically at discrete time intervals. Otherwise, the analogy between real and artificial neurons, synapses, and networks will be apparent.

In nature's neural networks, the input signals to each neuron are summed electrochemically. In artificial networks, the neuronal input signals are summed numerically. We begin with the LMS (least-mean squares) algorithm for adapting the weights of a linear combiner. This is a fundamental element of Hebbian-LMS.

A basic building block of all neural networks, living or artificial, and of adaptive signal processing and control systems, is the trainable linear combiner diagrammed in Fig. 3.1. During training, input vectors or patterns are presented at the input terminals, and each component of each input vector is weighted and a weighted sum is computed to provide an output. The kth input pattern or input vector is designated as X_k. The

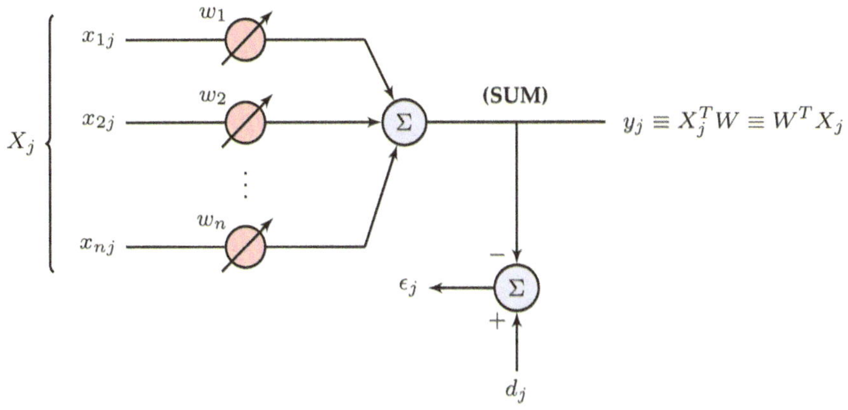

Fig. 3.1 A trainable linear combiner

output for that input is the (SUM), y_k. The weights are the components of the weight vector W. Thus,

$$y_k = X_k^T W = W^T X_k \tag{3.1}$$

During training, each input vector, X_k is supplied with a desired response d_k. The purpose of training is to cause the linear combiner to produce outputs for the entire set of training patterns that closest match the respective desired responses, in the least-squares sense.

For each input vector, the error ϵ_k is the difference between the desired response d_k and the actual output y_k.

$$\epsilon_k = d_k - y_k = d_k - X_k^T W. \tag{3.2}$$

The square of the error is

$$\epsilon_k^2 = d_k^2 - 2d_k X_k^T W + W^T X_k X_k^T W. \tag{3.3}$$

The mean square error, the MSE, is the mean value or expected value of (3.3)

$$\text{MSE} = E[\epsilon_k^2] = E[d_k^2] - 2E[d_k X_k^T]W + W^T E[X_k X_k^T]W. \tag{3.4}$$

The cross-correlation between the desired response and the input vector can be represented by the vector P

$$E[d_k X_k^T] \triangleq P \tag{3.5}$$

The input autocorrelation matrix can be represented by the matrix R

$$E[X_k X_k^T] \triangleq R \tag{3.6}$$

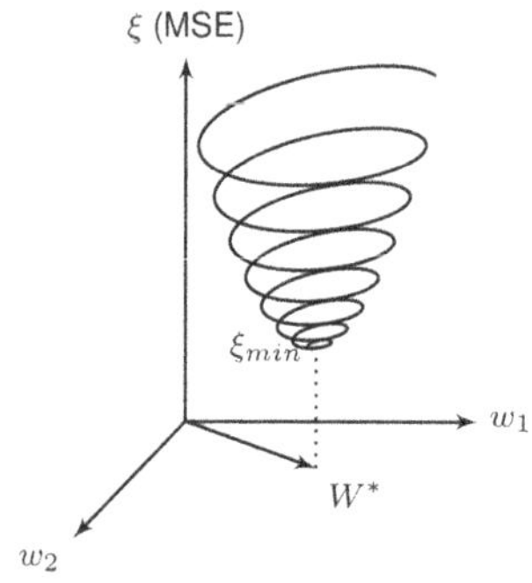

Fig. 3.2 A mean square error surface. A paraboloid

If the input vectors have zero mean, the matrix R is also called the input covariance matrix.

Substituting (3.5) and (3.6) in (3.4), we have

$$\text{MSE} = E[\epsilon_k^2] = E[d_k^2] - 2PW + W^T RW. \tag{3.7}$$

Thus, the mean square error is a quadratic function of the weights. A sketch of a typical MSE function is shown in Fig. 3.2. The MSE surface is a paraboloid. With many weights, this surface is a hyperparaboloid.

The MSE surface is bowl-shaped. The minimum MSE is the value of the MSE function at the bottom of the bowl. The linear combiner has a set of weights that, if chosen, will minimize MSE. The optimal set of weights that minimize MSE are components of a vector labeled W^*. Refer to Fig. 3.2.

During adaptation, the weights are varied to minimize the mean square error. One way of doing this makes use of the method of steepest descent. A series of steps is taken starting from an initial guess for the initial weight vector and following the negative gradient, step by step, approaching the bottom of the quadratic bowl. Accordingly

$$W_{k+1} = W_k + \mu(-\nabla_k) \tag{3.8}$$

The next weight vector equals the present weight vector plus a constant μ multiplied by the negative gradient. The parameter μ determines step size, stability, and speed of convergence.

The gradient at the kth iteration is defined as

$$\nabla_k \triangleq \frac{\partial E[\epsilon_k^2]}{\partial w_k} = \begin{bmatrix} \frac{\partial E[\epsilon_k^2]}{\partial w_{1k}} \\ \vdots \\ \frac{\partial E[\epsilon_k^2]}{\partial w_{nk}} \end{bmatrix} \tag{3.9}$$

The components of the gradient vector are the partial derivative of MSE with respect to the individual weights. The number of weights is n.

Differentiating equation (3.7) yields the gradient,

$$\nabla_k = -2P + 2RW_k \tag{3.10}$$

The optimal solution W^* is obtained by setting the gradient ∇_k to zero. Accordingly,

$$-2P + 2RW^* = 0. \tag{3.11}$$

The optimal weight vector is

$$W^* = R^{-1}P \tag{3.12}$$

The optimal weight vector W^* is called the Wiener solution.

The Wiener solution (3.12) often cannot be used in practice because one only has input data and would not have the correlation functions, P and R. In practice, the Wiener solution is an idealization, its performance a benchmark.

Minimizing the MSE is generally done step by step with the method of steepest descent,

$$W_{k+1} = W_k - \mu\nabla_k. \tag{3.13}$$

Following the negative gradient is like the action of a ball rolling down a hill.

An analytic expression for the gradient would not be available, only input data would be available. The idea of Widrow and Hoff was to take a single sample of error, associated with a single input vector, square it and call that the MSE, which, of course it is not, being only of a single error sample. Differentiating, it yields a crude "noisy" gradient estimate,

$$\hat{\nabla}_k = \frac{\partial\epsilon_k^2}{\partial w_k} = 2\epsilon_k\frac{\partial\epsilon_k}{\partial w_k}. \tag{3.14}$$

The expression for the error ϵ_k is given by Eq. (3.2). Differentiating, one obtains the crude gradient estimate as

$$\hat{\nabla}_k = 2\epsilon_k\frac{\partial\epsilon_k}{\partial w_k} = 2\epsilon_k(-X_k) \tag{3.15}$$

The expected value of this gradient estimate can be proven to be the true gradient. The crude estimate is unbiased. Using the gradient estimate with steepest descent,

$$W_{k+1} = W_k - \mu\hat{\nabla}_k \tag{3.16}$$

From this we have the LMS algorithm:

$$W_{k+1} = W_k + 2\mu\epsilon_k X_k \tag{3.17}$$

$$\epsilon_k = d_k - x_k^T W_k \tag{3.18}$$

The LMS algorithm can be described in words: "*With the presentation of each input pattern vector and its associated desired response, the weight vector is changed slightly by adding the pattern vector to the weight vector, making the sum more positive, or subtracting the pattern vector from the weight vector, making the sum more negative, changing the sum in proportion to the error in a direction to make the error smaller*" The parameter μ is generally chosen to be small. Many small steps on average follow the gradient toward the bottom of the bowl. In the limit, as the number of iterations increases, the expected value of the weight vector exponentially approaches the Wiener solution (3.12). The properties of the LMS algorithm are discussed in detail in the text of Widrow and Stearns [1], and in many other sources [2]. The original 1959 Widrow-Hoff paper is reference [3].

The LMS algorithm is used in adaptive filters and artificial neural networks. This algorithm is used in one form or another for channel equalization and echo cancelling in telecommunication systems. It is used in every MODEM in the world. It is one of the enabling technologies of the internet. It is the foundation of modern day neural networks based on the backpropagation learning algorithm, a gradient descent algorithm. The crude gradient estimate was a "crazy idea," but it turned out to be a pretty good crazy idea.

3.4 Steepest Descent, A Feedback Algorithm

The method of steepest descent is a feedback algorithm. A block diagram of the method of steepest descent applied to minimization of mean square error when MSE is a quadratic function of the weights is shown in Fig. 3.3.

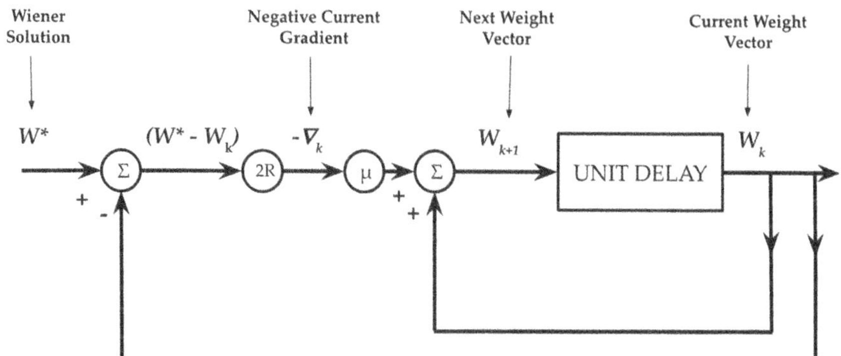

Fig. 3.3 Feedback diagram for a quadratic MSE surface, controlling the weight vector

To verify that this feedback system does indeed represent the method of steepest descent, begin the verification at the point marked, W_{k+1}. Delaying W_{k+1} by one cycle time gives W_k. From the diagram, W_{k+1} is

$$W_{k+1} = W_k + \mu(-\nabla_k) \tag{3.19}$$

This agrees with Eq. (3.13) and does represent steepest descent. Referring again to the diagram, the gradient is seen to be

$$\nabla_k = 2R(W_k - W^*) \tag{3.20}$$

Substituting the expression for the Wiener solution,

$$\nabla_k = 2R(W_k - R^{-1}P) = -2P + 2RW_k. \tag{3.21}$$

This confirms Eq. (3.10).

Referring again to the diagram, when the system converges, $W_{k+1} = W_k$ and accordingly the gradient $\nabla_k = 0$. Therefore, $W_k = W^*$. The weight vector W_k has converged on the Wiener solution W^*.

When employing the method of steepest descent, one does not implement the feedback system per se. One simply follows the gradient, starting from an initial condition, step by step toward the bottom of the quadratic bowl, like a ball rolling down a hill. The initial value of W_k is set to be a best initial guess of W^* which the iterative process of steepest descent will ultimately find.

The feedback diagram is a model of what happens and gives insight to the process. Its "error" signal is the difference $(W^* - W_k)$. Negative feedback is used to minimize this error. The feedback system is linear and its transient response is a sum of exponentials. The number of exponential components is equal to the number of distinct eigenvalues of the input autocorrelation matrix R. If all of the eigenvalues are equal, the quadratic mean square error surface is circular. An expression for the single time constant under this condition can be shown to be

$$\tau = \frac{n}{4\mu \text{trace} R} \tag{3.22}$$

The number of weights is n. Time is measured in number of adapt cycles.

The LMS algorithm does not perfectly perform the method of steepest descent since the LMS gradient is not the true gradient but an estimate. The feedback diagram can be used to represent this by adding noise to the gradient. The noise propagates in the feedback system and the result is noise in the weight vector W_k. At convergence, the weight vector does not go to the Wiener solution but varies randomly about it. This creates excess mean square error. The average excess mean square error divided by the minimum mean square error of the Wiener solution is a dimensionless measure of how far on average the LMS solution is from the ideal Wiener solution. This ratio is called the misadjustment. A formula for misadjustment for the LMS algorithm can be shown to be

$$M = \mu \mathrm{trace} R \tag{3.23}$$

All adaptive systems that learn from input data must have misadjustment. This is because the number of learning samples is finite, not infinite, that would be needed for perfect steepest descent.

It is important to note that increasing μ makes convergence faster (inversely related to learning time constant) but increases the misadjustment. In engineering practice, there is a tradeoff between speed of convergence and misadjustment. A good starting point would be to choose μ so that misadjustment would be something like 10%. That would result in a learning time, four time contants of learning, to be a number of training cycles equal to ten times the number of weights for a linear combiner or for a single artificial neuron. Combining equations (3.22) and (3.23), a relation between learning time and misadjustment is obtained:

$$4\tau = \frac{n}{M} \tag{3.24}$$

The learning time, 4τ, equals $\frac{n}{M}$.

For a given level of misadjustment, the learning time increases in proportion to the number of weights. A complete discussion of this subject is given in Widrow and Stearns "Adaptive signal processing" [1] and Widrow and Wallach "Adaptive inverse control" [4].

The simplicity of the LMS algorithm is unrivaled. On the face of it, who would expect such a simple algorithm to be using stochastic gradient descent to minimize mean square error, a quadratic function of the weights? Who would expect such a simple algorithm to play a major role in learning in the animal brain?

The LMS algorithm has been used worldwide for years. Its performance is "bullet proof." This reflects favorably on the solid stable behavior of living neural systems.

3.5 Summary

The LMS(Least Mean Squares) algorithm was intended for the adaptation of the coefficients of a linear combiner, which is a key element of a single neuron. Training patterns(vectors) and their associated desired responses are presented and the difference between the output, the sum (SUM) and the desired response for a given training pattern is an error. The LMS algorithm, based on the method of steepest descent, seeks an optimal solution, an optimal set of coefficients or weights that minimize the mean square error, averaged over the set of training patterns. The optimal solution is called the Wiener solution. The basic LMS algorithm is "supervised," in the sense that a desired response is specified for each input training pattern. An "unsupervised" version of LMS has input training patterns, but desired responses are not given. The unsupervised version is the basis for the Hebbian-LMS algorithm.

3.6 Questions and Experiments

1. Generate ten 4×4 patterns, 16 pixels each. Assign desired responses of either $+1$ or -1 to each of these patterns. Code an adaptive linear combiner with 16 weights. Apply the patterns to the inputs of the linear combiner and train iteratively with the LMS algorithm. What you have is a trainable pattern classifier. Experiment with the value that you choose for μ. As you train, plot the square of the error as a function of the number of training cycles. This should look like a noisy exponential. This is the learning curve. What effect does μ have on the learning curve, on stability, and rate of convergence?

References

1. Widrow, B., Stearns, S.D.: Adaptive Signal Processing. Prentice Hall, Englewood Cliffs, NJ (1985)
2. Haykin, S.S.: Adaptive Filter Theory. Pearson Education India (2008)
3. Widrow, B., Hoff, M.E.: Adaptive switching circuits, technical report, Stanford University CA Stanford Electronics Laboratory (1960)
4. Widrow, B.: Adaptive inverse control. In: Adaptive Systems in Control and Signal Processing 1986, pp. 1–5. Elsevier (1987)

Chapter 4
The Hebbian-LMS Algorithm

Abstract In order to effect adaptation and learning, a neuron must have inputs and a desired response. The objective is the difference between the desired response and the sum (SUM). With unsupervised learning, no external desired response is given. Bootstrap learning manufactures a desired response from the sum (SUM), from its own output. Bootstrap learning is the basis of the Hebbian-LMS algorithm. It is shown how the Hebbian-LMS algorithm implements the extended Hebbian learning rules and in addition, exhibits homeostasis.

4.1 Introduction

Although the fields of artificial neural networks and neurobiology have progressed independently, connections between the fields have been attempted. This chapter makes a small step toward closing the gap. There is a connection.

An unsupervised learning algorithm, the Hebbian-LMS algorithm [1], has been devised to have engineering applications, but it may likely be a good model for natural synaptic plasticity. This algorithm fits the extended Hebbian learning rules. It is an implementation of Hebbian learning by means of the LMS algorithm. Verification of this will be demonstrated.

In this chapter, we will introduce and explain the Hebbian-LMS algorithm, how it works, how it implements Hebbian learning, how it is used in practical applications. From this, one can gain additional insight into the working and purposes of living neural networks. We begin with the LMS algorithm for adapting the weights of a linear combiner.

4.2 The LMS Algorithm

A fundamental building block of all neural networks, living or artificial, and of adaptive signal processing and control systems, is the trainable linear combiner of Fig. 4.1. During training, input vectors or patterns are presented at the input terminals,

B. Widrow, *Cybernetics 2.0*, Springer Series on Bio- and Neurosystems 14,
https://doi.org/10.1007/978-3-030-98140-2_4

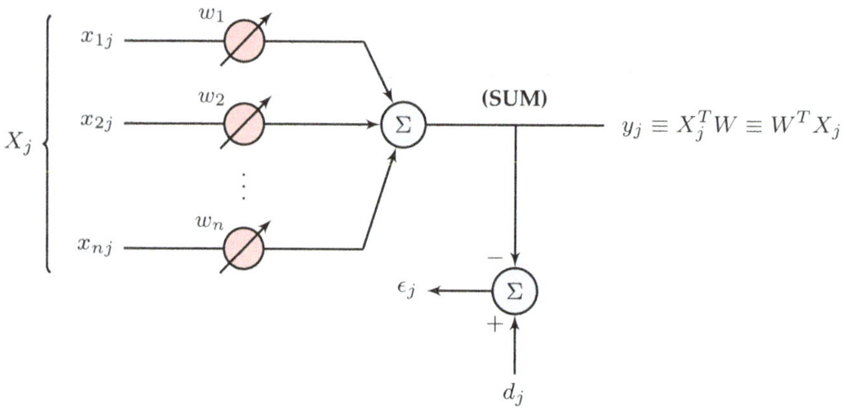

Fig. 4.1 A trainable linear combiner

and each component of each input vector is weighted and a weighted sum is computed to provide an output. The kth input pattern or input vector is designated as X_k. The output for that input is the (SUM), y_k. The weights are the components of the weight vector W. Thus,

$$y_k = X_k^T W = W^T X_k \tag{4.1}$$

During training, each input vector, X_k is supplied with a desired response d_k. The purpose of training is to cause the linear combiner to produce outputs for the entire set of training patterns that match closest to the respective desired responses in the least-squares sense.

For each input vector, the error ϵ_k is the difference between the desired response d_k and the actual output y_k.

$$\epsilon_k = d_k - y_k = d_k - X_k^T W. \tag{4.2}$$

The mean square error is a quadratic function of the weights, a hyperparaboloid. During adaptation, the weights are varied to minimize mean square error averaged over the set of training patterns. The LMS algorithm does this by the method of steepest descent. A series of steps are taken following the negative gradient, step by step, approaching the bottom of the quadratic "bowl". The LMS algorithm is

$$W_{k+1} = W_k + 2\mu\epsilon_k X_k \tag{4.3}$$

$$\epsilon_k = d_k - x_k^T W_k \tag{4.4}$$

The LMS algorithm is used in adaptive filters and artificial neural networks. This algorithm is used for channel equalization and echo cancelling in telecommunication systems. It is used in every MODEM in the world. It is one of the enabling technologies of the internet. You use it every time you use your cell phone, every time you log onto the internet, every time you use Wi-Fi.

4.3 ADALINE

ADALINE is an acronym for adaptive linear neuron. ADALINE is the simplest of all learning paradigms. It is a linear combiner whose output, y_k, the $(\mathrm{SUM})_k$, is applied to a nonlinear activation function to provide a neuron output. A diagram of ADALINE is shown in Fig. 4.2. ADALINE was adaptive but not really linear. It was more than a neuron since it also included the weights or synapses. Nevertheless, ADALINE was the name it was given in 1959.

ADALINE is a trainable classifier. The input patterns, the vectors X_k, $k = 1, 2, \ldots$, were weighted by the weight vector $W_k = \left[\, w_{1k}, \, w_{2k}, \, \ldots, \, w_{nk} \,\right]^T$, and their inner product was the sum $y_k = X_k^T W_k$. Each input pattern X_k was to be classified as a $+1$ or a -1 in accord with its assigned class, the "desired response." ADALINE was trained to accomplish this by adjusting the weights to minimize the mean square error. The error was the difference between the desired response d_k and the sum y_k. The error is $e_k = d_k - y_k$. ADALINE's output q_k was the sign of the sum y_k, i.e. $q_k = \mathrm{SGN}(y_k)$, where the function $\mathrm{SGN}(\cdot)$ is the signum, take the sign of. The sum y_k will henceforth be referred to as $(\mathrm{SUM})_k$. The weights of ADALINE were trained with the LMS algorithm.

A photograph of a physical ADALINE made by Widrow and Hoff in 1960 is shown in Fig. 4.3. The input patterns of this ADALINE were binary, 4×4 arrays of pixels, each pixel having a value of $+1$ or -1, set by the 4×4 array of toggle switches. Each toggle switch was connected to a weight, implemented by a potentiometer. The knobs of the potentiometers, seen in the photo, were manually rotated during the training process in accordance with the LMS algorithm. The sum was displayed by the meter. Once trained, output decisions were $+1$ if the meter reading was positive

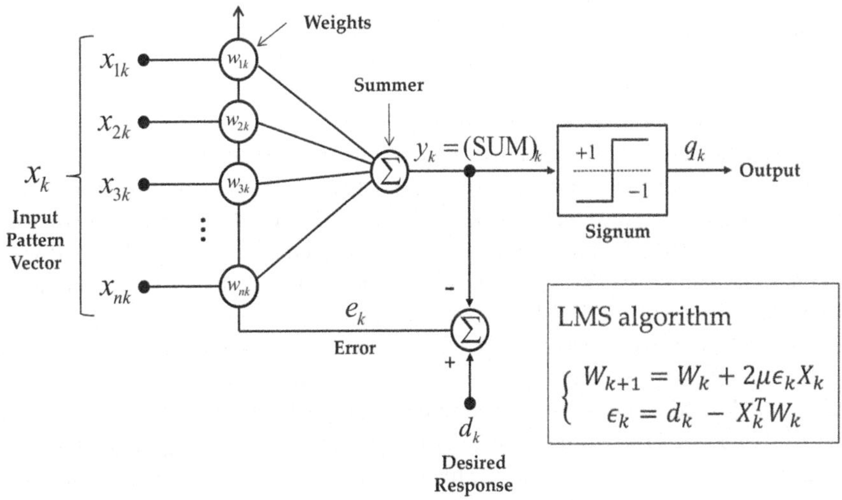

Fig. 4.2 ADALINE (Adaptive linear neuron)

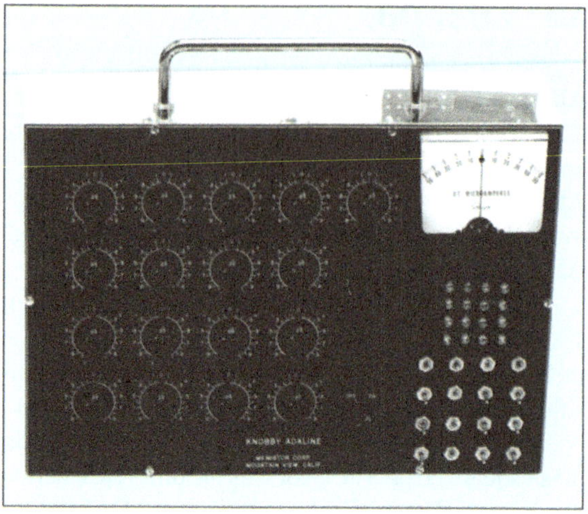

Fig. 4.3 Knobby ADALINE. Reproduced from B. Widrow, Y. Kim, and D. Park, The Hebbian-LMS Learning Algorithm, IEEE Computational Intelligence Magazine, vol. 10, no. 4, pp. 37–53, Nov. 2015 with permission from IEEE @ IEEE 2021

and -1 if the meter reading was negative. The earliest learning experiments were done with this ADALINE, training it as a pattern classifier. This was supervised learning, as the desired response for each input training pattern was given. A video showing Prof. Widrow training ADALINE can be seen online [2].

4.4 Bootstrap Learning

In order to train ADALINE, it was necessary to have a desired response for each input training pattern. The desired response indicated the class of the pattern. But what if one had only input patterns and did not know their desired responses, their classes? Could learning still take place? If this were possible, this would be unsupervised learning.

In 1960, unsupervised learning experiments were made with the ADALINE of Fig. 4.3 as follows. Initial conditions for the weights were randomly set and input patterns were presented without desired responses. If the response to a given input pattern was already positive (the meter reading to the right of zero), the desired response was taken to be exactly $+1$. A response of $+1$ was indicated by a meter reading half way on the right-hand side of the scale. If the response was less than $+1$, adaptation by LMS was performed to bring the response up toward $+1$. If the response was greater than $+1$, adaptation was performed by LMS to bring the response down toward $+1$.

If the response to another input pattern was negative (meter reading to the left of zero), the desired response was taken to be exactly -1 (meter reading half way on the left-hand side of the scale). If the negative response was more positive than -1, adaptation was performed to bring the response down toward -1. If the response was more negative than -1, adaptation was performed to bring the response up toward -1. With adaptation taking place over many input patterns, some patterns that initially responded as positive could ultimately reverse and give negative responses, and vice versa. However, patterns that were initially responding as positive were more likely to remain positive, and vice versa. When the process converges and the responses stabilize, some responses would cluster about +1 and the rest would cluster about −1. The objective was to achieve unsupervised learning with the analog responses at the output of the summer (SUM) clustered at +1 or −1. Perfect clustering could be achieved if the training patterns were linearly independent vectors whose number were less than or equal to the number of weights. Otherwise, clustering to +1 or −1 would be done as well as possible in the least squares sense. The result was that similar patterns were similarly classified, and this simple unsupervised learning algorithm was an automatic clustering algorithm. It was called "bootstrap learning" because ADALINE's quantized output was used as its desired response. This idea is embodied in the block diagram in Fig. 4.4.

Research done on bootstrap learning was reported in the paper "Bootstrap Learning in Threshold Logic Systems," presented by Bernard Widrow at an International Federation of Automatic Control (IFAC) conference in 1966 [3]. This work led to the 1967 Ph.D. thesis of William C. Miller, at the time a student of Professor Widrow, entitled "A Modified Mean Square Error Criterion for use in Unsupervised Learning" [4]. These papers described and analyzed bootstrap learning.

Figure 4.5 illustrates the formation of the error signal of bootstrap learning. The error is the difference between the quantized output q_k and the sum $(\text{SUM})_k$. The quantized output is the desired response. In Fig. 4.5a, the horizontal axis represents the (SUM), the quantized output, either +1 or −1, is a function of (SUM), represented by the step function. The straight line with slope of 1 represents (SUM).

Algebraically, the error is

$$e_k = d_k - (\text{SUM})_k \tag{4.5}$$
$$= q_k - (\text{SUM})_k \tag{4.6}$$

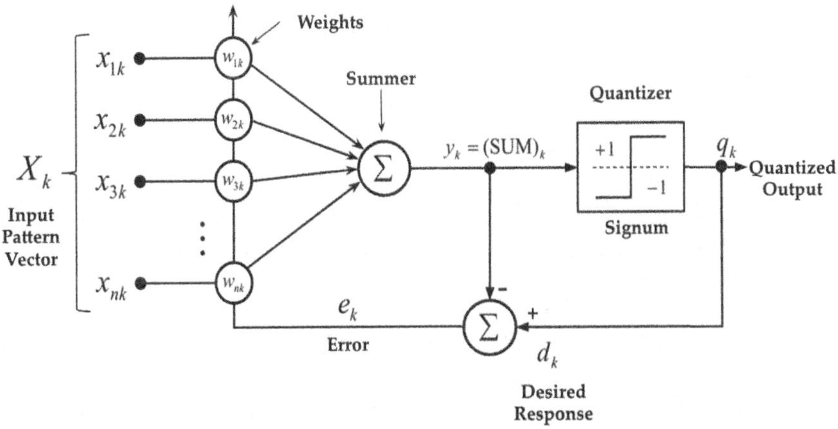

Fig. 4.4 ADALINE with bootstrap learning. Reproduced from B. Widrow, Y. Kim, and D. Park, The Hebbian-LMS Learning Algorithm, IEEE Computational Intelligence Magazine, vol. 10, no. 4, pp. 37–53, Nov. 2015 with permission from IEEE @ IEEE 2021

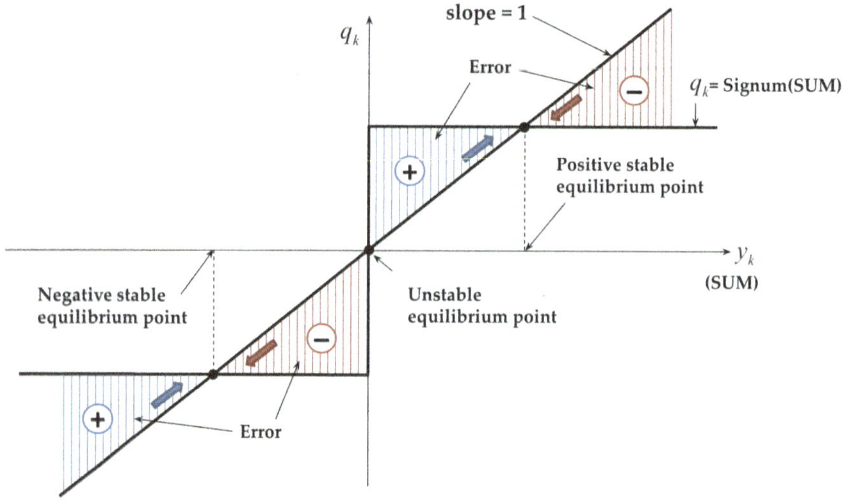

(a) Bootstrap learning with ADALINE.

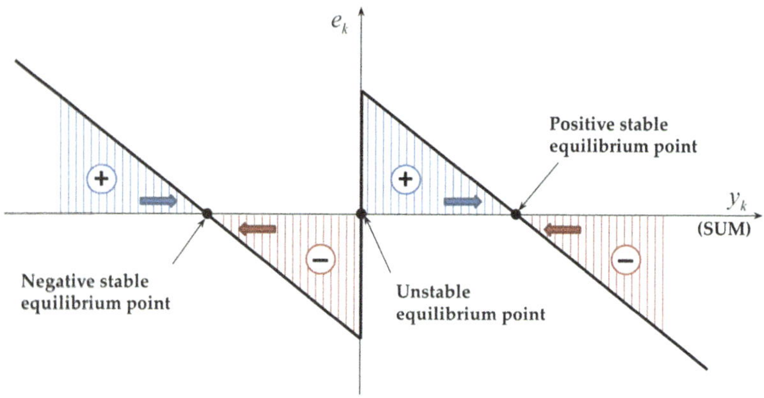

(b) Error function for Hebbian-LMS.

Fig. 4.5 Formation of the error signal for bootstrap LMS. Reproduced from B. Widrow, Y. Kim, and D. Park, The Hebbian-LMS Learning Algorithm, IEEE Computational Intelligence Magazine, vol. 10, no. 4, pp. 37–53, Nov. 2015 with permission from IEEE @ IEEE 2021

$$= \mathrm{SGN}((\mathrm{SUM})_k) - (\mathrm{SUM})_k \tag{4.7}$$

The polarities of the error are indicated in Fig. 4.5. Bootstrap learning comprised of the LMS algorithm of Eq. (4.3) and the error of Eq. (4.7) is unsupervised.

When the error is zero, no adaptation takes place. In Fig. 4.5a, one can see that there are three different values of (SUM) where the error is zero. These are the three equilibrium points. The point at the origin is an unstable equilibrium point. The other two equilibrium points are stable. Some of the input patterns will produce (SUM)s that gravitate toward the positive stable equilibrium point, while the other input patterns produce (SUM)s that gravitate toward the negative stable equilibrium point. The arrows in Fig. 4.5a indicate the directions of change to the (SUM) that would occur as a result of adaptation. Perfect values of (SUM), either $+1$ or -1, will result upon convergence if the training patterns are linearly independent. All input patterns will become classified as either positive or negative at the quantized output when the adaptation process converges.

ADALINE with bootstrap learning is a simple form of the Hebbian-LMS algorithm.

4.5 Bootstrap Learning with a Sigmoidal Neuron, the Hebbian-LMS Algorithm

The sharp quantizer of ADALINE in Fig. 4.2 could be replaced by a sigmoidal function in order to be more "biological." The sharp edges of the quantizer would be rounded. Figure 4.6 is a diagram of a sigmoidal neuron whose weights are trained with bootstrap learning. The learning process of Fig. 4.6 is characterized by the following error signal:

$$\text{error} = e_k = \text{SGM}((\text{SUM})_k) - (\gamma)(\text{SUM})_k. \tag{4.8}$$

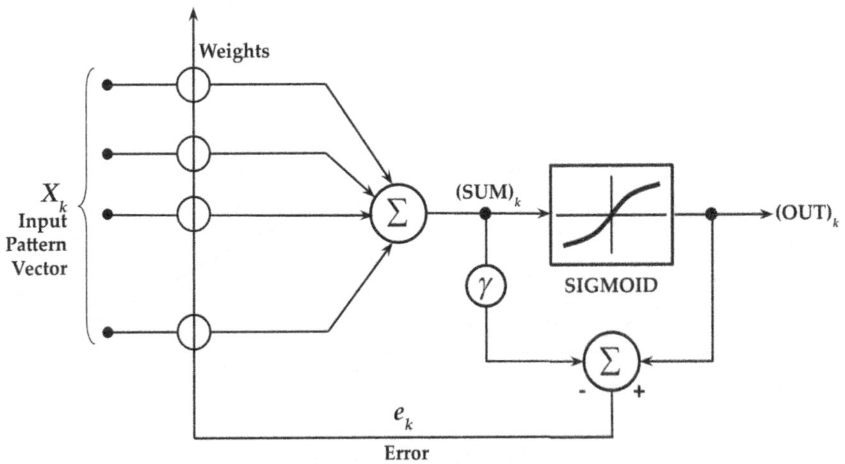

Fig. 4.6 A sigmoidal neuron trained with bootstrap learning. Reproduced from B. Widrow, Y. Kim, and D. Park, The Hebbian-LMS Learning Algorithm, IEEE Computational Intelligence Magazine, vol. 10, no. 4, pp. 37–53, Nov. 2015 with permission from IEEE @ IEEE 2021

The sigmoidal function is represented by $SGM(\cdot)$. Input pattern vectors are weighted, summed, and then applied to the sigmoidal function to provide the output signal, $(\mathrm{OUT})_k$. The weights are initially randomized, then adaptation is performed using the LMS algorithm (4.3), with an error signal given by (4.8). An example of a sigmoidal function is hyperbolic tangent

$$\mathrm{SGM}(x) = tanh(x)$$

Insight into the behavior of the form of bootstrap learning with a sigmoidal quantizer can be gained by inspection of Fig. 4.7. The error is the difference between the

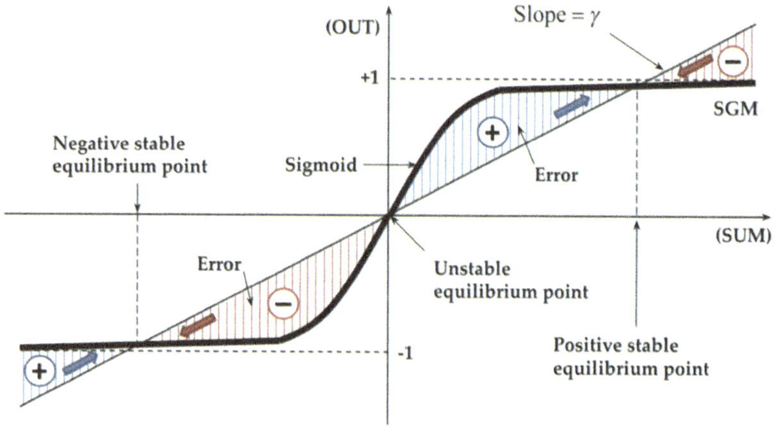

(a) Bootstrap learning with sigmoidal neuron.

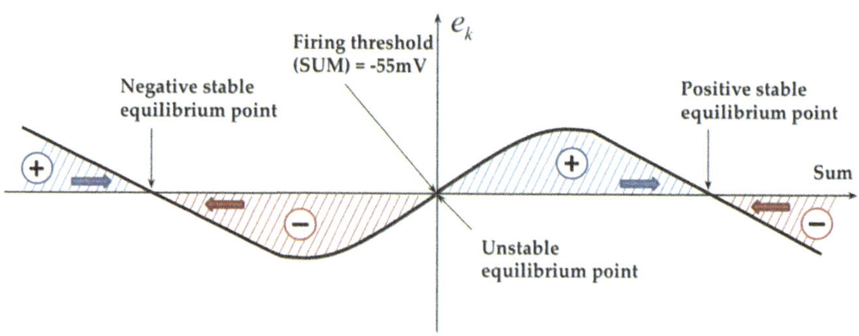

(b) The error function.

Fig. 4.7 Formation of the error signal with a sigmoidal quantizer. Reproduced from B. Widrow, Y. Kim, and D. Park, The Hebbian-LMS Learning Algorithm, IEEE Computational Intelligence Magazine, vol. 10, no. 4, pp. 37–53, Nov. 2015 with permission from IEEE @ IEEE 2021

sigmoidal output and the (SUM) multiplied by the constant γ, in accordance with Eq. (4.8). γ is a positive coefficient chosen by the algorithm designer. As illustrated in the figure, the slope of the sigmoid at the origin has a value of 1, and the straight line has a slope of γ. These values are not critical, as long as the slope of the straight line is less than the initial slope of the sigmoid. The polarity of the error signal is indicated as + or − in Fig. 4.7a and b. The arrows in the figures indicate directions of adaptivity, the directions the algorithm would push (SUM) during adaptation. There are two stable equilibrium points, a positive one and a negative one, where the error is zero,

$$\text{SGM}((\text{SUM})) = \gamma \cdot (\text{SUM}), \tag{4.9}$$

The directions of the arrows confirm this. An unstable equilibrium point exists where (SUM) = 0. This equilibrium point is of no physical significance.

When (SUM) is positive and (γ)(SUM) is less than SGM((SUM)), the error will be positive. The LMS algorithm will adapt the weights in order to increase (SUM) toward the stable positive equilibrium point. If (SUM) is positive and (γ)(SUM) is greater than SGM((SUM)), the error will reverse and will be negative. The LMS algorithm will adapt the weights in order to decrease (SUM) toward the stable positive equilibrium point. The opposite of all this will take place when (SUM) is negative.

Training begins with random initial weights. When the training patterns are linearly independent and their number is less than or equal to the number of weights, all input patterns will have (SUM) outputs exactly at either the positive or negative equilibrium point, upon convergence of the LMS algorithm. The "LMS capacity" or "capacity" of the single neuron can be defined as being equal to the maximum number of patterns that can be trained in, with zero error. This is equal to the number of weights.

When the number of training patterns is greater than capacity, the LMS algorithm will cause the pattern responses to cluster, some near the positive stable equilibrium point and some near the negative stable equilibrium point. The error corresponding to each input pattern will generally be small but not zero, and the mean square of the errors averaged over the training patterns will be minimized by LMS. The LMS algorithm maintains stable control and prevents saturation of the sigmoid and of the weights. The training patterns divide themselves into two classes without supervision. Clustering of the values of (SUM) at the positive and negative equilibrium points as a result of LMS training will prevent the values of (SUM) from increasing without bound. This algorithm is another form of Hebbian-LMS. Error reduction is accomplished by means of stable negative feedback.

The crossover of the error signal at the positive equilibrium point gives homeostasis to the Hebbian-LMS algorithm which will be related to homeostasis in living systems.

4.6 Bootstrap Learning with a More "Biologically Correct" Sigmoidal Neuron

The inputs to the weights of the sigmoidal neuron in Fig. 4.6 could be positive or negative and the weights themselves could be positive or negative. As a biological model, this would not be satisfactory. In the biological world, an input signal coming from a presynaptic neuron must have positive values (presynaptic neuron firing at a given rate) or have a value of zero (presynaptic neuron not firing). Some synapses are excitatory, some inhibitory. They have different neurotransmitter chemistries. The inhibitory inputs to the postsynaptic neuron are subtracted from the excitatory inputs to form (SUM) in the cell body of the postsynaptic neuron. The adding of the excitatory inputs and the subtraction of the inhibitory inputs is indicated in Fig. 4.8. The (SUM) is proportional to the membrane potential. A bias of −70 mV is added to (SUM) to insure that with zero inputs to the weights, the resting membrane potential will be −70 mV.

Biological weights or synapses behave like variable attenuators and can only have positive weight values. The output of the postsynaptic neuron can only be zero (neuron not firing) or positive (neuron firing).

The postsynaptic neuron and its synapses diagrammed in Fig. 4.6 have the indicated properties and are capable of learning as indicated in Fig. 4.7. The LMS algorithm of Eq. (4.4) will operate as usual with positive excitatory inputs or negative

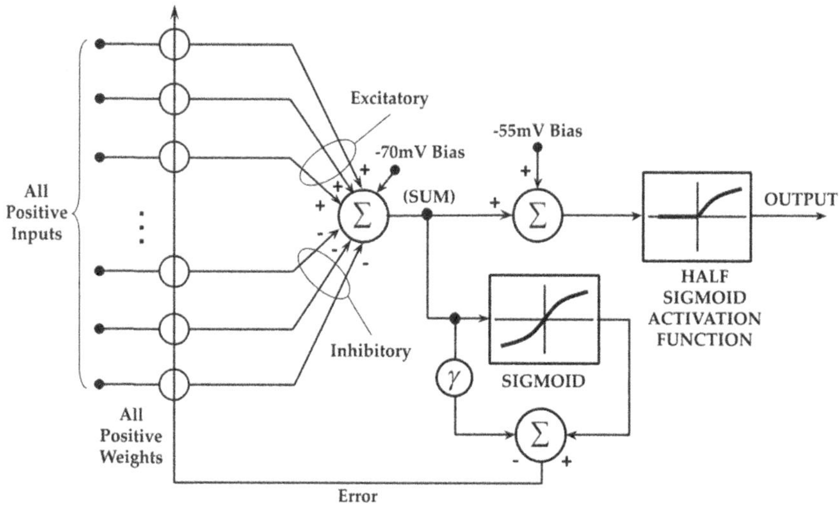

Fig. 4.8 A postsynaptic neuron with excitatory and inhibitory inputs and all positive weights trained with Hebbian-LMS learning. All outputs are positive after rectification. The (SUM) could be positive or negative. A 55 mV bias is included at the input to the activation function Reproduced from B. Widrow, Y. Kim, and D. Park, The Hebbian-LMS Learning Algorithm, IEEE Computational Intelligence Magazine, vol. 10, no. 4, pp. 37–53, Nov. 2015 with permission from IEEE @ IEEE 2021

inhibitory inputs. For LMS, these are equivalents of positive or negative components of the input pattern vector.

LMS will allow the weight values to remain within their natural positive range even if adaptation caused a weight value to be pushed to one of its limits. Subsequent adaptation could bring the weight value away from the limit and into its more normal range, or it could remain saturated. Saturation would not necessarily be permanent (as would occur with Hebb's original learning rule). Note that LMS will keep (SUM) within its normal dynamic range, away from saturation, hovering near one of its stable equilibrium points. This is homeostasis.

The neuron and its synapses in Fig. 4.8 are identical to those of Fig. 4.6, except that the final output is obtained from an activation function that is a "half sigmoid." The neuron output will be positive, the weights will be positive, and some of the weighted inputs will be excitatory, some inhibitory, equivalent to positive or negative inputs. The (SUM) could be negative or positive. The neuron fires and produces an output when (SUM) is greater than the firing potential of -55mV. Accordingly, a bias of +55mV is added to the input of the half sigmoid. The training processes for the neurons and their synapses of Figs. 4.6 and 4.8 are identical, with identical stabilization points. The training algorithm for the neuron of Fig. 4.8 is Hebbian-LMS, given by equations

$$\begin{aligned} W_{k+1} &= W_k + 2\mu e_k X_k, \\ e_k &= \mathrm{SGM}(X_k^T W_k) - \gamma X_k^T W_k = \mathrm{SGM}((\mathrm{SUM})_k) - \gamma \cdot (\mathrm{SUM})_k \end{aligned} \tag{4.10}$$

With this algorithm, there is no externally inputted desired response with each input pattern X_k. Learning is unsupervised. The positive parameter μ is chosen by the algorithm designer and it controls stability and speed of convergence. Increasing μ speeds convergence but increases misadjustment.

4.7 Hebbian Learning and the Hebbian-LMS Algorithm

The Hebbian-LMS algorithm implements the extended Hebbian learning rules. A diagram of the Hebbian-LMS neuron and its synapses is shown in Fig. 4.8. The error function is shown in Fig. 4.7. The Hebbian rules will be reviewed and then checked against Figs. 4.7 and 4.8 in order to confirm that the Hebbian-LMS algorithm does or perhaps does not implement the Hebbian learning rules.

Two bias inputs are shown in Fig. 4.8. The first bias insures that when the inputs are all zero, the (SUM), the membrane potential, will be the resting potential of −70 mv. The second bias will cause the neuron to begin firing and producing an output when the membrane potential is at the firing threshold of −55 mV. The neuron will fire at −55 mV and will fire faster as the voltage rises above the threshold.

In Fig. 4.8, the excitatory signals add and contribute positively to the (SUM). The inhibitory signals contribute negatives to the (SUM). The excitatory weights adapt while learning in order to reduce the magnitude of the error ϵ. The inhibitory weights at the same time adapt to minimize the magnitude of the error, ϵ. Adaptation of both the excitatory and inhibitory weights contribute in concert to reduce the magnitude of the error.

When the excitatory weights become larger, the inhibitory weights become smaller, and vice versa. Together, they adapt to reduce the error magnitude. When learning with a set of input training patterns, they adapt to minimize mean square error. Now, back to the Hebbian rules.

The first Hebbian rule is, "when a presynaptic neuron is not firing, there will be no weight change." Check: If any of the inputs have values of zero, the connected weights will not change. During an adaptation cycle, the LMS algorithm will not change weights with zero inputs. This applies of course to both excitatory and inhibitory weights.

A second Hebbian rule is, "when the presynaptic neuron is firing and the postsynaptic neuron is also firing, the excitatory synaptic weights increase at a rate proportional to the product of the presynaptic and postsynaptic firing rates while the inhibitory synaptic weights decrease in the same proportion." Check: It is apparent from Fig. 4.8 that the excitatory and inhibitory weights will change in opposite directions with LMS adaptation. Furthermore, in accord with the LMS algorithm, the rate of change of a given synaptic weight will be proportional to the product of its input signal and the error signal. The input signal is proportional to the presynaptic firing rate. The error signal is related to the postsynaptic neuron's firing rate.

So far, so good. But the second Hebbian rule will only be satisfied by the Hebbian-LMS algorithm if its error signal is directly proportional to the postsynaptic neuron's firing rate. The error is plotted as a function of the postsynaptic (SUM) in Fig. 4.7b. For small values of the (SUM) above the firing threshold, the error is linearly related to (SUM), and this is fine. But for larger values of (SUM), the linearity breaks down and the error is no longer linearly related to the (SUM). The error curve peaks and then bends down into a line toward the equilibrium point. This is not fine. Something is amiss.

Something is incorrect with the Hebbian-LMS algorithm or something is incorrect with the second Hebbian learning rule. Which is it?

Fire together, wire together. The "rate of wiring," the rate of change of the synaptic weight, is proportional to the product of the presynaptic firing rate and the postsynaptic firing rate, according to the second Hebbian rule. If the postsynaptic neuron begins to fire slowly, there will be weight change and it is reasonable to assume that the rate of change will be proportional to the postsynaptic neuron's firing rate. But what happens when the postsynaptic firing rate becomes larger? Does the rate of synaptic weight change also get larger? What happens when the postsynaptic firing rate becomes even larger, comes up to the homeostatic rate, does the rate of weight change become even larger? No! At the homeostatic rate, there must be no weight change at all. Otherwise, there will be no homeostasis.

There is trouble with the second Hebbian learning rule. It needs to be modified. What about the Hebbian-LMS algorithm? Does it need adjustment? No!

The Hebbian-LMS algorithm will cause positive excitatory weight change at a rate proportional to the value of (SUM) above the firing threshold for small firing rates.

The rate of weight change will peak for a certain value of (SUM) between the firing threshold and the equilibrium point. Higher values of (SUM) will result in a slowing of the rate of weight change. At the equilibrium point, weight change will cease, resulting in homeostasis. Higher values of (SUM) will cause a reversal of the error and a negative weight change as the algorithm pushes (SUM) back to the homeostatic point. Note that if the value of the (SUM) were below the firing threshold, the error reverses and this would cause weight reduction. See Fig. 4.7b. This agrees with the third Hebbian rule that "if the presynaptic neuron is firing and the postsynaptic neuron is not firing, the excitatory weight decreases." Note that excitatory and inhibitory weight changes are always reverses of each other, and firing rate is related to (SUM).

The Hebbian-LMS algorithm does everything right. It seems to be doing what nature intends, and gives homeostasis. The second Hebbian learning rule has been fixed and it now reads as follows: "When the presynaptic neuron is firing and the postsynaptic neuron is also firing, the excitatory synaptic weights increase at a rate proportional to the product of the presynaptic and postsynaptic firing rates for small postsynaptic firings rates, but the excitatory synaptic weights decrease for postsynaptic firing rates above homeostatic, and the inhibitory weights change oppositely in like proportion." The complexity of this rule is unavoidable. It is what it is. The extended Hebbian rules of previous chapters work well for small magnitudes of (SUM). For larger magnitudes of (SUM), the nonlinearity and crossover of the error function is natural and gives rise to homeostasis.

Inspection of the error function of Fig. 4.5 shows a form of symmetry in the response of the Hebbian-LMS algorithm to training patterns that cause (SUM) values above the threshold and training patterns that cause (SUM) values less than the threshold. The algorithm adapts and pushes (SUM) responses toward the positive equilibrium point and other responses toward the negative equilibrium point. Upon convergence, some responses cluster near the positive equilibrium point, others cluster near the negative equilibrium point. At the neuron output, the negative responses will give zero outputs. The positive responses will give neuron outputs close to 1. The neuron output will be essentially binary. The Hebbian learning rules have very little to say about this, so there is nothing to check with the Hebbian-LMS rules. It seems that the brain in maintaining homeostasis computes with 1 and 0's.

4.8 A More General Learning Algorithm

The diagram of Fig. 4.8 represents a model of neuronal and synaptic learning with the Hebbian-LMS algorithm. It is likely that nature is performing this algorithm but it is highly unlikely that nature is performing this algorithm exactly in accord with the diagram of Fig. 4.8. The idea that nature would generate an error signal as a function of (SUM) by taking the difference between a sigmoidal function of (SUM) and a scaled version of the (SUM) itself seems unlikely. Performing Hebbian-LMS as in Fig. 4.8 is fine for computer simulation. Nature generates an error signal from the membrane potential, the (SUM), but not exactly like the method presented in Fig. 4.8.

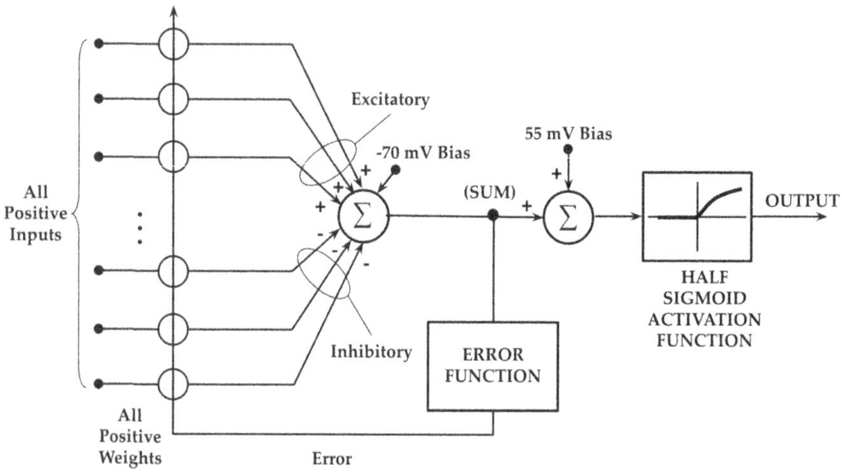

Fig. 4.9 A general form of Hebbian-LMS. Reproduced from B. Widrow, Y. Kim, and D. Park, The Hebbian-LMS Learning Algorithm, IEEE Computational Intelligence Magazine, vol. 10, no. 4, pp. 37–53, Nov. 2015 with permission from IEEE @ IEEE 2021

A more general form of Hebbian-LMS with the error signal simply being a nonlinear function of (SUM) is shown in Fig. 4.9. This is a better representation of what nature actually does, generating an error signal as a nonlinear function of the membrane potential. The function is not easily expressed algebraically.

Nature's error signal would need to be like that of Fig. 4.7, but not exactly the same. It would need to have two stable equilibrium points with crossover of error polarity at the equilibrium points. The positive equilibrium point is of interest, the negative one not so much, since the neuron would not be firing with the (SUM) at the negative equilibrium point. The positive equilibrium point with the error polarity having reversal or crossover is essential for nature's homeostasis.

4.9 An Algebraic Description of the Neuron of Fig. 4.8 Including the Biases

The neuron of Fig. 4.8 can be described algebraically, with its biases. Let −70 mV bias be called B_1 and the +55 mV be B_2. Equations (4.10) can be modified to include the biases, as

$$\begin{aligned} W_{k+1} &= W_k + 2\mu\epsilon_k X_k \\ \epsilon_k &= \text{SGM}(X_k^T W_k + B_1) - \gamma(X_k^T W_k + B_1) \end{aligned} \tag{4.11}$$

Equations 4.11 comprise the adaptive algorithm. The neuron output is

$$(\text{OUTPUT}) = \begin{cases} \text{SGM}(X_k^T W_k + B_1 + B_2), & \text{for } (.) \geq 0 \\ 0, & \text{otherwise} \end{cases}$$

Equations (4.11) could be defined as the Hebbian-LMS algorithm. These equations perfectly describe the Hebbian rules for small values of (SUM). They exhibit homeostasis at the positive equilibrium point.

4.10 Nature's Learning Algorithm

In the next three chapters, a deeper look into synaptic learning will be presented. Nature's learning process, increasing or decreasing the neuroreceptor population, is implemented within the membrane of the postsynaptic side of the synapse. Looking inside the membrane, it becomes evident that further additions to the Hebbian-LMS algorithm are necessary in order to model actual synaptic learning. This is the subject of the next chapter and the succeeding two chapters.

4.11 Summary

To summarize, for low firing rates, the Hebbian learning rules and the Hebbian-LMS algorithm fit each other perfectly. At higher firing rates, they begin to differ.

At the homeostatic rate, the Hebbian rules predict rapid weight changing, while the Hebbian-LMS algorithm predicts a cessation of weight changing. When firing faster than homeostatic, the Hebbian learning rules predict even faster weight changing and the Hebbian-LMS algorithm predicts a reversal in weight changing. Hebbian-LMS always pulls the firing rate toward the homeostatic rate. On the other hand, the Hebbian learning rules per se do not permit homeostasis.

Although the Hebbian-LMS algorithm was born from extended Hebbian learning, it supersedes Hebbian learning and accounts for the phenomenon of homeostasis.

4.12 Questions and Experiments

1. Repeat experiment 3.6, but initialize the weights randomly and do not assign desired responses for each input pattern. Train with bootstrap learning, unsupervised learning. The input patterns will be classified as in the $+1$ class or the -1 class. What you have is the simplest form of Hebbian-LMS. How similar are the patterns in each class? Repeat for different sets of random initial values.

2. Create 100 times the number of input training patterns by adding independent Gaussian noise to the components of the original patterns. The variance of the noise can be varied with different experiments. Now examine the patterns in each class and see if there is similarity among the patterns in each class. Repeat for different sets of random initial weight values.
3. How does nature form the error signal as a function of membrane potential? Examine the literature, then try to imagine how this might work. This is essential for homeostasis.

References

1. Widrow, B., Kim, Y., Park, D.: The Hebbian-LMS learning algorithm. IEEE Comput. Intell. Mag. **10**(4), 37–53 (2015)
2. (32) The LMS algorithm and ADALINE. Part ii—ADALINE and memistor ADALINE—YouTube. https://www.youtube.com/watch?v=skfNlwEbqck. block Accessed on 04 Oct 2020
3. Widrow, B.: Bootstrap learning in threshold logic systems. In: Proceedings International Federation Automatic Control, pp. 96–104 (1966)
4. Miller, W.C.: A modified mean-square-error criterion for use in unsupervised learning. Ph.D. thesis, ProQuest Information & Learning (1968)

Chapter 5
Inside the Membrane; The Percent Hebbian-LMS Algorithm

Abstract A presynaptic neuron and postsynaptic neuron are coupled by a synapse. The synaptic weight is proportional to the number of receptors of the synapse. Learning is effected by increase and decrease in the number of receptors under the control of an adaptive algorithm, very likely the Hebbian-LMS algorithm. A "mechanism" implements the algorithm and is located in the membrane and cell body of the postsynaptic neuron. In the synaptic cleft, increase and decrease is accomplished by the mechanism undergoing "receptor traffic." To do its work, the mechanism implementing Hebbian-LMS needs an input X-vector component and an error signal. It senses an error signal from the membrane potential. It senses its X-component from the neurotransmitter in the cleft via the receptors. The X-component is proportional to the product of the presynaptic neuron's signal as reflected in the neurotransmitter concentration in the cleft, and the number of receptors that act like antennas sticking into the fluid of the cleft. Accounting for this, the Hebbian-LMS algorithm is modified and becomes the percent Hebbian-LMS, making the weights with the largest values adapt fastest.

5.1 Introduction

A neuron receives as many as thousands of input signals, forms a composite by summation, and transmits the composite signal down its axon to an axon terminal, and thereupon it enervates other neurons by synaptic connection. A main function of the neuron is signal consolidation and transmission.

The neuron's inputs come from presynaptic neurons via synaptic connection. The synapses are adjustable. Hebb said that their "efficiencies" (weights) increase depending on the firing of the presynaptic neurons and the postsynaptic neurons. In Chap. 2, Hebb's learning rule was extended in a natural way, and the connection to the LMS algorithm became apparent. The result was the implementation of Hebbian learning by means of the LMS algorithm. Several forms of Hebbian-LMS were described in Chap. 4.

The neuron and its synapses is an adaptive linear combiner. If adaptation of the synaptic weights were to be done with Hebbian-LMS, each synapse, in order to adapt,

B. Widrow, *Cybernetics 2.0*, Springer Series on Bio- and Neurosystems 14,
https://doi.org/10.1007/978-3-030-98140-2_5

would need to know its input signal and the error signal, which is common to all the synapses. The error signal is derived from the linear combiner's (SUM), which is proportional to the membrane potential. Knowledge of the membrane potential is available to all the synapses. Their neuroreceptors are embedded in the membrane. Since the neuroreceptors are exposed to the fluid and neurotransmitters in the synaptic clefts, they can sense the input signals. Therefore the input signals and the error are available to the receptors. So, the synapses and their receptors have all the information they need to decide to increase or decrease the number of their neuroreceptors and at what rate.

The purpose of this chapter is to look deeper into the membrane, as well as can be done, to try to understand the mechanism that causes the membrane and the cell to augment or to diminish the number of neuroreceptors in each synapse when implementing a learning algorithm. The focus is on the membrane.

5.2 The Membrane and the Neuroreceptors

Every cell in the body is surrounded with a membrane that has neuroreceptors embedded within. When the appropriate neurotransmitter is present, its molecules can bind to neuroreceptors and open co-located gates. These gates allow ions and fluids to flow through the membrane in either direction, as dictated by concentration gradients. The flow of ions affects the membrane potential. Rapid changes in membrane potential can result from ion flow. The long-term average of the membrane potential is regulated by increasing or decreasing the number of receptors, the number of gates.

The increasing and decreasing of the number of receptors can be accounted for by the Hebbian—LMS algorithm. The rate of change in the number of receptors is proportional to the product of the concentration of neurotransmitter and the error signal.

The error signal is a nonlinear function of the membrane potential. There is an equilibrium point, a set point, such that if the membrane potential is below the set point the error is positive and if the membrane potential is above the set point, the error is negative. The objective of changing the number of receptors is to keep the long-term average of the membrane potential near the set point. This is homeostasis. Good health and good functioning of the cell requires keeping the membrane potential near the set point.

The mechanism for creating and for destroying receptors is located in the membrane itself. It is clear what the signals are that are needed by the mechanism to cause the number of receptors to increase or decrease and at what rate. How this mechanism is controlled is not yet understood. How this mechanism receives its input signals, its control signals, can be surmised.

Figure 5.1 is a diagram of a membrane and receptors, and a mechanism that creates and destroys receptors. The inputs to the mechanism are the neurotransmitter concentration and the membrane potential (from which the error is derived). The mechanism, embedded in the membrane, cannot directly measure membrane poten-

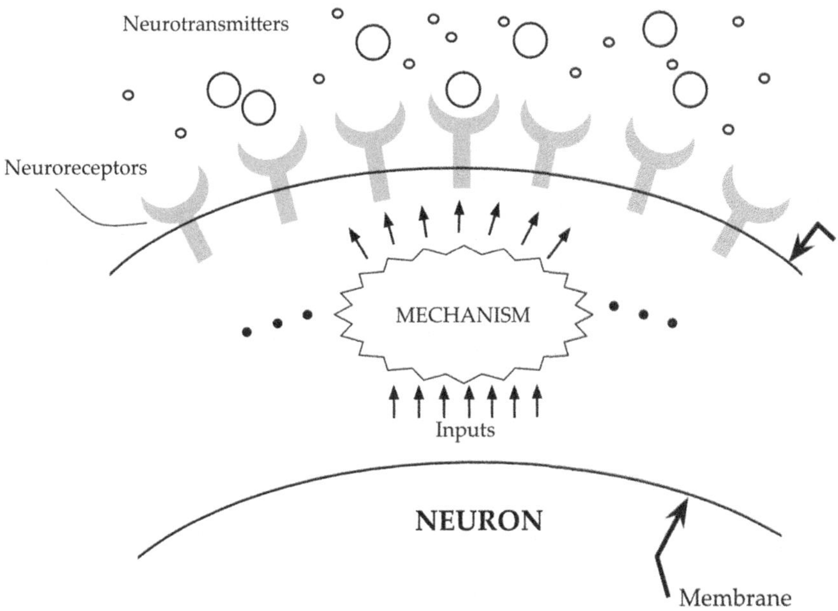

Fig. 5.1 A membrane with a mechanism that creates and destroys neuroreceptors

tial. It does not have a voltmeter with probes attached to the inner surface and the outer surface of the membrane. The membrane is made of lipid protein and although not highly electrically conductive, it will conduct and act as a shunt to the membrane potential. Ohmic current in accord with Ohm's law will flow through the membrane and its amplitude will be proportional to the membrane potential. This ohmic current can be sensed by the embedded mechanism, thereby providing information about the membrane potential and the error signal.

The neurotransmitter is located in the fluid of the synaptic cleft outside the membrane. Its concentration cannot be directly sensed by the embedded mechanism. The only way that the mechanism could receive information about the neurotransmitter concentration would be from the receptors that reach outside to the neurotransmitter. A receptor bound to a neurotransmitter molecule opens its gate to allow an ionic flow into and out of the membrane. The ionic flow is a movement of ions and because they are charged, there results an electric current. Because this is an ionic current, it does not get confused with the ohmic current which is a flow of electrons. The ionic current is proportional to the product of the neurotransmitter concentration in the cleft and the number of receptors.

This is a product of the synaptic input signal, i.e. the neurotransmitter concentration, and the weight value, i.e. the number of receptors. The ionic current is sensed by the embedded mechanism and is one of its inputs.

Figure 5.2 is a drawing that illustrates the ionic current and the ohmic current which are the two inputs to the embedded mechanism.

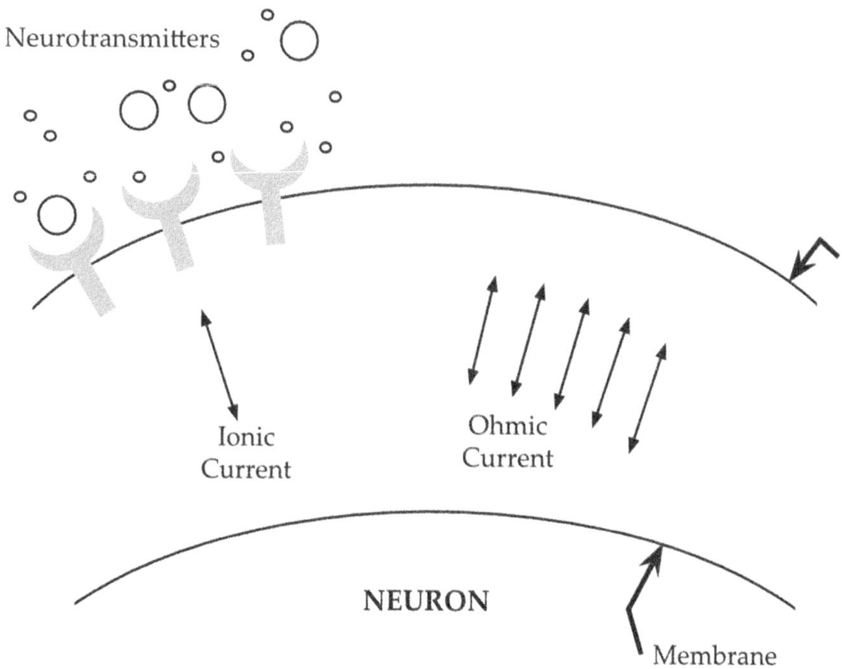

Fig. 5.2 Input currents to mechanism

Creating and destroying receptors is really a metaphor for what actually occurs. This subject is presented in more detail in Sect. 5.5 on receptor traffic.

5.3 A New Learning Algorithm

In order to effect adaptation and learning, each synapse requires knowledge of its input signal and the error signal. The error signal is derived from the membrane potential, reflected in the ohmic current. The embedded mechanism senses the ohmic current. The synaptic input signal is reflected in the ionic current. The embedded mechanism senses the ionic current and the ohmic current.

The input signals to the synapses are the outputs of the presynaptic neurons. The input signals are the components of the X vector. These components are proportional to the time-average concentrations of neurotransmitter in the respective synaptic clefts.

In an individual synapse, the ionic current is proportional to the neurotransmitter concentration and to the number of receptors. Accordingly, among all the synapses, the "X-vector" fed to the embedded mechanism has the components of the original

X-vector, but now weighted by the weight values, proportional to the number of receptors.

$$\{\text{“X- VECTOR” FED TO THE EMBEDDED MECHANISM}\}_k = \begin{bmatrix} x_1 w_1 \\ x_2 w_2 \\ \vdots \\ x_n w_n \end{bmatrix}_k \tag{5.1}$$

This is an unusual product of two vectors, the original X-vector and the weight vector. It is not a classical inner product or outer product. It is called a Hadamard product. It is represented symbolically as

$$\begin{bmatrix} x_1 w_1 \\ x_2 w_2 \\ \vdots \\ x_n w_n \end{bmatrix}_k = (X_k \circ W_k) \tag{5.2}$$

Among all the synapses, the ionic current is proportional to $(X_k \circ W_k)$.

The original Hebbian-LMS algorithm is

$$\begin{aligned} W_{k+1} &= W_k + 2\mu\epsilon_k X_k \\ \epsilon_k &= f(X_k^T W_k) \end{aligned} \tag{5.3}$$

A new form of this algorithm believed to be actually performed by the embedded mechanisms is

$$\begin{aligned} W_{k+1} &= W_k + 2\mu\epsilon_k (X_k \circ W_k) \\ \epsilon_k &= f(X_k^T W_k) \end{aligned} \tag{5.4}$$

Equations (5.4) **are proposed herein to be nature's learning algorithm.**

The adaptive algorithm of Eqs. (5.4) takes iterative steps along the components of the negative gradient, except that each gradient component is now weighted by the respective current weight value. Bigger weights get adapted faster.

The algorithm of Eqs. (5.4) makes weight changes that are proportional to the sizes of the weights. A name for the new algorithm is percent Hebbian-LMS, represented by % Hebbian-LMS.

The original Hebbian-LMS algorithm is

$$\begin{aligned} W_{k+1} &= W_k + 2\mu\epsilon_k X_k \\ \epsilon_k &= \text{SGM}(X_k^T W_k) - \gamma X_K^T W_k \end{aligned} \tag{5.5}$$

The activation function for this algorithm is sigmoidal. The new version of this algorithm is % Hebbian-LMS, given by Eqs. (5.6).

$$
\begin{aligned}
W_{k+1} &= W_k + 2\mu\epsilon_k(X_k \circ W_k) \\
\epsilon_k &= \mathrm{SGM}(X_k^T W_k) - \gamma X_K^T W_k
\end{aligned} \tag{5.6}
$$

The ionic current is proportional to $(W_k \circ X_k)$. The ohmic current is proportional to $X_K^T W_k = W_k^T X_k$ which is proportional to the membrane potential. In order to implement Eqs. (5.6) or to implement Eqs. (5.4), the embedded mechanism only needs to know the ionic current and the ohmic current. To control the number of receptors, the rate of change, up or down, depends on the ohmic current compared to a mid value, the homeostatic set point built into the embedded mechanism. With ohmic current less than the set point, increase the number of receptors. With ohmic current greater than the set point, reduce the number of receptors. The error is proportional to the difference between the ohmic current and the set point.

The objective of the Hebbian-LMS algorithm is to minimize error. The error is a function of membrane potential. Equations (5.4) and (5.6) reflect this. A certain membrane potential is the "sweet spot" that is good for the health of the neuron. This is the homeostatic potential, the equilibrium point, the crossover point of the error function. At homeostasis, the error is zero. With membrane potential above homeostatic the error is negative, below homeostatic the error is positive. Hebbian-LMS keeps the membrane potential close to the homeostatic level.

Hebbian-LMS regulates the membrane potential. Inside the membrane, the mechanism senses the ohmic current which is proportional to the membrane potential. Therefore, Hebbian-LMS regulates ohmic current. The ohmic current is an input to the mechanism that implements Hebbian-LMS. The error signal is used by the mechanism to reduce the error signal. This is the feedback loop of homeostasis.

The rate of change in number of receptors is proportional to the product of the error and the amplitude of the ionic current.

The ohmic current is the variable which is regulated by adding and subtracting receptors. Its value determines the direction of adaptive change and the associated error is one factor in determining the rate of change. The other factor is the ionic current which is probably the power supply for the operation of the embedded mechanism. The speed of convergence is proportional to the ionic current. The ionic current is the enabler, and the ohmic current provides a sense of direction. In order to do Hebbian-LMS, the mechanism only needs to know the ionic current and the ohmic current. This is the basis of homeostasis.

The arguments presented above are based on the assumption that the membrane is of a neuron. This assumption is not necessary, and it is clear that the above arguments will apply to the receptors of any cell in the body. All forms of homeostasis are explainable in terms of the Hebbian-LMS theory, whether applied to neurons or to any other cell type. Hebbian-LMS theory, although plausible, is a theory only and not yet proven.

5.4 Simulation of a Hebbian-LMS Neuron Trained with the % Hebbian-LMS Algorithm

Computer simulation was performed to demonstrate learning and clustering by the neuron and synapses of Fig. 4.8 now trained by the % Hebbian-LMS algorithm. Initial values for the weights were chosen randomly, independently, with uniform probability between 0 and 1. There were 50 excitatory and 50 inhibitory weights. There were 50 training patterns whose vector components were chosen randomly, independently, with uniform probability between 0 and 1. Initially, some of the input patterns produced positive (SUM) values, designated in Fig. 5.3a by blue crosses, and the remaining patterns produced negative (SUM) values, designated in Fig. 5.3a by red crosses. After 50 iterations, some of the crosses have changed sides, as seen in Fig. 5.3b. After 250 iterations, as seen in Fig. 5.3c, clusters have begun to form and membership of the clusters has stabilized. There are no responses near zero. After 800 iterations, tight clusters have formed as shown in Fig. 5.3d. At the neuron output, the output of the half sigmoid, the responses will be binary, approximate 0's and approximate 1's. Upon convergence, the patterns selected to become 1's or those selected to become 0's are strongly influenced by the random initial conditions, but not absolutely determined by initial conditions. The patterns would be classified very differently with different initial weights. A learning curve, mean square error as a function of the number of iterations, is shown in Fig. 5.3e. This curve demonstrates learning and convergence of the % Hebbian-LMS algorithm.

When using a supervised LMS algorithm, the learning curve is known to be a sum of exponential components [1]. With unsupervised LMS, the theory has not yet been developed. The nature of this learning curve and the speed of convergence have only been studied empirically.

5.5 Receptor Traffic

It is not accurate to say that receptors are created and destroyed with upregulation and downregulation. The receptors are subject to traffic, physical movement in both the membrane and in the cell itself. This movement has an effect on the availability of neuroreceptors for binding to neurotransmitters, thus upregulation and downregulation are possible.

There are complex intracellular pathways for receptor trafficking. Various proteins move receptors around, like trucks shipping goods to many different locations. The "delivery" process is quite complex. Sometimes things are moved to a location nearby first, and then moved to the exact location, in the neuron near the synapse. This process has been estimated to require 15–30 min to deliver receptors. The speed of adaptation is therefore limited to have time constants of no less than 15–30 min.

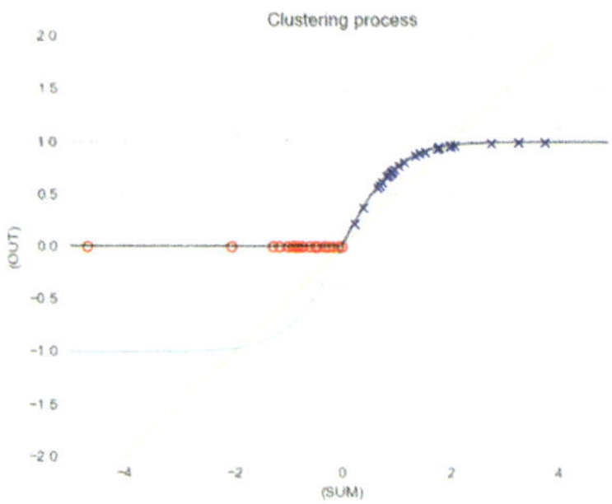

(a) Initial responses

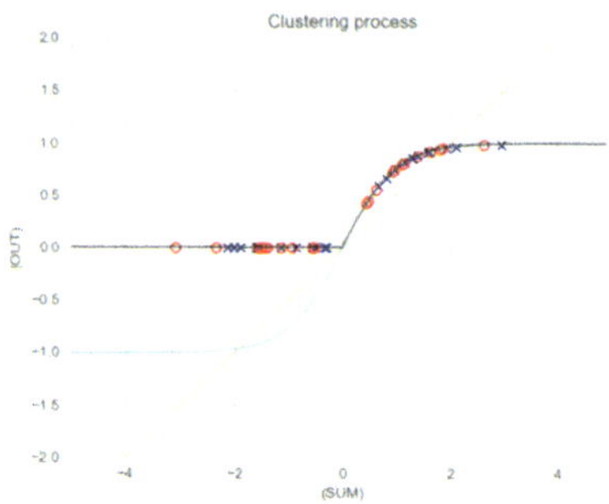

(b) Responses after 50 iterations

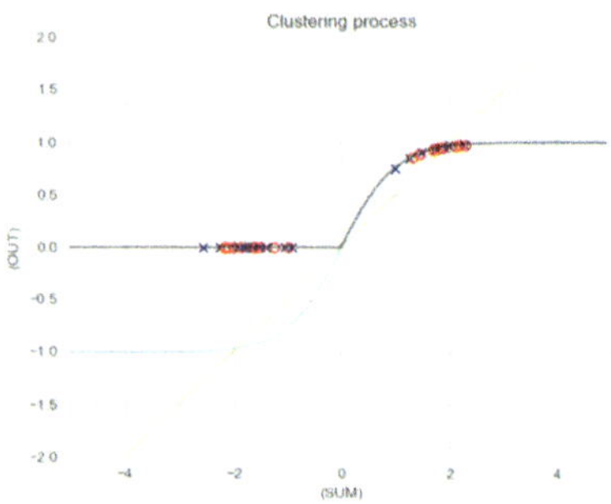

(c) Responses after 250 iterations

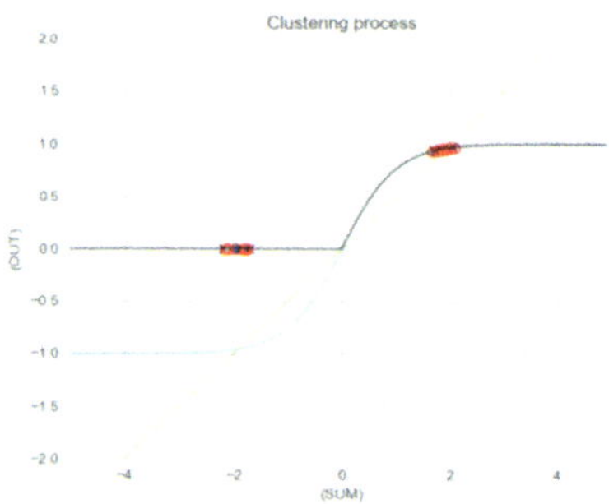

(d) Responses after 800 iterations

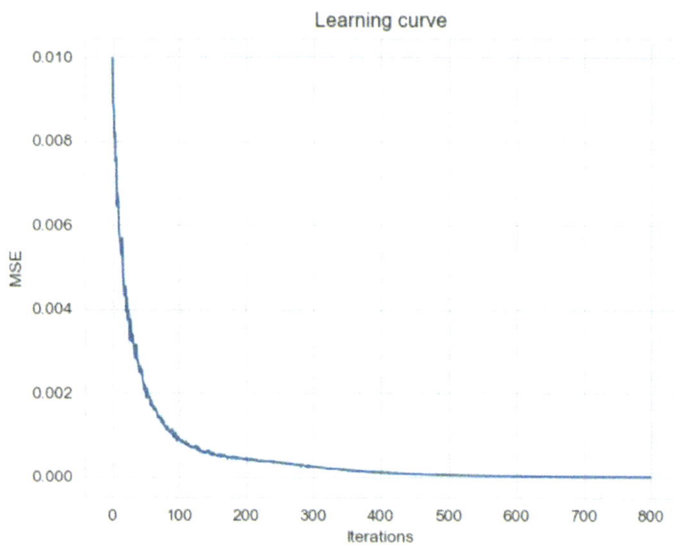

(e) A learning curve

Fig. 5.3 A learning experiment with a single neuron trained with % Hebbian-LMS, hyperbolic tangent activation function

The delivery trucks park receptors in "garages" (actually vesicles near the synapse) when not needed, during downregulation, shielding them from binding to neurotransmitter molecules. The receptors emerge from the garages during upregulation and appear in the synapse.

Receptors get manufactured and moved to distribution centers, and from there to specific locations as needed. Recycling is common. The garage is in the neuron, close to the synapse, not in the membrane itself. The "mechanism" described in the sections above is not strictly in the membrane, but is in addition in the neuron. A good reference is Victor Anggono and Richard L. Huganir, 2012, entitled "Regulation of AMPA trafficking and synaptic plasticity."

The full answer to trafficking and up and down regulation is unknown. This issue is technically challenging and people have spent many years trying to understand receptor movement. What is known however does not contradict the model of up and down regulation of the previous sections, although creating and destroying receptors is only a metaphor, but it does correspond to manufacturing and recycling.

5.6 Summary

A presynaptic neuron communicates with a postsynaptic neuron by means of a synapse. The strength of the connection, the synaptic weight, is proportional to the number of neuroreceptors within the synapse. The presynaptic signals are carried across the cleft of the synapse by neurotransmitter molecules which bind to the neuroreceptors and impart signals. The average concentration of neurotransmitter in the cleft is proportional to the average firing rate of the presynaptic neuron. Inside the postsynaptic neuron's membrane at the synapse resides a mechanism that controls the receptors. Upregulating and downregulating them in accord with an adaptive algorithm. The input signal to the mechanism is not simply the neurotransmitter concentration but this concentration multiplied by the number of receptors. To model this adaptive process, Hebbian-LMS is modified to account for the multiplication and the result is the % Hebbian-LMS algorithm where big weights converge faster than small weights. The mechanism that physically controls the number of receptors is based on "receptor traffic," moving receptors in the membrane and the cell body as needed to increase or decrease the number of receptors in the cleft. It seems that % Hebbian-LMS controls the traffic that controls the synaptic weight. Over thousands of such synapses, the linear combiner's weights change in order to accomplish learning.

5.7 Questions and Experiments

1. The ideas of ohmic current, ionic current, and the mechanism that exercises control over the number of receptors are hypothetical. There is no description of these elements in the literature. This hypothesis, although intuitive and reasonable, is unproven. It would be a major biomedical effort to undertake proof or disproof. Can you think of how you would go about making a proof? Would in-vitro preparation like that of Bi and Poo be useful in this quest? Would (dc) electric currents in solution have an effect on Bi and Poo wight change? Could you predict this effect? Actually this would be a good Ph.D. thesis project.

Reference

1. Widrow, B., Stearns, S.D.: Adaptive Signal Processing. Prentice Hall, Englewood Cliffs, NJ (1985)

Chapter 6
Synaptic Scaling and Homeostasis

Abstract Synaptic scaling is a hypothesis that is accepted by many neuroscientists. Its main premise is that both excitatory and inhibitory synaptic weights that feed inputs to a neuron are scaled up or down together as needed to bring the neuron's membrane potential to its homeostatic level. Adjusting the weights directly affects the sum (SUM) which is proportional to the membrane potential. With synaptic scaling, large weights adapt faster than small weights. This also happens with % Hebbian-LMS. Both adaptive algorithms have the same objective, to restore the neuron to homeostasis. Both algorithms require the same signals for adaptation, namely the X-vector and the error. Synaptic scaling operates with a common mode, excitatory and inhibitory weights scale up or down together. % Hebbian-LMS implements a differential mode so that when excitatory weights scale up, the inhibitory weights scale down, etc. Nature could implement either algorithm, but % Hebbian-LMS has the advantage of requiring less weight change to achieve error correction. Physical evidence supporting synaptic scaling also supports % Hebbian-LMS. The algorithms are identical with excitatory synapses but not so with inhibitory synapses.

6.1 Introduction

Among neuroscientists there is a community of researchers who have been studying the phenomenon of synaptic scaling over a period of as many as 20 years. A leader in this field is Dr. Gina Turrigiano of Brandeis University. A researcher in this field at Stanford University is Dr. Robert C. Malenka, who has been consulted by Dr. Widrow on synaptic plasticity and synaptic scaling.

Nature's purpose seems to be regulation of a neuron's firing rate, keeping it close to a homeostatic rate, by controlling its membrane potential. This is the objective of the % Hebbian-LMS algorithm and it is also the objective of synaptic scaling. These two adaptive algorithms are similar in many ways yet they are very different. The question is, which algorithm is nature employing? Both algorithms seem to be supported by the same physical evidence.

B. Widrow, *Cybernetics 2.0*, Springer Series on Bio- and Neurosystems 14,
https://doi.org/10.1007/978-3-030-98140-2_6

6.2 Synaptic Scaling

Googling the subject of "multiplicative synaptic scaling," one finds the following definition: "synaptic scaling is a slow process in which the strengths of synapses impinging onto a neuron are scaled by a common multiplicative factor in a way that adjusts the total to match the neuron's dynamic range while preserving the relative strengths of the synapses and thus, presumably, the memory traces."

What this definition is saying is that all the synapses of a neuron are scaled up or down to bring the (SUM) close to the positive equilibrium point, the homeostatic set point. The same scale factor applies to all synapses, both excitatory and inhibitory. All the synaptic weights go up and down together by the same percentage. Contrasting with synaptic scaling, the % Hebbian-LMS algorithm raises the excitatory weights and lowers the inhibitory weights by the same percentage, or lowers the excitatory weights and raises the inhibitory weights by the same percentage, as required to raise or lower the (SUM) to match the homeostatic level. The excitatory and inhibitory weights go up and down differentially with % Hebbian-LMS. They go up and down together with synaptic scaling.

The % Hebbian-LMS algorithm of Chap. 5 is given by

$$\begin{aligned} W_{k+1} &= W_k + 2\mu\epsilon_k(X_k \circ W_k) \\ \epsilon_k &= f(X_k^T W_k) \end{aligned} \tag{6.1}$$

The Eq. 6.1 apply to excitatory and inhibitory synapses with the understanding that the components of the X-vector are positive for excitatory and negative for inhibitory synapses.

If a neuron only had excitatory synapses, no inhibitory synapses, synaptic scaling and the % Hebbian-LMS algorithm would be identical. Equation 6.1 would describe synaptic scaling.

If the neuron had both excitatory and inhibitory synapses, Eq. 6.1 would still describe synaptic scaling with the understanding that the components of the X-vector are positive for excitatory and negative for inhibitory synapses. However, the polarity of the inhibitory weight changes needs to be reversed, so that excitatory and inhibitory weights would increase and decrease together rather than in opposition.

6.3 Comparison of Synaptic Scaling with % Hebbian-LMS

The objective of these algorithms is to adjust, change, or adapt the synaptic weights to bring the (SUM) close to the positive stable equilibrium point, the homeostatic level. If the voltage of the (SUM) is below that of the equilibrium point, raise it up. If the voltage of the (SUM) is higher than that of the equilibrium point, lower it.

The % Hebbian-LMS algorithm raises the (SUM) by increasing the excitatory synaptic weights and at the same time decreasing the inhibitory synaptic weights. Reversing this will lower the (SUM). Excitatory and inhibitory synaptic weights adapt differentially, when one goes up the other goes down. With % Hebbian-LMS, large weights adapt faster than small weights.

Synaptic scaling has a similar purpose, but it adapts the weights in a different way. All synaptic weights, both excitatory and inhibitory, are increased or decreased in unison. The weights do not adapt differentially. Large weights adapt faster than small weights.

Comparing % Hebbian-LMS and synaptic scaling, there are similarities and differences. They both have independent adaptation of the individual synapses. Each adapts independently of all the others. In order to adapt, % Hebbian-LMS requires knowledge for each individual synapse of its neurotransmitter concentration in the cleft and the membrane potential, which is needed to create an error signal. The error signal goes to zero at the equilibrium point.

In order to adapt, synaptic scaling requires knowledge of the neurotransmitter concentration in the cleft of the individual synapse, and the membrane potential, which is needed to create an error signal. These requirements are the same as for % Hebbian-LMS. Nature could have implemented either algorithm as easily as the other when creating the membrane and the mechanism of adaptation incorporated in the synapse. Which algorithm did nature choose?

6.4 Papers of Turrigiano and Stellwagen and Malenka on Synaptic Scaling

A reflection on the thinking of researchers in the field of synaptic scaling can be gained by review of a paper by Dr. Turrigiano and another paper by Drs. David Stellwagen and R. C. Malenka.

The abstract of the paper "The dialectic of Hebb and homeostasis" by Dr. Turrigiano, Philosophical Transactions B, Royal Society, London, March 5, 2017, [1] is the following:

Abstract

It has become widely accepted that homeostatic and Hebbian plasticity mechanisms work hand in glove to refine neural circuit function. Nonetheless, our understanding of how these fundamentally distinct forms of plasticity complement (and under some circumstances interfere with) each other remains rudimentary. Here, I describe some of the recent progress of the field, as well as some of the deep puzzles that remain. These include unravelling the spatial and temporal scales of different homeostatic and Hebbian mechanisms, determining which aspects of network function are

> under homeostatic control, and understanding when and how homeostatic and Hebbian mechanisms must be segregated within neural circuits to prevent interference. This article is part of the themed issue 'Integrating Hebbian and homeostatic plasticity'.

Dr. Turrigiano's abstract acknowledges homeostatic and Hebbian plasticity's contributions to neural circuit function. The abstract indicates that homeostasis and Hebbian plasticity are separate mechanisms that have separate locations and different operational time scales. These are deep puzzles regarding how these two mechanisms cooperate and knowledge of how they operate remains rudimentary.

The % Hebbian-LMS algorithm seems to solve the puzzle. According to the % Hebbian-LMS hypotheses, homeostasis is a natural byproduct. Hebbian learning and homeostasis are not separate mechanisms but are both incorporated in the % Hebbian-LMS algorithm. % Hebbian-LMS also provides something like synaptic scaling. This algorithm clarifies many of the outstanding puzzles, as will be seen.

A summary of the Stellwagen, Malenka paper [2] is:

> **Abstract**
> Two general forms of synaptic plasticity that operate on different timescales are thought to contribute to the activity-dependent refinement of neural circuitry during development: (1) long-term potentiation (LTP) and long-term depression (LTD), which involve rapid adjustments in the strengths of individual synapses in response to specific patterns of correlated synaptic activity, and (2) homeostatic synaptic scaling, which entails uniform adjustments in the strength of all synapses on a cell in response to prolonged changes in the cell's electrical activity. Without homeostatic synaptic scaling, neural networks can become unstable and perform sub-optimally. Although much is known about the mechanisms underlying LTP and LTD, little is known about the mechanisms responsible for synaptic scaling except that such scaling is due, at least in part, to alterations in receptor content at synapses.

The Stellwagen and Malenka paper is especially interesting. They state that homeostatic synaptic scaling entails uniform adjustments in the strength of all synapses on a cell in response to prolonged changes in the cell's electrical activity. This is exactly what % Hebbian-LMS does. They also state that "little is known about the mechanisms responsible for synaptic scaling except that such scaling is due, at least in part, to alterations in receptor content at synapses."

The % Hebbian-LMS hypothesis answers many related questions about synaptic plasticity. However, % Hebbian-LMS theory does raise deeper questions, such as how does the embedded mechanism in the membrane work? The embedded mechanism is an hypothesis that has as yet not been observed directly. Its effects are observable, but that still does not explain how it works. The embedded mechanism hypothesis is an idea that might provide a sense of direction leading to deeper research in the field.

6.5 Selected Quotes From The Turrigiano Paper

Dr. Turrigiano is a distinguished researcher in the fields of Hebbian plasticity and synaptic scaling. Her work has been highly cited over the past 20 years. The following quotes reflect not only her work but also the thinking of other researchers in the field. We will comment on her individual quotes.

The following are numbered quotes from the Turriagiano paper and our correspondingly numbered responses and comments.

1. *"there has been a growing acceptance of the idea that stabilizing plasticity mechanisms are critical for many aspects of proper circuit function"*

   We agree, stabilization of firing rate is critical.
2. *"we still do not fully understand how homeostatic and Hebbian mechanisms cooperate to enable, shape and constrain microcircuit plasticity"*

   We disagree that Hebbian learning and homeostasis are two separated mechanisms. According to % Hebbian-LMS, they are both parts of the same mechanism.
3. *"there is solid evidence for the existence of a form of synaptic plasticity, synaptic scaling, that operates in a global manner to homeostatically adjust the postsynaptic weights of excitatory synapses"*

   We agree about global, but synaptic scaling takes place individually in each synapse. What makes it global is that the synapses have a common error signal.
4. *"the formal definition of a homeostatic system is that it operates in a negative feedback manner, so that when the system deviates from a set point value, an ensuing error signal triggers compensatory mechanisms that bring the system precisely back to this set point"*

   This is classical control theory. The set point is what we call the equilibrium point. The % Hebbian-LMS algorithm provides negative feedback to maintain homeostasis.
5. *"In the case of synaptic scaling, there is considerable evidence that the variable under control is some function of average neuronal firing rate"*

   We agree. Synaptic scaling intends to control the (SUM), the membrane potential, which is related through a sigmoidal activation function to the firing rate.
6. *"The timescale over which perturbations in firing are sensed and integrated, and the speed of the resulting homeostatic compensation, are still not entirely clear"*

   Adaptation toward homeostasis has a learning curve that is similar to exponential, probably similar to LMS learning. Nature chooses the parameter, μ.
7. *"on average, firing rates returned to within 15% of their initial value, even though neurons started from widely different mean firing rates"*

According to % Hebbian-LMS, convergence takes place from random initial weights.

8. *"FRH (firing rate homeostasis) is largely achieved through a set of cell-autonomous homeostatic mechanisms"*

   We agree. Each neuron adapts and converges individually.

9. *"so the changes one can measure after a given manipulation will be owing to a complex mixture of local and global, Hebbian and homeostatic processes"*

   We agree. In a network of neurons and synapses, a change of input will cause adaptation and convergence, restoring homeostasis. % Hebbian-LMS adaptation within the synapse is local. Homeostasis is global, being a property derived from actions of thousands of synapses.

10. *"Why build the system this way, with large (or fast) Hebbian mechanism and small (or slow) homeostatic mechanisms?"*

    Since Hebbian learning and homeostasis are of the same mechanism, the phenomena will take place within the same time frame. It may seem that there are two different time frames. Hebbian learning, adaptation of the weights of a synapse, is local and relatively fast. Homeostasis, resulting from the % Hebbian-LMS algorithm, is global and is averaged over thousands of synapses and may seem to change slowly. It has a lot more inertia.

11. *"While computational approaches to understanding the interplay between Hebbian and homeostatic mechanisms have been enormously useful for outlining a landscape of possible functions, our understanding of where, when and how various learning rules operate in real circuits is still rudimentary enough that it is hard to draw firm conclusions"*

    Hebbian and homeostatic mechanisms are all the same mechanism, according to the % Hebbian-LMS algorithm. This is a theory, based on physical evidence.

The biologist, experimenting with living neurons in vitro and vivo, are seeing two different phenomena, Hebbian learning and homeostasis. This is complicated. There are questions of how they cooperate in achieving stable results and questions of their relative speeds of operation. Thinking in terms of the % Hebbian-LMS algorithm, these issues become much easier to understand and model. The % Hebbian-LMS algorithm is in a sense, a unified theory of Hebbian learning and homeostasis. It is a simple algorithm and the information that nature would need to execute it, namely the neurotransmitter concentration and the membrane potential, are available at the synapse.

6.6 Advantages of the % Hebbian-LMS Hypothesis over the Synaptic Scaling Hypothesis

If a neuron has only excitatory synapses, no inhibitory synapses, % Hebbian-LMS and synaptic scaling are identical. When the neuron has both excitatory and inhibitory synapses, there will be big differences in the learning algorithms and their effectiveness. The latter is the case of greatest interest.

The objective is to change the synaptic weights in order to bring the (SUM) closer to the homeostatic level. One would like to get the largest change in the (SUM) from the smallest percent change in the synaptic weights, up or down, excitatory or inhibitory. Changing the weights to achieve the current objective disturbs already learned experience. Adaptation with minimal disturbance is of the essence.

Suppose, for example, that a neuron of interest has 60 excitatory synapses and 40 inhibitory synapses. Suppose that a 10% change in the synaptic weights is made: adapting with % Hebbian-LMS, the change in the (SUM) will be proportional to

$$60(0.1) + 40(0.1) = 10$$

The excitatory weights were increased by 10% and the inhibitory weights were decreased by 10%, contributing to an increase in the (SUM).

Adapting now with synaptic scaling, the change in the (SUM) will be proportional to

$$60(0.1) - 40(0.1) = 2$$

The excitatory weights were increased by 10% and the inhibitory weights were increased by 10%, contributing an increase in the (SUM) from the excitatory weight change and a negative increase from the inhibitory weight change. For the same disturbance to the weights, % Hebbian-LMS has a 5x advantage over synaptic scaling for this example.

Suppose for example that the neuron has 80 excitatory and 20 inhibitory synapses. With a 10% weight change, the change in (SUM) will be proportional to

$$80(0.1) + 20(0.1) = 10$$

for % Hebbian-LMS, and

$$80(0.1) - 20(0.1) = 6$$

for synaptic scaling. The % Hebbian-LMS algorithm will have an advantage of 1.6x over synaptic scaling. With less inhibitory weights, the advantage of % Hebbian-LMS becomes smaller.

On the other hand, suppose the neuron has an equal number of inhibitory and excitatory synapses. For this case, the advantage of % Hebbian-LMS over synaptic scaling would be infinite. Changing the weight values would have zero effect on the (SUM) with synaptic scaling.

The takeaway result is that % Hebbian-LMS is always effective but synaptic scaling will work in most cases but not very effectively in some cases. Since both algorithms would be just as easy for nature to implement, it is highly likely that % Hebbian-LMS would be nature's choice.

6.7 Memory Traces

An argument in favor of synaptic scaling often made by researchers in the field is that synaptic scaling achieves homeostasis, but at the same time, does not change the ratio of excitatory weight to inhibitory weight. The idea is that memory traces stored in the brain's neural networks would be preserved, or so they say.

The argument could be turned around in favor of % Hebbian-LMS. It might be more likely that to preserve memory traces, adaptation with % Hebbian-LMS would retain the ratios among the excitatory weights, retain the ratios among the inhibitory weights, and retain the same ratios between the excitatory weights and the reciprocals of the inhibitory weights. This seems a little more complicated, but it makes more sense.

This argument brings up the whole subject of memory traces, accepted widely in neuroscience. The memory trace hypothesis is based on the idea that long-term memory is stored in the brain's neural networks. This is highly unlikely and in addition no one has ever seen a memory trace. The brain's neural networks are too volatile for storage over a lifetime. In the very young, circuits are added constantly. In the middle-aged and elderly, circuits die out constantly.

So where is long-term memory stored? Probably in the massive amount of DNA in the neurons and glia. This is digital memory. The DNA cannot be altered? Not so. People have built small computers with DNA. The computers perform logic and memory operations.

Where would the idea of memory storage in DNA come from? A fertilized egg destined to become a human being has DNA from both parents. All inherited information is stored in the nucleus of this single cell, intrinsic information such as how to cry, pee, poop, suck, etc. There are no neural networks in the egg, no memory traces, but DNA is there.

If long-term memory is stored in DNA and not in the neural networks, what is the purpose of the neural networks. Alzheimer's patients experience memory loss because their neural networks are being destroyed by the disease. Neural networks must have something to do with memory. But what?

The answer is that neural networks in the brain are the mechanisms of memory retrieval, not storage. This is associative retrieval. Incoming patterns prompt the memory to retrieve stored patterns that are related, in the same cluster. Memory retrieval and pattern recognition are really the same thing. But all this is the subject of another book on both human and artificial memory.

6.8 Summary

Synaptic scaling, like % Hebbian-LMS, is an adaptive algorithm. The purpose of adaptation is to adjust synaptic weights to bring a neuron's membrane potential closer to homeostatic. Synaptic scaling increases and decreases the synaptic coupling by changing the synaptic weights, excitatory and inhibitory, together. % Hebbian-LMS changes the weights differentially so that when the excitatory weights are increased, the inhibitory weights are decreased. When the inhibitory weights are increased, the excitatory weights are decreased. % Hebbian-LMS follows the "principle of minimal disturbance" since its differential mode minimally disturbs the synaptic weights during adaptation. The most widely quoted neuroscientist in the fields of Hebbian plasticity and synaptic scaling is Dr. Gina Turrigiano. Her 2017 paper titled, "The dialectic of Hebb and homeostasis" cites Hebbian learning and homeostasis as mechanisms of synaptic plasticity that are separate and seemingly incompatible. The theory of % Hebbian-LMS solves this dilemma. Both Hebbian learning and homeostasis are of the same learning algorithm.

6.9 Questions and Experiments

1. Study the literature of homeostasis. Record your references.
2. Study the literature of multiplicative synaptic scaling. Record your references and observations.
3. Do scientific articles presenting physical evidence in support of the synaptic scaling hypothesis inadvertently support the % Hebbian-LMS hypothesis? Cite the articles and explain your reasoning.

References

1. Turrigiano, G.G.: The dialectic of hebb and homeostasis. Philos. Trans. R. Soc. B: Biol. Sci. **372**(1715), 20160258 (2017)
2. Stellwagen, D., Malenka, R.C.: Synaptic scaling mediated by glial tnf-α. Nature **440**(7087), 1054–1059 (2006)

Chapter 7
Synaptic Plasticity

Abstract This chapter summarizes the teaching of Chaps. 1, 2, 3, 4, 5 and 6. The Hebbian "wire together fire together" is extended to apply to situations where the presynaptic neuron is firing and the postsynaptic neuron is firing or not firing, and extended to apply to both excitatory and inhibitory synapses. The % Hebbian-LMS algorithm implements the extended Hebbian learning rules for small values of the postsynaptic membrane potential, but at larger values, the Hebbian rules begin to break down. The % Hebbian-LMS algorithm applies at small and large values of the membrane potential and predicts the phenomenon of homeostasis.

7.1 Introduction

Hebbian learning rules, their extensions, and their various aspects have been presented in the past chapters. The purpose here is to bring all of this together in one place, to summarize it, and in some sense, to establish a set of laws of synaptic plasticity and homeostasis. Another objective is to define the % Hebbian-LMS neuron, a neuron that is believed to exist and function in accord with the derived laws of synaptic plasticity.

7.2 Hebbian Learning, Synaptic Plasticity, Homeostasis

A neuron and its synapses perform two basic functions, signal propagation and learning. The neuron and its synapses work together and communicate with each other. What are they saying? A human observer can learn about this from many sources. Readily available from Google are innumerable websites on the subjects "neuron", "synapses", "synaptic plasticity," etc. Excellent sources are Wikipedia and Khan Academy. On any one topic, corroboration from many sources is necessary to have a clear picture.

Does the neuron and its synapses have a code book with a long set of executable rules that must be followed with mathematical precision every time it would be

B. Widrow, *Cybernetics 2.0*, Springer Series on Bio- and Neurosystems 14,
https://doi.org/10.1007/978-3-030-98140-2_7

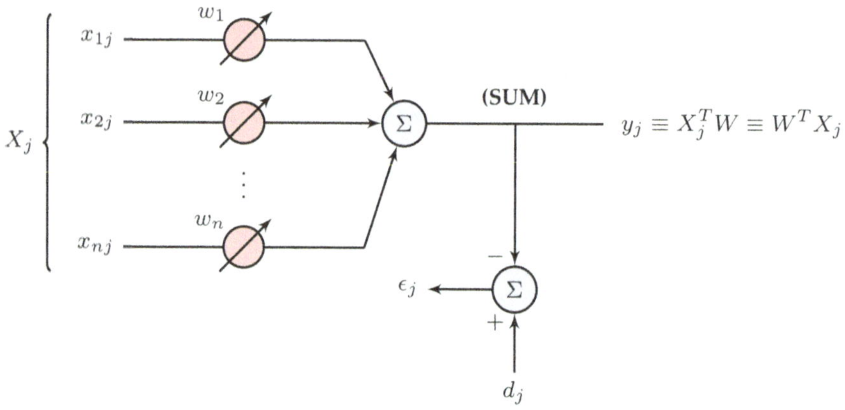

Fig. 7.1 A linear combiner

necessary to make a move? No. Of course not. The neuron and its synapses simply do what they do. Nature does what it does. She/he knows nothing about mathematics.

The human observer makes up a set of rules, for example Hebbian rules, in attempting to understand nature. Different observers with different backgrounds will establish their own descriptions or models of nature's behavior.

From an electrical engineer's standpoint, pictorial diagrams are very helpful. For example, Fig. 7.1 shows a linear combiner. With this diagram, signal flow is easily visualized. A (SUM) is created by weighting the input signals. The (SUM) is compared with a desired response to produce an error signal that will be used by an adaptive algorithm to control the weights. The weights are circles, and arrows through them indicate variability. The input signals comprise a vector X_k, the weight values comprise a vector W_k and the (SUM) is the inner product of these vectors. The membrane potential is proportional to (SUM). This is the main idea of synaptic and neuronal signal flow.

The (SUM) is proportional to the membrane potential.

$$(\text{SUM})_k = X_k^T W_k = W_k^T X_k. \tag{7.1}$$

A representation of nature's neuron and synapses is diagrammed in Fig. 7.2. This neuron is believed to incorporate the most general form of % Hebbian-LMS [1, 2]. This neuron is a linear combiner with a half sigmoid activation function. The neuron's output signal is a half sigmoidal function of (SUM) plus 55mV. The firing rate is proportional to this signal.

The (SUM) has a bias of –70 mV, the resting potential. The +55mV bias is part of the activation function. It insures that the neuron does not fire unless the (SUM) is greater than –55 mV. In % Hebbian-LMS fashion, the error signal is generated from the (SUM). The error function could be like the one that will be seen in Fig. 7.3, but it need not be exactly the same. It needs to have the same general shape with a

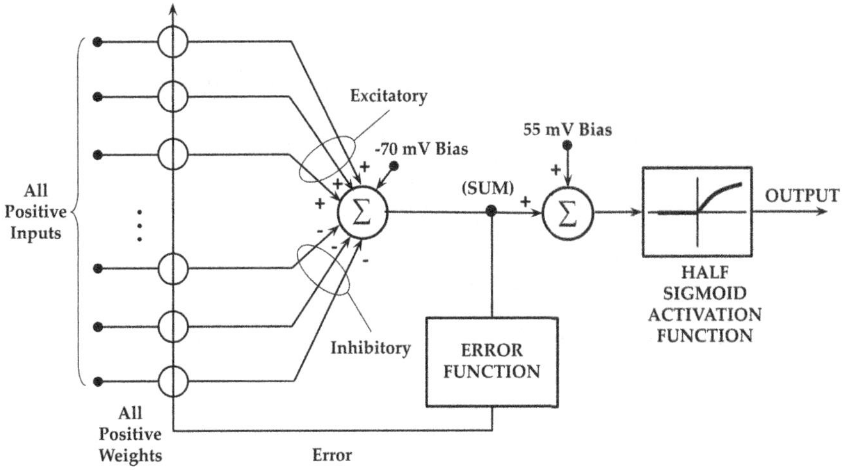

Fig. 7.2 The % Hebbian-LMS neuron

crossover at a positive equilibrium point. The negative equilibrium point of Fig. 7.3 is of less concern because the corresponding negative (SUM) would not allow the neuron to fire.

Nature's objective is to maintain the membrane potential, the (SUM), as close as possible to the homeostatic voltage, the homeostatic set point or positive equilibrium point. If the membrane potential is higher than the set point, the error, the difference between the membrane potential and the set point will be negative and approximately proportional to the difference. If the membrane potential is lower than the set point, the error will be positive. This is in accord with the error function of Fig. 7.3.

Nature will adapt the weights slowly, both excitatory and inhibitory, in their respective directions to reduce the magnitude of the error. That is what nature does. It's very simple.

Figure 4.7 is repeated here as Fig. 7.3 for ready reference. Observe the arrows pointing at the positive stable equilibrium point. The adaptive algorithm adjusts the weights to push the (SUM) to this homeostatic equilibrium point.

7.3 Hebbian Learning Rules

What nature does can be codified into a set of rules. What nature does seems to be simple, however representing this with a set of rules reveals subtlety and complexity.

The most succinct expression of % Hebbian-LMS learning in its most general form is given by Eq. (7.2)

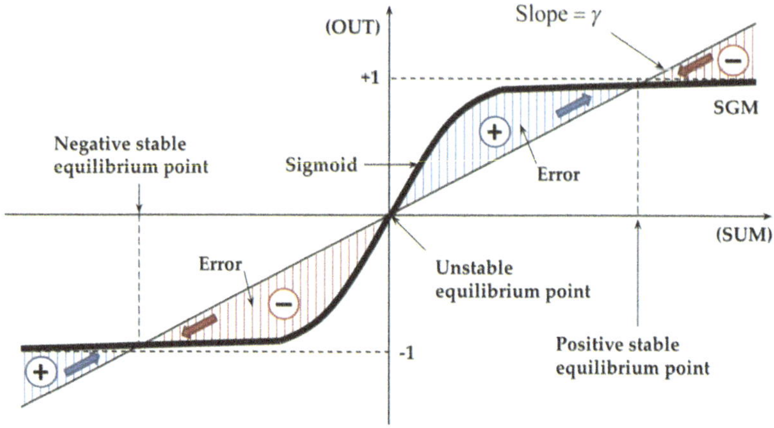

(a) The Output and Error vs (SUM)

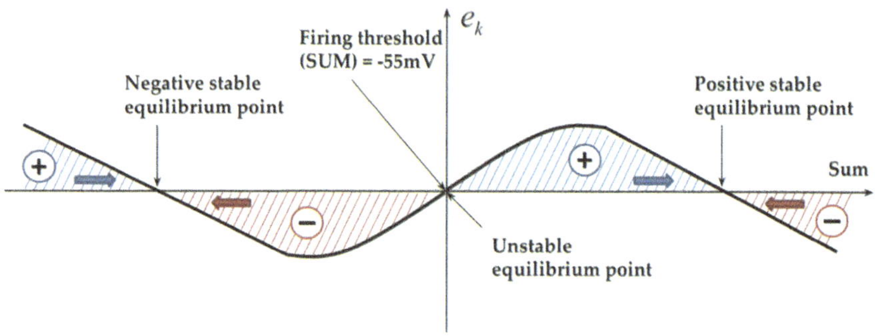

(b) The error function.

Fig. 7.3 Formation of the error function

$$
\begin{aligned}
W_{k+1} &= W_k + 2\mu\epsilon_k(X_k \circ W_k) \\
\epsilon_k &= f(X_k^T W_k) \\
(\mathrm{SUM})_k &= X_k^T W_k \\
(\mathrm{OUTPUT})_k &= \begin{cases} \mathrm{SGM}[(\mathrm{SUM})_k + 55\,\mathrm{mV}] & \text{for } [.] \geq 0 \\ 0, & \text{otherwise} \end{cases}
\end{aligned}
\tag{7.2}
$$

These equations comprise the % Hebbian-LMS algorithm. They incorporate all of the features of extended Hebbian learning and beyond. The parameter μ controls rate of adaptation and stability.

Equation (7.2) are written as they would be for digital simulation. The time index k represents the iteration cycle number. Digital implementation takes many finite small steps along the negative gradient of the mean square error surface. Such adaptive steps would not be taken by nature. Natural adaptivity is a continuous process, not discrete in time.

The Hebbian rules start with "neurons that fire together wire together." The extended rules encompass that, but go much further. The % Hebbian-LMS algorithm encompasses all of that and goes very much further.

There is one rule that applies under all conditions: When the presynaptic neuron is not firing, there will be no synaptic weight change. Another general condition is the following: The learning rules for excitatory and inhibitory synapses are opposites, regarding synaptic weight change.

The % Hebbian-LMS learning rules are the following, but keep in mind that the postsynaptic firing rate is related to postsynaptic membrane potential:

1. Excitatory synaptic adaptation
   (a) Postsynaptic firing rate lower than the homeostatic firing rate:
      i. When the presynaptic neuron is firing and the postsynaptic neuron is firing, the synaptic weight increases. (This is Hebb's original rule).
   (b) Postsynaptic firing rate greater than the homeostatic firing rate:
      i. When the presynaptic neuron is firing and the postsynaptic neuron is firing, the synaptic weight decreases.
2. Inhibitory synaptic adaptation
   All increases and decreases in synaptic weight values are the opposite of the above.

The extended Hebbian learning rules apply when the postsynaptic firing rate is low. Then Hebbian learning and % Hebbian-LMS learning are identical in behaviour. At higher firing rates, the two learning rules depart. At the homeostatic firing rate, Hebbian learning, Hebbian learning rules call for very rapid synaptic weight adaptation. At the homeostatic firing rate, % Hebbian-LMS calls for a cessation of adaptation. The Hebbian learning rules do not permit homeostasis at any firing rate. The % Hebbian-LMS algorithm predicts homeostasis when the (SUM) is at the crossover of the error function. (Fig. 7.3 shows the crossover). At this point, the membrane potential is at its homeostatic level. The firing rate is homeostatic.
At low firing rates, the Hebbian learning rules work perfectly, but at higher firing rates, Hebbian learning breaks down. % Hebbian-LMS learning takes over and supersedes Hebbian learning.
The perplexing problems that Dr. Turrigiano pointed out in her paper when trying to reconcile homeostatic plasticity with Hebbian plasticity seem to be solved with % Hebbian-LMS theory. Hebbian learning and homeostasis are

not mutually incompatible. We propose that the % Hebbian-LMS algorithm is the key to synaptic plasticity. We believe that the % Hebbian-LMS algorithm of equations (7.2) is most likely nature's learning algorithm. This is plausible, but not yet proven.

7.4 Summary

This chapter summarizes the ideas of Chaps. 1, 2, 3, 4, 5 and 6. Synaptic plasticity, the control of the number of receptors in a neuron's input synapses, is determined by a learning algorithm. Hebbian learning and homeostasis are at issue. The Hebbian-LMS algorithm fits Hebbian rules and predicts homeostasis. A general algorithm for synaptic plasticity is not yet known. % Hebbian-LMS theory is a good candidate for a general theory of synaptic plasticity.

7.5 Questions and Experiments

1. Refer to Fig. 7.2. Is there a general way, mathematically or otherwise, to describe a generic error function for % Hebbian-LMS? What properties must it have to create homeostasis?

References

1. Widrow, B., Kim, Y., Park, D.: The Hebbian-LMS learning algorithm. IEEE Comput. Intell. Mag. **10**(4), 37–53 (2015)
2. Widrow, B., Stearns, S.D.: Adaptive Signal Processing. Prentice Hall, Englewood Cliffs (1985)

Part II
Addiction and Mood Disorders

Introduction to Part II

A question exists regarding the effect of a therapeutic drug on the synapses of the brain. With life experience, the synaptic weights will have adapted and learned to perform usefully. Suppose one were to take aspirin or tylenol to alleviate the pain of a headache. Could these drugs cause changes to the synaptic weights and destroy learned functions? If one were to take one of these drugs, would 4 plus 3 still equal 7? What effect in general would taking a substance that crosses the blood-brain barrier have on learned experience? What if this substance were a neurotransmitter?

A recent article in the New York Times on opioid addiction became very interesting when it began to discuss neurotransmitters and neuroreceptors. That seemed to fall into the domain of synaptic plasticity. This is an example of a drug that permeates the brain and has neurological effects. What does it do to the knowledge learned and stored in the synapses? These questions led to a deeper study of addiction and its effects.

Another article in a Stanford University publication on research on pain perception brought up neurotransmitters and neuroreceptors. Going deeper into the subject led to a study of both pain and pleasure that turned out to be related to the study of addiction. One thing led to another and soon anxiety, depression, and bi-polar disorder became of interest. Hebbian-LMS theory turns out to be very helpful for explaining various aspects of these mental phenomena. Other cognitive processing may also be involved, it should be noted.

Chapter 8
Addiction

Abstract Endorphin receptors are involved with opioid addiction. In the brain, opioid masquerades as the neurotransmitter endorphin and binds to endorphin receptors. Under normal conditions, without opioid, neurons in the part of the brain having synapses with endorphin receptors fire at their homeostatic rate, likely regulated by the % Hebbian-LMS algorithm. Endorphin is inhibitory, so when opioid is injected into the bloodstream, it crosses the blood-brain barrier, and binds to endorphin receptors; it is inhibitory and the neurons' firing rates go below their homeostatic level. The brain perceives this as pleasure, euphoria. % Hebbian-LMS "fights" against the opioid trying to raise the firing rate back toward homeostatic by reducing the population of endorphin receptors, down-regulating them. When the drug wears off, the reduced population is still pulling the firing rate upward, and the rate overshoots above homeostatic and this is perceived as pain or agony, the opposite of euphoria. More opioid is taken to stop the pain, and this goes on and on. The result is addiction.

Keywords Addiction · Opioid, endorphin receptors · Habituation · Blood-brain barrier

8.1 Introduction

There are many substances, that, inhaled, taken orally, or by injection, enter the bloodstream, cross the blood-brain barrier, and become present in every synapse of the brain. Some of these substances are neurotransmitters and some, while not neurotransmitters, are capable of imitating neurotransmitters or causing an increase in the natural production of neurotransmitters. One way or the other, these substances can effect an increase in neurotransmitter molecules that can bind with their respective neuroreceptors. Specific substances affect only specific neuroreceptors and none other. Some substances are excitatory and others are inhibitory. Thinking of the brain's many trillions of synapses, each having learned weight values from normal living, a question arises: what happens to the brain and brain function when suddenly one of these substances is ingested, inhaled, or injected and it causes an increase in neurotransmitter in the synapses that respond to that substance? This will affect

B. Widrow, *Cybernetics 2.0*, Springer Series on Bio- and Neurosystems 14,
https://doi.org/10.1007/978-3-030-98140-2_8

signals sent throughout the parts of the brain that contain synapses with the specific neuroreceptors. This can have profound effects on brain function. This is clearly not what nature intended. When this does happen and strange behavior results, how can this behavior be explained? In some cases, this results in addiction. Activity of the neuroreceptors is involved, the same neuroreceptors that are involved in normal learning activity. The % Hebbian-LMS algorithm gives insight into normal learning, and it also can give insight into what happens when the entire brain is flooded with an active substance, even when this substance or drug causes addiction. How does addiction work is a good question. The theory of the % Hebbian-LMS algorithm offers a possible explanation.

8.2 Neurotransmitters

There are dozens, perhaps a hundred or more different neurotransmitters that are involved with brain activity. The most common neurotransmitter by far is glutamate, an excitatory neurotransmitter. The next most common neurotransmitter is GABA (gamma-aminobutyric acid), an inhibitory neurotransmitter. Another common neurotransmitter is dopamine that, depending on location and function, can be either excitatory or inhibitory. Glutamate is associated with learning and memory. GABA contributes to motor control, vision, and other control functions. Since most neurons in the brain have GABA and glutamate receptors, these neurotransmitters affect almost all sensory and behavioral functions.

Dopamine plays a large role in motivation and reward. Motivation is important for memory and interest in things. Melatonin is an inhibitory neurotransmitter that is associated with sleepiness. Seratonin is also an inhibitory neurotransmitter that has an effect on emotions, mood, appetite, and pain. Norepinephrine and histamine are neurotransmitters that can be either excitatory or inhibitory, depending on function. Norepinephrine is associated with arousal, learning, and mood. Histamine plays a role in metabolism, temperature control, and the sleep/wake cycle. Many other neurotransmitters exist that have a variety of cognitive effects. Many of the neurotransmitters have overlapping functions, affecting sleepiness, arousal, pain, etc. Each neurotransmitter seems to affect a number of functions. The literature is not completely clear regarding these functions. One might think that a specific neurotransmitter would have only a specific cognitive effect, but this is not the case. For a specific neurotransmitter, neuroreceptors exist in synapses connected to neurons located in various parts of the brain, each brain part having its own cognitive function. Hence, each neurotransmitter can be associated with a variety of cognitive effects.

8.3 Drugs

There are many addictive drugs that unfortunately exist, many of which can be devastating to health and well being. Some of the more common ones are nicotine, alcohol, heroin, cocaine, amphetamines, cannabis, and so forth. These drugs are not neurotransmitters but in the brain, some of them masquerade as neurotransmitters. Nicotine mimics the neurotransmitter acetylcholine. Heroin and morphine molecules, opioids, bind to endorphin receptors. Cocaine and amphetamines cause an abundance of dopamine. They do not mimic dopamine but they enhance its natural production. The effect of all these drugs is to increase the amount of their respective neurotransmitters [1, 2].

A case in point is that of opioid that attaches to endorphin receptors. Endorphin is an inhibitory neurotransmitter, so the use of opioid causes the synapses with endorphin receptors to apply inhibitory effects to the attached post-synaptic neurons, lowering their membrane potentials and making them less likely to fire, or to fire at a lower rate. Lowering the firing rate of these specific neurons creates a feeling of euphoria. Increasing the firing rate causes a feeling of pain, discomfort, agony, and fear, the opposite of euphoria. If one accepts the idea that nature uses the % Hebbian-LMS algorithm to control the synaptic weights, it is easy to explain how addiction works, how one can get "hooked," what happens to the synaptic weights initially and when using these drugs steadily as an addict, and what happens when the use of the drug is stopped, resulting in withdrawal symptoms.

8.4 % Hebbian-LMS and Addiction

The effect of heroin intake and how this drug works is fairly typical of the addictive drugs, although not universal. Using heroin as an example, the mechanism of addiction can be explained. One should keep in mind Figs. 2.1, 2.2, 4.8 and 5.3. Therein lies the secret of addiction, it seems.

Start with a normal brain with no drugs. The usual signal traffic will flow throughout the neural networks. Some of the neurons may be undergoing learning and slow synaptic weight changing. For a typical neuron after convergence, Fig. 5.3d illustrates values of (SUM) clustering about the two equilibrium points of Fig. 4.8a and b. The cases where the (SUM) clusters at the negative equilibrium point will not be of interest since they relate to the neuron not firing. The interesting cases are where the (SUM) is at or in the vicinity of the positive equilibrium point, where the value of the (SUM) corresponds to the homeostatic firing rate.

For convenience in this work, the neurons with synapses with endorphin receptors will be called endorphin neurons. Synapses with endorphin receptors will be called endorphin synapses. The endorphin neurons have both excitatory and inhibitory synapses. The inhibitory synapses are the endorphin synapses.

Like all neurons, the endorphin neurons have homeostatic set points, equilibrium points. A continuous adaptive process prevails whose purpose is to have these neurons firing at their homeostatic rate. Their membrane potentials need to be close to their homeostatic level, and the membrane potentials are affected by the excitatory and inhibitory synaptic weights.

When the endorphin neurons are firing at a rate lower than the homeostatic set point, the experience is pleasurable. When the endorphin neurons are firing at a rate higher than the homeostatic set point, the experience is unpleasant, painful. When the endorphin neurons are firing at the homeostatic rate, the experience is normalcy, no euphoria, no agony. The synaptic weights are constantly being adjusted in accord with the % Hebbian-LMS algorithm to keep the (SUM) values of the endorphin neurons close to their homeostatic level, and this balance keeps the individual in a normal state.

Now suppose heroin is suddenly injected. Every synapse with endorphin receptors will be fooled, "thinking" that the opioid is the neurotransmitter endorphin, and every neuron with attached endorphin synapses will experience an inhibitory kick. These neurons' (SUM) values will suddenly drop with negative voltage increments. The amplitude of a neuron's voltage drop will be proportional to the amount of opioid injected and to the number of its endorphin synapses. Millions of endorphin neurons will have their firing rates reduced, producing a feeling of euphoria. How that sensation happens is not clear, but the perception does indeed occur.

The euphoria lasts for five to ten hours and then gradually wears off. The heroin disappears and that contributes primarily to the loss of euphoria. Another contribution to the loss of euphoria is effected by the % Hebbian-LMS algorithm. Refer to Fig. 4.8a and b. When the value of (SUM) is lower than the positive equilibrium value, % Hebbian-LMS will adapt the weights to push the value of (SUM) up toward the positive equilibrium value. In Fig. 4.8a and b the arrows indicate this. A more positive (SUM) gives less euphoria. When % Hebbian-LMS adapts the weights to make the (SUM) more positive, it does this by making the endorphin weights smaller, less positive, and making all other weights somewhat more positive. This is synaptic plasticity. The total number of endorphin receptors in the endorphin synapses is reduced, and the total number of receptors in all other synapses is somewhat increased.

This explanation of pain and euphoria is somewhat simple. It does not take into account the contributions of other brain circuits to observed human behavior. These contributions can be significant.

8.5 Overshoot and Addiction

A shot of opioid gives the endorphin synapses a very powerful input signal. The other synapses of the endorphin neurons have no such inputs. They ignore the opioid. Their inputs are at a much lower level, due only to normal brain signal traffic. Consequently, when % Hebbian-LMS adaptation occurs, the rate of adaptation of the endorphin synapses will be much higher than that of the other synapses of the endorphin neurons.

Given a shot of opioid, the endorphin synapses experience rapid adaptation as long as the opioid is present. As it wears off, the rate of adaptation slows substantially. The purpose of adaptation is to fight against the unnatural opioid effects.

The synaptic weights were adapting to push the membrane potential up toward the homeostatic level. The opioid is pushing in the opposite direction. The membrane potential can get close to the homeostatic level but remains below. There will be an error, employed by the adaptive algorithm to push toward homeostasis.

The cumulative effect of the weight changes is to cause the endorphin synapses' weights to be much smaller than normal. The other weights of the endorphin neurons will be somewhat greater than normal.

As the opioid is present and gradually wears off, the membrane potentials rise toward the equilibrium level. Because of the cumulative weight changes pushing the membrane potential in the positive direction, the membrane potential will not rest at the homeostatic level when the opioid wears off. The membrane potentials will rise well above the equilibrium point and the result will be a sensation of pain. This overshoot of the membrane potential causes agony known as withdrawal.

With the opioid gone and the individual having withdrawal, the membrane potentials of the opioid neurons will be well above the homeostatic level. % Hebbian-LMS will then be adapting the synaptic weights in the opposite direction to bring the membrane potentials back toward the homeostatic level. This adaptation will be much slower in the absence of the strong synaptic input signals due to opioid. Withdrawal symptoms will persist over many hours, perhaps days, as % Hebbian-LMS gradually returns all of the synapses to their normal values.

Withdrawal is so awful that a person will do almost anything to stop the pain. The way to stop it is to take another shot of opioid. Once that happens, the person is addicted. Addiction results from the dire need to relieve pain. The desire for euphoria has a small effect on addiction, but the principal motivation is pain relief.

8.6 Habituation

An addicted person usually takes a new dose of opioid before severe withdrawal sets in. The weights are still adapted to counteract the opioid. A larger dose of opioid is needed to achieve the original euphoric effect. After many cycles of dosing, the synaptic weights settle into a pattern that on average has them adapted more strongly in the direction to counteract the effects of the opioid. This is a cumulative phenomenon that will require much more opioid each cycle until reaching a steady state, getting the original level of euphoria. Taking more and more of the drug is called habituation. This happens with all addictive substances.

8.7 Graphical Interpretations

When a person takes a drug such as heroin for the first time, or has been taking it steadily, or suddenly stops, the level of heroin in the brain varies as a function of time. Among the neurons affected by heroin, the (SUM) voltages, i.e. the membrane potentials, vary as functions of time. For those neurons, the values of their inhibitory weights, the endorphin weights are also functions of time. The average value of their excitatory weights, all non-endorphin weights, is another function of time. It is instructive to examine these time waveforms to get another perspective on addiction and the possible role of the % Hebbian-LMS algorithm.

Figures 8.1 and 8.2 have plots of these time functions. These plots do not represent real data. Obtaining such data from living subjects would be difficult if not impossible. These plots represent a thought experiment based on knowledge and experience with the % Hebbian-LMS algorithm. They are predictions based on the idea that nature is performing the % Hebbian-LMS algorithm and thereby is controlling synaptic response to drugs. The results account for observed human response to addictive drugs.

In a normal state, never having taken heroin, the neurons with both endorphin and non- endorphin synapses have weights and values of (SUM) that differ somewhat from neuron to neuron. It is neither practical nor insightful to attempt to follow the effects of subsequent heroin taking on individual synapses and (SUM) values. It makes more sense to examine and study these effects averaged over the (SUM) values of many neurons, perhaps millions of them, and to study the effects on the

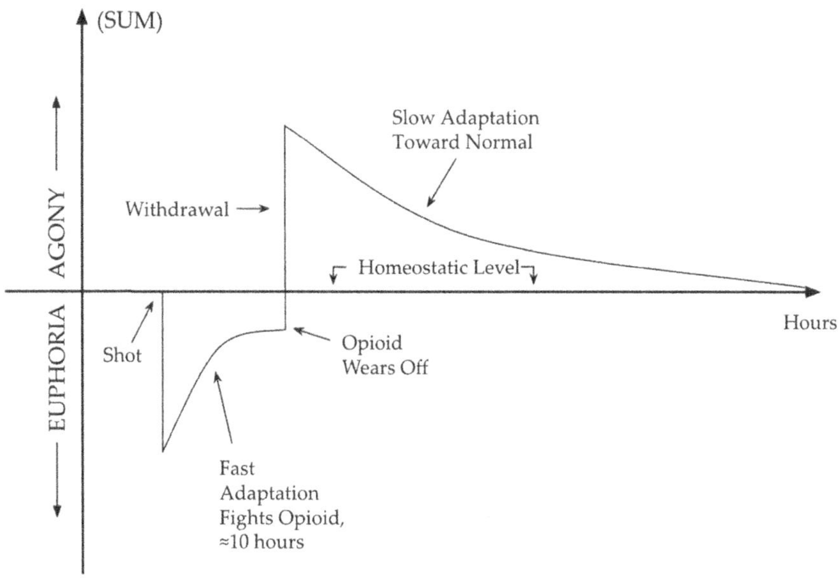

Fig. 8.1 Membrane potential of an endorphin neuron after a single shot of opioid

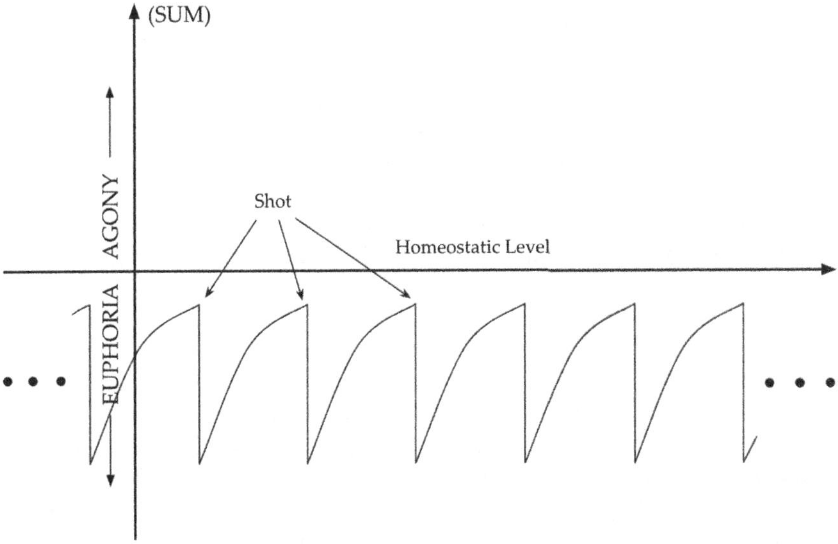

Fig. 8.2 Membrane potential of an endorphin neuron in addictive steady-state

endorphin synapses by averaging over all these synapses of a given neuron and then averaging over all neurons with endorphin synapses. The same averaging is done for the non-endorphin synapses. The endorphin synapses are inhibitory. The non-endorphin synapses are generally excitatory.

Figure 8.1 is a plot of membrane potential of an endorphin neuron after receiving a single shot of opioid. The membrane potential starts at the homeostatic level, the normal state. Receiving the shot, the membrane potential drops very rapidly. With such a strong signal inputted to the endorphin synapses, they begin to adapt rapidly, decreasing in amplitude in order to fight the opioid and restore the membrane potential to the homeostatic level. With much smaller input signals, the excitatory weights adapt more slowly to increase their magnitudes. All synaptic weights adapt to bring the membrane potential toward the homeostatic level. The net effect is rapid adaptation.

After about 10 h, the opioid wears off and with its affect shut down, the membrane potential rises by an amount similar to the amount of depression when first receiving the shot. The result is withdrawal with voltage overshoot. The plot of Fig. 8.1 shows this. The membrane potential is now well above the homeostatic level, into the range of agony.

With the opioid gone, the input signals to the endorphin synapses are now at the level of normal neuron-to-neuron traffic, much lower than with opioid. Adaptation becomes much slower. % Hebbian-LMS reverses its direction and

now raises the magnitudes of the endorphin synaptic weights and reduces that of the excitatory synaptic weights in order to bring the membrane potential back to the homeostatic level.
The plot of Fig. 8.1 was drawn assuming the effect of the opioid was uniform over time and suddenly dropped to zero when the opioid wore off. Affectation by the opioid will in nature not be uniform over time and will not stop suddenly when the drug wears off. Realistically, the plot of Fig. 8.1 should have its sharp features rounded somewhat, but the general idea is there.

An addicted person would be taking doses of opioid on a regular basis, motivated by the threat of the pain and agony of withdrawal. Figure 8.2 is a plot of membrane potential of an addict, assuming that the opioid doses are uniform in amplitude and are taken uniformly in time. This would not happen exactly in practice, but the general idea is represented by the plot. Each dose is assumed to be taken before the opioid from the previous dose wears off.
The ups and downs of the membrane potential is a "sawtooth" waveform over time. It undergoes a sharp drop when the shot is administered, and from then it undergoes an exponential-like rise as % Hebbian-LMS works to raise the membrane potential toward the homeostatic level. The plot shows a steady-state periodic waveform. Habituation is assumed to be established. The individual doses would be strong. Euphoria is experienced when the shot is taken. This gradually wears off due to the action of % Hebbian-LMS. Another shot is taken to bolster the feeling of euphoria, before reaching homeostasis and beyond with withdrawal symptoms. Withdrawal with agony is the threat that keeps addiction going. The addict is in a state of euphoria that goes up and down in intensity.

8.8 Opioid Detox

A program to break an opioid addiction is called detox. The idea is to rid the body of a toxic substance. With opioid addiction, the toxic substance is the opioid. Breaking addiction to nicotine, alcohol, cocaine, etc., is also called detox. Most often, opioid detox is performed in a 24 h per day clinical setting where the addicted person is helped by a trained staff that includes nurses and doctors. The simplest form of detox is "cold turkey." This is a sudden stop of opioid intake. The experience of withdrawal is brutal, painful, with fear and agony. Withdrawal symptoms begin about 12 h after the last dose of opioid. This is the beginning of what is called acute withdrawal. The symptoms peak in about 3 to 5 days, and last for 1 to 4 weeks. At the end of the acute phase, the opioid is completely out of the body and the brain has adjusted to the condition of no opioid. Addiction has ceased.

The post-acute phase can last for as much as two years. The post-acute phase is a psychological battle against the craving for the drug. This is not addiction, it is fighting to suppress the memory of the euphoria of the drug and fighting to prevent action in response to the memory. A lot of psychological support from family, friends, and professionals is needed to prevent resumption of drug use. After the post-acute phase, the person is "cured" but is always in danger of a drug relapse.

During the acute phase with no drug input, the endorphin weights gradually increase from near zero to their nominal values. The non-endorphin weights decrease gradually from large positive values to their nominal values. The neurons' membrane potentials will relax from large positive values and converge on the equilibrium point, whereupon there will be no perception of either pain or pleasure. Convergence is reached in about one week or so.

The brutal experience of cold turkey can be mollified by a continuation of drug input, gradually reducing the dose until the dose reaches zero. Gradual tapering will require a longer period of time than cold turkey, but the withdrawal symptoms that accompany tapering will be much less severe than with cold turkey.

A method of tapering is described herein that will essentially eliminate withdrawal symptoms. The method is based on simple reasoning in accord with the behavior of the % Hebbian-LMS algorithm. The method would require a clinical setting with trained attendants available on a 24 h per day basis.

The method works as follows. The addicted person is initially given no drug until experiencing the first indication of withdrawal symptoms. Without delay, this person is given 90% of the usual dosage of the opioid and no more. When the first signs of withdrawal appear again, an 80% dose is given. Waiting once again until withdrawal begins, the subject person is given a 70% dose, and so forth until the final dose of 10%. Each dose is given first at the onset of withdrawal symptoms. The amplitude of the dosage is tapered down towards zero. Stopping at 10%, some withdrawal will be felt, but it will be very much less than would be felt with cold turkey. A final dose of 5% could be added to the procedure to further reduce the last of the withdrawal effects.

8.9 Summary

After a shot of opioid, % Hebbian-LMS fights against the opioid, attempting to restore homeostasis. Weights have been changed in order to raise the membrane potential. If not for % Hebbian-LMS, the membrane potential would return to the homeostatic level as the opioid wears off and there would be no withdrawal. With % Hebbian-LMS the changed weights cause the membrane potential to overshoot as the opioid wears off. The result is withdrawal that persists as the synaptic weights slowly adapt back toward their homeostatic values. The % Hebbian-LMS algorithm is responsible for withdrawal. Withdrawal is responsible for addiction.

% Hebbian-LMS seems to be the "bad guy" here, until one stops to realize its true purpose. The purpose of % Hebbian-LMS is to maintain normalcy, to keep the membrane potentials of the endorphin neurons close to homeostatic. Opioid intake is not normal and upsets normalcy.

8.10 Questions and Experiments

1. Opioids bind to endorphin receptors and the result is euphoria. When the opioid wears off, withdrawal results unless another dose of opioid is taken. Avoidance of the agony of withdrawal is addictive behavior.
   Nicotine is addictive. What are the receptors for this drug? How does addiction work in this case?
2. Cocaine is addictive. What are the receptors for this drug? How does addiction work with cocaine?
3. Cannabis is potentially addictive. What are the receptors and how does addiction work with cannabis? How is its use sometimes addictive and sometimes not addictive?
4. Alcohol is potentially addictive. What are the receptors involved? How does alcohol addiction work? Why is alcohol use sometimes not addictive, and sometimes addictive?
5. Some persons with chronic pain use opioids for pain relief but do not want to be addicted. One possibility would be to use the drug sparingly, some background level of pain remaining. The other possibility would be to use the drug at a higher dose level to totally relieve the pain. Would the lower dose usage be less addictive? Explain your reasoning. Think of withdrawal effects.

References

1. Heroin addiction explained: How opioids hijack the brain - The New York Times. https://www.nytimes.com/interactive/2018/us/addiction-heroin-opioids.html, December 2018. (Accessed 10 April 2020)
2. "Researchers discover the brain cells that make pain unpleasant | news center | Stanford Medicine." http://med.stanford.edu/news/all-news/2019/01/researchers-discover-the-brain-cells-that-make-pain-unpleasant.html, January 2019. (Accessed 10 April 2020)

Chapter 9
Pain and Pleasure

Abstract Neurons in a certain part of the brain have synapses with endorphin receptors. These synapses are inhibitory. The same neurons also have synapses that are excitatory. This part of the brain is the focus of both pain and pleasure. Pain sensors all over the body transmit signals to the excitatory synapses. A painful accident causes the firing rate of neurons to climb well above homeostatic. This is perceived as pain. A sudden injection of heroin causes the firing rate to go well below homeostatic and this is perceived as pleasure. Having pleasure first, when it stops, overshoot causes pain. This is addictive since one would do anything to stop the pain. Having pain first that suddenly stops, overshoot causes euphoria and this is not addictive. Long distance runners have great pain in their legs that stops when they stop running. They get "runner's high." Natural childbirth is another example. The pain is great but stops suddenly when the baby is born. The overshoot causes what is called "postpartum euphoria." Having babies is not addictive. Eating food is. This is nature's calculation for survival of the species.

Keywords Pain, pleasure · The body's pain sensors · Runner's high · Postpartum euphoria

9.1 Introduction

You were resting, then suddenly stood up to walk, tripped and fell, and landed on the floor with a broken arm. The pain is severe. Neuronal pain sensors in the arm are firing, screaming from the trauma, sending axonal signals to the spinal cord and thence to the brain. Before the fall, parts of the brain containing neurons with endorphin synapses and non-endorphin synapses, inhibitory and excitatory inputs, had been experiencing normal nominal synaptic and neuronal signal traffic. After the fall, the signal traffic in this part of the brain increases dramatically. Axons carrying the pain signals terminate and innervate neurons distributed over this part of the brain, a part that contains endorphin neurons. Excitatory neurotransmitter is released. This traffic delivers extraordinary inputs to the excitatory synapses. Among the endorphin neurons, the membrane potential is increased by the action of the excitatory synapses.

B. Widrow, *Cybernetics 2.0*, Springer Series on Bio- and Neurosystems 14,
https://doi.org/10.1007/978-3-030-98140-2_9

The effect is modulated by the synaptic weight values. The average effect will be a positive change in membrane potential over and above the equilibrium point. The psychological effect will therefore be a feeling of pain and agony. This is analogous to the effect that one gets from opioid withdrawal.

Suffering from a broken arm, one could take an opioid such as heroin that will reduce the membrane potential of the neurons and thus will eliminate the sensation of pain and could even reverse it to euphoria, depending on dosage.

The % Hebbian-LMS algorithm will try to bring the neurons' membrane potentials to the equilibrium point, but the continual pain input signals will cause the membrane potentials to be raised above the equilibrium point and pain will be perceived. On the other hand, using opioids, the pain can be made to go away. % Hebbian-LMS will as always try to bring the membrane potentials up to the equilibrium point but with sufficient opioid input, the membrane potentials will remain somewhat below the equilibrium point and no pain will be felt.

If no painkillers were used, there would be pain. The initial pain would be intense. Gradually the pain will subside to a stable low level or to vanish altogether as % Hebbian-LMS brings the membrane potentials closer to the equilibrium point. This is homeostasis.

An injury anywhere on the body will cause pain signals to be transmitted to the brain where the endorphin neurons reside, causing the perception of pain. The sense of touch allows one to localize a point of contact where something touches the body. The sense of touch will allow one to locate the source of pain. Feeling the pain and being able to locate the injury allows one to take defensive action, critical for survival.

9.2 Transient Pain and Euphoria

Transient pain is the type that is hurtful for a relatively shot period of time and then stops, and the period might be seconds, minutes, or hours. There are many examples of transient pain. Several examples will be described next.

Vigorous running causes pain in the leg muscles that stops soon after running stops. While running, the pain signals go to the brain and provide strong inputs to the endorphin synapses and the non-endorphin synapses. Inputs to these synapses sharply reduce as soon as the pain signals stop soon after running stops.

While running, the pain signals cause the membrane potentials of the endorphin neurons to rise and % Hebbian-LMS adapts the synaptic weights, strongly increasing the endorphin weights while slowly reducing the non-endorphin weights, in order to push the membrane potentials down toward the equilibrium point. A long, hard run will cause the non-endorphin weights to be reduced to values close to zero, and the endorphin weights to take large values. When running stops and the pain signals stop, the normal synaptic and neural signal traffic is restored. The non-endorphin weights are very small and the endorphin weights are very large. The signal traffic applied to these weights causes the membrane potential to be below the equilibrium point and euphoria is felt. This is overshoot. % Hebbian-LMS now works in the opposite direction to bring the membrane potentials up to the equilibrium point over a period of 2 or 3 h, while euphoria subsides. This is homeostasis.

Euphoria following intense pain is known as "runner's high." The stronger the pain, the longer and deeper will be the euphoria that follows. Generally a mild buzz from the endorphin neurons will be felt for up to 24 h after initial euphoria. Since pain does not return after euphoria subsides, running is not addictive. Running is voluntary, not compulsory.

Hot chili peppers contain capsaicinoids which cause pain in the mouth when eaten. Pain is perceived when eating the chillies and persists for perhaps a half hour after finishing eating them. When the pain stops, a mood boost is felt that lasts for 2 to 3 h. Pain followed by mild euphoria is desirable to many people. This is perhaps one reason why hot spicy food is widely appreciated. Eating hot spicy food is voluntary, not compulsory. Spicy food is not addictive.

Deep tissue massage is painful but people submit to this for the mild euphoria that follows after the pain stops. The euphoria lasts for 2 to 3 h, and it can be followed by a mood boost for a day or so. Massage is not addictive, but often will be repeated periodically as motivated by the memory of the mild euphoria.

There are all sorts of self-harm that follow the pattern of first pain then 2 to 3 h of pleasure. Acupuncture is an example. Sadomasochism (S and M) is another example. A dictionary definition is the following: "Sadomasochism is the giving or receiving pleasure from acts involving the receipt or infliction of pain or humiliation." This is generally performed by two people and it is not addictive, although memory of the pleasure prompts repetition.

Another example of self-harm is the practice of "cutting," cutting oneself to inflict pain to be followed by pleasure. This is crazy but some unfortunate people do it.

Eating food represents another interesting form of pain and pleasure. When the stomach is empty, one experiences hunger pains. Consuming food puts a quick end to hunger pains and this results in mild euphoria for 2 to 3 h. When hunger pains return, the only way to make them stop is to eat. Eating is addictive. This type of addiction is critically needed for survival.

Many forms of experience fit the pattern of pain and pleasure. There is an expression "no pain, no gain." In all these cases, the % Hebbian-LMS algorithm tries to force homeostasis to keep the endorphin neurons at their equilibrium points. The % Hebbian-LMS algorithm cannot do this perfectly, since the error signal goes to zero at the equilibrium point and therefore the correcting force goes to zero there. Nevertheless, % Hebbian-LMS moderates the perceived effects of pain and pleasure, which would be much stronger without % Hebbian-LMS. The main "goal in life" of % Hebbian-LMS is to maintain homeostasis, working against the sensation of either pain or pleasure.

9.3 Conclusion: Addiction and Pain-Pleasure/Pleasure-Pain

Assuming that the % Hebbian-LMS algorithm is nature's method for controlling the brain's synaptic weights for Hebbian learning and synaptic plasticity, this same algorithm is involved with addiction and the perception of pain and pleasure. The fact that this algorithm explains so many aspects of behavior and pain-pleasure sensation without any known contra-indications, its credibility as nature's learning algorithm is enhanced.

Neurons with endorphin synapses are of interest here. There are perhaps millions of these. When their membrane potentials are as usual at or near the equilibrium point, one does not feel pain or pleasure. If the average of their membrane potentials is greater than the equilibrium point, pain is perceived, and if lower, pleasure is experienced. In the event of either pain or pleasure, the % Hebbian-LMS algorithm is activated with the goal of restoring the membrane potentials to the equilibrium point. This is homeostasis, a natural regulator. Over the time needed for the synaptic weights to adapt, the regulator moderates the sensation of pain or pleasure. Addiction and the perception of pain or pleasure can all be explained in light of knowledge of homeostatic regulation.

What the % Hebbian-LMS algorithm does is unsupervised synaptic learning and homeostasis. Homeostasis and synaptic learning, synaptic plasticity, have a profound effect on drug addiction and the perception of pain and pleasure.

Pain and pleasure are closely related experiences. Having pleasure first and then pain is a requisite sequence for addiction. An example is the injection of an opioid to create pleasure, then pain when the drug wears off. To quench the pain, it is necessary to have another injection of opioid. Another example of addiction is eating food to get pleasure, which will be followed later by the hunger pains of the empty stomach. To stop the pain, one must eat.

Having pain first, then pleasure does not lead to addiction. An example is runner's high. Pain in the leg muscles from vigorous running followed by euphoria after running stops is not addictive. Memory of the euphoria can prompt repetition however, but this is voluntary and not compulsory. Another example is self-harm. A few people will cut themselves to feel pain, to be followed by euphoria. This is not addictive, but voluntary.

All of these experiences can be accounted for by the % Hebbian-LMS hypothesis. A final example of pain and pleasure is postpartum euphoria, which is experienced by mothers during and just after giving birth. Googling these two words gives a great deal of information about this phenomenon.

Natural childbirth is done without any painkilling drugs. This is generally very painful. Once the baby is born, the pain stops. The % Hebbian-LMS hypothesis predicts that the sudden cessation of this intense pain will be followed by euphoria. Overshoot!

The google websites give testimony of mothers who had natural childbirth that was followed by euphoria. However, not all of the mothers felt euphoria. It is not very clear why this is. Perhaps tissue damage creates a significant level of pain that persists for some time after the birth and that may prevent euphoria. About half of the mothers experienced euphoria, a real high that is above and beyond the great pleasure of their new baby being presented to them.

Professor Widrow has some stories from his own family experience. He is the oldest of four children. When he was a teenager, he asked his mother if childbirth was painful. There was no such thing as epidural in those days. His mother said that childbirth was painful. He asked her why she had more kids after he was born. She said that "one forgets the pain soon after the baby is born." Perhaps she had euphoria after each birth and was able to forget the pain. His father had an explanation for all of this. He said "we had four kids in five years and then found out what was causing it."

Having babies is not addictive. The recollection of the pleasant part is what motivates mommy, however. Daddy is motivated by you know what. The net effect is that families generally have more than one child, and this is important for survival of the species.

When taking a drug and you suddenly stop, as with opioids, the result is overshoot, withdrawal. When performing a physical activity, such as running, and you suddenly stop, the result is overshoot, runner's high. When you have a baby and the baby is born, you have overshoot, postpartum euphoria. What is involved is the endorphin neurons and most likely the % Hebbian-LMS algorithm.

9.4 Summary

Pleasure first then pain is addictive. To alleviate the pain, the pleasure part must be enlisted. This is compulsory, not optional. An example is heroin addiction.

Pain first, then pleasure is not addictive. Memory of the pleasure encourages repetition but does not compel it. An example is runner's high. Another example is natural childbirth without drugs.

Pain sensors all over the body connect to the endorphin neurons. Pain perception is not localized. Touch sensing is highly localized. The combination allows one to respond to an injury.

An injury is most painful initially. Then the pain subsides somewhat. % Hebbian-LMS is in action. Other brain function might also be involved, however. There is evidence that endorphins are released during the pain experience to help reduce pain severity [1].

9.5 Questions and Experiments

1. Every six months, I visit my dermatologist to be inspected for keratosis spots on my skin and have them frozen with liquid nitrogen. They can be pre-cancerous and need to be removed. When a cotton swab that was dipped into liquid nitrogen is applied to the skin, the pain is intense, but within a few minutes it subsides and drops to a steady much more tolerable level and gradually disappears over several hours. Can you explain what is happening here?
2. When you eat ice cream too fast, you can get a severe headache that lasts for a minute or two. What is going on here? The pain from the super cold ice cream creates pain in the stomach. Why is the pain felt as a headache? There are pain sensing neurons in the stomach, but are there are also touch sensing neurons there? Would any pain in the body where there are pain sensing neurons but no touch sensing neurons raise the firing rate of the endorphin neurons in the brain and be perceived as a generic pain with no location, and the generic pain would be perceived as a headache? Since there are no pain sensors in the brain, how could you have a headache? With a head ache, what is the source of the pain?
3. Find a woman who just gave birth, the old fashioned way without anesthesia or pain killers. Ask about postpartum euphoria. Find another woman who had given birth, but with anesthesia and pain killers. Ask about postpartum euphoria. Note the experiences of the two women regarding euphoria.
4. Do pain sensing neurons have receptors? If they do, then the number of receptors can be increased or decreased to regulate the firing rate and to try to maintain homeostasis. Does the % Hebbian-LMS algorithm apply to pain sensing neurons? If so, would that contribute to pain subsiding after sensing the extreme pain of sudden trauma?

Reference

1. Boecker, H., Sprenger, T., Spilker, M.E., Henriksen, G., Koppenhoefer, M., Wagner, K.J., Valet, M., Berthele, A., Tolle, T.R.: The runner's high: opioidergic mechanisms in the human brain. Cereb. Cortex **18**(11), 2523–2531 (2008)

Chapter 10
Anxiety, Depression, Bipolar Disorder, Schizophrenia and Parkinson's Disease

Abstract In this chapter, several different brain disorders are addressed. The common thread is upregulation and downregulation of critical neuroreceptors. Each of these brain maladies is associated with a specific brain area whose neurons' firing rates determine normalcy or affliction. For example, anxiety is caused by the relevant neurons having an abnormally low homeostatic firing rate. Excitatory drugs can provide relief by raising the rate above homeostatic. Another example is depression caused by an abnormally elevated level of glutamate, which is excitatory. A higher than homeostatic firing rate can be lowered by use of the drug Ketamine, an NMDA antagonist that diminishes the binding of glutamate to the NMDA receptors. Without this drug, nature's % Hebbian-LMS algorithm alone is not strong enough to enforce homeostasis, but with Ketamine together, relief is possible. Another example is bipolar disorder. When the relevant neurons fire at the homeostatic rate, the mood is normal. Above the rate, there is mania, below this rate gives depression. Lithium sulfate treats this disorder. It does not seem to raise or lower the rate. How it works is not understood. It is speculated here that it increases the parameter μ of the % Hebbian-LMS allowing it to pull the firing rate closer to homeostatic with smaller error. Schizophrenia and Parkinson's disease are further examples of disorders that can be mitigated by control of receptor populations.

Keywords Bipolar disorder · Parkinson's disease · Substantia Nigra · Dopamine · Striatum · Basal Ganglia · Neuromodulation, CPCR's

10.1 Introduction

Human anxiety and depression are two different psychological afflictions. Like synaptic learning, these afflictions are affected and also controlled by synaptic weight variations and their neuronal responses. These variations take place in the parts of the brain that are configured to regulate mood, fear, etc., and self-confidence, interest in things, etc. With the involvement of neurotransmitters, neuroreceptors, and synaptic weights, it is possible to model this kind of behavior with the % Hebbian-LMS algorithm and see if the expected behavior of % Hebbian-LMS will account

B. Widrow, *Cybernetics 2.0*, Springer Series on Bio- and Neurosystems 14,
https://doi.org/10.1007/978-3-030-98140-2_10

for observations of anxiety and depression, and the effects of therapeutic drugs. If it fits, this will give further support to the % Hebbian-LMS model.

10.2 Anxiety

Anxiety is a disorder that involves several neurotransmitters in the brain. It is not clear if anxiety is related to the concentration level of only one or of several of these neurotransmitters. The neurotransmitters in question are serotonin, GABA, dopamine, and norepinephrine. It is not possible yet to measure these concentrations and to determine their effect on anxiety.

A certain part of the brain is involved with anxiety. Its various synapses have receptors for serotonin, GABA, or norepinephrine. The firing rates of the neurons in this part of the brain yield a perception of anxiety or no anxiety. Low firing rates are associated with anxiety. It is assumed that the firing rates are controlled by the % Hebbian-LMS algorithm, which establishes homeostasis. A person having anxiety symptoms has abnormal equilibrium points for one or more of the neurotransmitters. They are set at levels well below normal. The homeostatic firing rates are too low. Relief from anxiety can be had by raising the firing rate.

An anti-anxiety drug that is effective for many people with GAD (generalized anxiety disorder) is Buspirone. This drug binds to serotonin and dopamine receptors, and it has an excitatory effect on neurons having synapses with these receptors. With steady use, relief can be achieved within about six weeks. The firing rate is brought and held above the equilibrium point by this drug.

Many other neurons with synapses having these same receptors, neurons that are not associated with anxiety, will also have their firing rates raised by this drug. The result of this could be side effects such as nausea or headache from the use of Buspirone. These side effects could in some cases be serious and are then called "serotonin syndrome."

When the first dose of buspirone is ingested, within an hour or so the neuron firing rate is increased above the equilibrium point and this has an effect on anxiety but it is too soon to be perceived. The excitatory weights are gradually decreased and the inhibitory weights are gradually increased as % Hebbian-LMS "fights" to decrease the firing rate to maintain homeostasis at the abnormal equilibrium point, thus trying to defeat the drug. With daily dosage over many weeks of treatment, the reduced excitatory weight values and the elevated inhibitory weight values are sustained. The firing rate remains somewhat above the equilibrium point, enough to relieve the perceived anxiety.

If the treatment is suddenly stopped, there will be withdrawal. The anxiety will return stronger than ever because of the effects of % Hebbian-LMS on the weights but will soon subside to baseline as % Hebbian-LMS reverses and gradually returns the firing rate to the homeostatic equilibrium point. Withdrawal symptoms may not be noticed since only the usual kind of anxiety is sensed at withdrawal and beyond.

The patient does not observe withdrawal as being unusual, and this drug is therefore not addictive. Withdrawal does not compel addiction.

Treatment with Buspirone begins to be sensed after a few weeks and comes to full value in about six weeks. It is not clear why the response takes so long. Relief occurs much quicker, but for some unknown reason the perception of relief seems to take much longer. The treatment is not a cure for anxiety. As soon as the treatment is stopped, the anxiety returns.

10.3 Depression

Depression is quite different from anxiety. The firing rate of neurons in another part of the brain play a controlling role in the perceived ailment of depression. The firing rate is determined by neurotransmitters and neuroreceptors. It is not clear which neurotransmitters are involved. Most antidepressant drugs target serotonin, norepinephrine, and dopamine. These drugs work slowly and often not at all for some depression patients. The newest drug that produces rapid and spectacular reduction in depression for most patients is Ketamine, and this drug targets glutamate receptors. From this experience, it seems that glutamate is the most important neurotransmitter involved with depression.

In this part of the brain, NMDA receptors are the receptors for glutamate. These receptors are unusual in that both glutamate and glycene must be present to activate them. The NMDA receptor is "two-headed," having binding sites for glutamate and glycene. Glycene is generally in good supply, so the critical neurotransmitter is glutamate. Glutamate is excitatory.

With a normal brain, the glutamate level is properly set and there is no depression. In the case of a depressed person, the glutamate level is elevated. Ketamine is an NMDA antagonist that, when administered, blocks the glutamate receptor of the NMDA receptors. Since glutamate is excitatory, blocking the NMDA receptor is inhibitory, the opposite of excitatory. The neuron's firing rate drops and the effect is that within an hour or two, the symptoms of depression are gone.

Ketamine is not a new drug. The use of it for the treatment of depression is new. This drug has been used as an anesthetic. It has been abused as a recreational drug. It can alter one's sense of sight and sound, can produce profound relaxation, hallucinations and delusions, like LSD. In high doses, it is addictive. With much smaller doses, it is effective at treating depression.

A sufferer with depression has an elevated level of glutamate, although the set of neurons involved with depression have normal homeostatic set points. The higher levels of glutamate cause the firing rate of these neurons to be above the homeostatic rate and the result is depression. % Hebbian-LMS tries to hold the firing rate at the homeostatic rate by reducing the excitatory weights and increasing the inhibitory weights, but cannot do this perfectly. It must have some error in order to push toward homeostasis. % Hebbian-LMS helps to reduce the level of depression, but it is not perfect. % Hebbian-LMS does the best that it can. Nevertheless, the afflicted person has depression without the additional help of drugs.

A sudden dose of Ketamine blocks the glutamate receptors and causes the membrane potential of the affected neurons to suddenly drop. Their firing rate suddenly drops below homeostatic. The effect of the drug is strongly sensed. Hallucinations and pain relief replace depression. % Hebbian-LMS now reverses and acts to restore the neurons' membrane potentials back up to the equilibrium point. The inhibitory non-glutamate synapses' weight values will gradually decrease. The glutamate synapses with their NMDA receptors blocked or partially blocked will tend to remain unchanged. The overall effect for the patient is an almost instantaneous intense sense of the drug. This will gradually subside over an hour or so to a steady state homeostatic level as the inhibitory weights adapt, leaving no hallucinations or depression. Now at the homeostatic level, the synaptic weights will retain their values.

After the dose of Ketamine is administered, its effectiveness will continue. If this were not so, % Hebbian-LMS would restore the weights to their original values as the drug level decreased, leaving the patient with depression. This does not happen. Depression does not return for a week or so. The conclusion is that Ketamine is a long-acting drug. In a treatment program, more Ketamine is taken before depression returns. Once a week seems to be about the right rate to prevent depression. The dosage is titrated between "feel no pain" and the symptoms of depression.

A much stronger dose, at the abuse level, will cause much stronger effects. Given a single dose, the immediate effect is delusional, and this will diminish over a few hours to a "feel no pain" level. After a few days, the effect of the drug will slowly subside and the weight values, not yet restored to normal, will cause withdrawal. Withdrawal symptoms would be depression. To get relief, another dose of Ketamine will be sought. This is addiction. If no dose is taken and withdrawal is endured, the addiction can be broken.

Treatment with this drug needs to be medically supervised. The dosage needs to be adjusted for the patient. Long term treatment is possible and will generally be successful. Stopping treatment will result in return of depression.

10.4 Bipolar Disorder

Manic-depressive disorder, now called bipolar disorder, is a serious mood disorder. The bipolar subject experiences periods of mania that may last from a fraction of a day to many months. The person's mood may then switch to normal for a period of time and then switch to depression. Switching from depression to normal to a manic state and vice versa is a characteristic of bipolar disorder.

It is generally accepted that an imbalance of neurotransmission is at the root of bipolar disorder. The neuropathology and the pathophysiology of bipolar disorder are not understood. There is no behavioral animal model for bipolar disorder.

Bipolar disorder is perplexing to researchers and to clinicians. Its cause is unknown. There is a general consensus however that synapses, neurotransmitters, an neuroreceptors are involved. Certain neurotransmitters are implicated, such as dopamine, norepinephrine, serotonin, GABA, glutamate, and acetylcholine. Perhaps the most important are norepinephrine, an excitatory neurotransmitter, and serotonin, and inhibitory neurotransmitter.

A portion of the brain is the seat of bipolar disorder, consisting of perhaps millions of neurons and their excitatory and inhibitory synapses. The average firing rate of these neurons results in mood perception. A normal firing rate is perceived as a normal mood. A higher firing rate is perceived as the mood of manic. A lower firing rate is perceived as depression. This idea is controversial, and the literature is not clear. Googling bipolar disorder and considering many references, carefully connecting the dots leads to this idea.

A simplified model of bipolar disorder would involve the excitatory neurotransmitter norepinephrine and the inhibitory neurotransmitter serotonin. These neurotransmitters are secreted at the presynaptic sides of their respective synapses and are present in the synaptic clefts. In addition, these neurotransmitters are present all over in the brain's cerebro-spinal fluid, and can enter the clefts, supplementing the neurotransmitters therein. There are normal levels of these neurotransmitters in the cerebro-spinal fluid. An elevated level of norepinephrine or a lower than normal level of of serotonin or both will cause an increase in the firing rates of the relevant neurons, and this will be perceived as mania. A lower than normal level of norepinephrine or a higher than normal level of serotonin or both will cause a lower than normal firing rate which will be perceived as depression. Normal levels of these neurotransmitters in the fluid will result in a normal average firing rate which will be perceived as a normal mood.

For some unknown reason, bipolar patients may experience a rise above normal of the excitatory neurotransmitter norepinephrine in the cerebro-spinal fluid, and they will enter a manic episode. Alternatively, the same patients could have a rise above normal of the inhibitory serotonin and then go into depression. Why this happens with bipolar patients is not understood.

Homeostasis tends to keep the average firing rate at normal. In the manic state, homeostasis is not strong enough to lower the firing rate all the way to normal. The residual elevation in the firing rate yields mania. Similarly, in the depressed state, homeostasis is not strong enough to raise the firing rate all the way to normal. The residual lowering of the firing rate yields depression.

With a normal subject, homeostasis keeps the average firing rate at a normal level. With a bipolar patient, homeostasis helps stabilize both manic and depressive episodes but is not strong enough to totally maintain firing rate normally. % Hebbian-LMS always tries to bring the firing rate to homeostatic. It does not do this perfectly, since there will be some error in steady state.

The manic state is brought on by an excess of an excitatory neurotransmitter such as norepinephrine in the cerebro-spinal fluid, or by a lower than normal concentration of an inhibitory neurotransmitter such as serotonin in the cerebro-spinal fluid. The depressed state is brought on by a reversal of these neurotransmitter concentrations.

What causes these imbalances in neurotransmitter concentration with the bipolar patient is unknown.

Bipolar is a very complex disorder. There is no generally accepted way to analyze it. Many theories have been proposed. They are often unclear and indeed contradictory. A novel theory, quite different than those of the literature, is proposed here. This theory is an unproven speculative hypothesis. It has an advantage of simplicity and it seems to fit the manifestation of bipolar.

The new theory is as follows. Homeostasis results from % Hebbian-LMS controlling the neuron's synaptic weights, both excitatory and inhibitory. The neuron's equilibrium points correspond to the normal homeostatic firing rate.

With the bipolar patient, the % Hebbian-LMS algorithm will be constantly fighting in the struggle to maintain the membrane potentials at the equilibrium point. This comes into play during manic or depressive episodes. In between, during normal periods, % Hebbian-LMS will be functioning but not struggling.

The oldest and still most commonly prescribed drug for the treatment of bipolar disorder is lithium carbonate. Researchers and clinicians report that for most patients, lithium is effective for the treatment of mania. Other researchers and clinicians describe the effectiveness of lithium for the treatment of bipolar depression. Lithium is often called a bipolar mood stabilizer. It is used to treat both mania and depression, and it is used as a maintenance drug during normal periods to sustain the status quo. Many antidepressant drugs are available for bipolar patients experiencing depression. These drugs reduce its severity. They raise the neurons' firing rates, bringing them closer to normal. Antipsychotic drugs taken during episodes of mania reduce the severity of the symptoms. These drugs reduce the firing rates, bringing them closer to normal.

Most of these drugs either raise or lower the firing rates. Lithium is different. It does both. It "knows" when to raise and when to lower the firing rates. How could this be? A smart drug? There are many theories proposed in the literature that explain how lithium works. These explanations differ from one another, are very complicated, and sometimes contradictory. The effects of lithium seem to be not well understood.

Another theory about the action of lithium is given here: Lithium is a mood stabilizer that seems to reinforce homeostasis. Lithium is not a neurotransmitter or an imitator of a neurotransmitter. As a stabilizer, lithium helps attain and maintain homeostasis. If one accepts the hypothesis that % Hebbian-LMS controls the synaptic weights and thereby establishes an equilibrium point, then lithium in some way works through the % Hebbian-LMS algorithm to bring the neuron's membrane potential closer to homeostasis at the equilibrium point. The membrane potential at the equilibrium point results in a neuronal firing rate that is perceived as normal mood.

With % Hebbian-LMS, the error signal goes to zero with the membrane potential at the equilibrium point. The algorithm creates a restoring force proportional to the error that pulls the membrane potential toward the equilibrium point. Near the equilibrium point, the error gets small, the restoring force gets small, but the loss of the restoring force results in an error that does not reach zero. The patient is left with mania or depression, depending on the neurotransmitter balance in the cerebro-spinal fluid. It is hypothesized that use of lithium either magnifies the error signal, or it magnifies the change in the synaptic weights, increase or decrease, in proportion to the error signal. In accord with the % Hebbian-LMS hypothesis, lithium may be increasing the size of the parameter μ, without causing instability. In this way, lithium helps the % Hebbian-LMS algorithm to be more effective, making the magnitude of the error signal converge to smaller values, thereby reducing the magnitude of the bipolar symptoms of either mania or depression.

Some bipolar patients transition directly from depressive states to manic states. This phenomenon can be explained by the % Hebbian-LMS model. An interesting paper on the subject is "Transition to mania during treatment of bipolar depression," published in Neuropsychopharmacology, 2010, December 35(13), pages 2545–52, by the Department of Psychiatry, Bipolar Clinic and Research Program, Mass General Hospital and Harvard Medical School. This article describes bipolar patients, 23% of them, who transition directly from depressive states to manic states. Some were being treated with antidepressants, some not. There is no given understanding of this phenomenon. An explanation may be the following. This phenomenon looks like overshoot predicted by % Hebbian-LMS.

These patients start in a depressed state caused by a neurotransmitter imbalance, for example an excess of serotonin, in the cerebro-spinal fluid. This creates a force that lowers the firing rate of the relevant neurons. % Hebbian-LMS will decrease the synaptic weights of the inhibitory synapses with serotonin receptors in an attempt to raise the neurons' firing rates up to the equilibrium point. If whatever causes the neurotransmitter imbalance suddenly ceases and the serotonin level in the fluid returns to normal, the firing rates will go up and overshoot beyond normal. The overshoot is caused by the serotonin weights having been adapted to below normal levels. This overshoot causes mania. Gradually, % Hebbian-LMS will adapt the serotonin weights back to normal and it is predicted that the mania will subside and a normal mood will prevail.

Neurotransmitter imbalance in the cerebro-spinal fluid pulls neuronal firing rates away from the equilibrium point. % Hebbian-LMS creates a restoring force, homeostasis, that attempts to pull the firing rates toward the equilibrium point. The restoring force is approximately proportional to the error of % Hebbian-LMS. There is an analogy here to feedback control systems that are very important in engineering applications. The % Hebbian-LMS equilibrium point is analogous to the setpoint of a feedback control systems. A type 0 control system will experience a static error when the plant being controlled is subject to a static disturbance. The type 0 control system has restoring force proportional to the error, and a static disturbance will cause a static error. A type 1 control system has integration in its feedback loop, and a static disturbance will result in zero error in steady state. Both types are present in the neuronal system.

The % Hebbian-LMS algorithm is a type 0 feedback control system. A static disturbance such as a neurotransmitter imbalance will result in a small but non-zero static error. The homeostasis of % Hebbian-LMS pulls the firing rate toward the equilibrium point, the control set point, but it does not reach the set point without error. Mania and depression occur in spite of the natural action of homeostasis trying to regulate and restore the firing rate to normal. Lithium seems to make the magnitude of the error smaller, mitigating the effects of both mania and depression.

10.5 Schizophrenia

Schizophrenia is a mental affliction that is associated with yet another small brain area containing perhaps a million neurons, possibly more, possibly less. These neurons have excitatory and inhibitory synapses. The excitatory synapses have dopamine receptors.

It was thought that the disease was caused by "too much dopamine." Subsequent research established that dopamine levels were normal, but that the number of dopamine receptors in each synapse was substantially greater than normal. The net result is too much dopamine. This is believed to be the cause of schizophrenia.

The disease schizophrenia is characterized by auditory hallucinations, hearing voices that do not exist. Thinking is delusional and disordered. This is probably the worst of all mental diseases. It is incurable. Relief from the symptoms can be had by using the antipyschotic drugs chlorpromazine or haloperidol. Newer ones are now available. These are dopamine blockers. They bind to dopamine receptors, do not activate the synapses, but prevent dopamine molecules from attaching to dopamine receptors. This reduces the effect of dopamine.

Without drugs, schizophrenia patients would have normal levels of dopamine but an excess of dopamine receptors. The large number of receptors would be established there by the action of % Hebbian-LMS in creating homeostasis. The homeostatic firing rate being higher than normal. The high inputs would cause higher than normal membrane potentials. Thus, the homeostatic firing rates would be higher than normal, resulting in schizophrenia.

The anti-psychotic drugs lower the membrane potentials and lower the neuron's firing rates. % Hebbian-LMS fights against the drugs in trying to maintain the homeostatic disease state. Enough drug can overcome % Hebbian-LMS and provide relief. The drug level would need to be adjusted to get the desired effect.

10.6 Parkinson's Disease

Parkinson's disease is primarily a muscular control disorder, but in many cases it leads to dementia. This is not a mood disorder, although it has a lot in common with some mood disorders. Parkinson's is a motor control disorder that worsens over time. Muscle tremor and slow muscular response are some of the manifestations of the disease.

The trouble begins in the Substantia Nigra Pars Compacta. Dopaminergic neurons herein project axons ito the striatum, which plays a major role in motor control.

These neurons manufacture dopamine for the striatum. These neurons in addition are a major source of dopamine for the entire brain.

For reasons that are not well understood, the dopaminergic neurons can gradually die, and this leads to Parkinson's disease. When these neurons die, they are not replaced. These neurons fire at their homeostatic rate, and do not fire faster to compensate for lost neurons. Whenever these neurons die, the amount of dopamine delivered to the striatum diminishes. The result is Parkinson's disease.

Within the striatum, these dopaminergic projections synapse to two separate groups of dopamine-receptive neurons. One group provides a "direct pathway" to the basal ganglia that is inhibitory, the other group connects to the basal ganglia and it provides an "indirect pathway," and it is excitatory. The balance with these pathways allows the basal ganglia to smoothly regulate voluntary muscle activity. A shortage of dopamine upsets the control process. Treatment with L-dopa, which converts to dopamine, often relieves the debilitating symptoms.

Elderly people often exhibit symptoms like Parkinson's disease. With ageing, neurons in the substantia nigra can die off slowly, and not be replaced. The difference between neuron die off in much younger Parkinson's patients is speed of occurrence. The rate of neuron die off with Parkinson's patients of any age is much higher than that from ageing. The dramatic rate of neuron die off with Parkinson's make this a disease. The much slower rate with normal ageing is not considered a disease and is generally not treated, although it could be. Old people drive their cars slowly. Young people have no idea why.

Learning goes on continuously day and night in the memory segments. Some researchers associate learning with growth of new synapses. It seems that evidence for this is scant. Slow changes in neural network circuitry due to cell death and due to addition of new synapses are disturbing but easily accommodated by the constant adaptation and learning that normally takes place.

10.7 Summary

A wide range of subjects is presented discussing a number of brain disorders. A common thread is adaptive control of upregulation and downregulation of critical neuroreceptors. Each of the brain disorders is associated with the synapses, the neurons, and the firing rate of the neurons in a specific brain area. Knowledge of the % Hebbian-LMS algorithm and homeostasis sheds light on the nature of these afflictions. The firing rate of neurons in the relevant brain areas determine normalcy or affliction.

Drugs are used to mitigate the symptoms of these disorders. The drugs do not cure these diseases, but they make it easier to live with them.

% Hebbian-LMS always tries to keep firing rates homeostatic. In some cases, homeostatic firing is perceived as normalcy, a desired state for the patient. In other cases, this is not so. For example, the homeostatic firing rate of certain neurons creates anxiety. For some unknown reason, this firing rate is well below normal. Untreated, a mood of anxiety persists. Relief comes from drugs that increase the firing rate to be higher than homeostatic. As such, % Hebbian-LMS fights the drug trying to restore homeostasis. The drug is stronger and relief is possible.

With depression, homeostatic firing is perceived as normalcy, the desired state. Without treatment, the firing rate is higher than homeostatic. This is caused by an abnormally high level of glutamate. Why this happens is not well understood. The remedy is to decrease the binding of glutamate to its receptors. An often used drug for this is Ketamine. In this case, % Hebbian-LMS works with the drug to bring the firing rate close to homeostatic.

Homeostasis is normalcy with bipolar disorder. % Hebbian-LMS fights against mania and against depression in attempting to establish homeostasis. The drug lithium sulfate probably helps % Hebbian-LMS keep tighter control of firing rate, pulling it toward homeostasis with smaller error.

The disorders Schizophrenia and Parkinson's disease also involve homeostasis and receptor regulation. Each of these disorders involves different brain areas and work in different ways. They all involve homeostasis, one way or the other.

10.8 Questions and Experiments

1. Among the neurons in the part of the brain associated with anxiety, the homeostatic firing rate is low and this low rate with these neurons is perceived as anxiety. Is there any explanation for why the equilibrium rate is too low to prevent anxiety? Search the literature to see if there is an answer to this natural defect.
2. A part of the brain is associated with depression. With a depressed subject, the neurons in this part of brain have a firing rate that is too high and the high firing rate is perceived as depression. Lowering the firing rate gives relief from depression. The high firing rate is caused by an elevated level of glutamate, an excitatory neurotransmitter. The question is, with a normal level of glutamate, would the firing rate be the homeostatic rate? What kind of experiment could be performed to find an answer? Another question is, what causes the elevated level of glutamate?
3. In yet another part of the brain resides a set of neurons that, in a bipolar subject, when firing at the homeostatic rate yields a perception of normalcy. When firing at a rate higher than this causes a perception of mania. When firing at a rate lower than this causes a perception of depression. Lithium is a drug that is able to reduce the symptoms of both mania and depression. Other drugs mitigate the effects of one or the other mood anomaly. Read the literature to discover how lithium works. Can you reconcile the literature on the action of lithium with the hypothesis of lithium as a mood stabilizer as presented in this chapter?
4. A project at Stanford University by Leanne Williams of Psychiatry and Behavioral Science and Zhenan Bao of Chemical Engineering is developing a sensor to measure cortisol, a hormone in sweat and saliva, with the idea that measuring cortisol is a measure of stress that leads to depression [1]. How does this correlate with the teaching of this chapter? Production of cortisol is related to stress, but

what else aside from stress is cortisol production related to? Save your result of this study for study of Chaps. 12, 13, and 14.

5. Separate areas of the brain are associated with anxiety, depression, and schizophrenia. Can these areas be located and identified?

Reference

1. Science seeks a better way to measure stress, anxiety and depression|stanford school of engineering. https://engineering.stanford.edu/magazine/article/science-seeks-better-way-measure-stress-anxiety-and-depression. (Accessed 01 July 2020)

Part III
Regulation and Control of Physiological Variables and Body Organs

Introduction to Part III

Blood pressure, blood salinity, blood flow rate, body temperature, and many other physiological variables throughout the body are normally controlled within narrow limits as required for optimal functioning. Homeostasis plays a major role in regulating these variables. Observations of homeostasis have been made over many years, but the seat of homeostasis, i.e. the location of the keeper of the set point, is often unknown, not well understood, and not discussed. It is the purpose of this writing to describe proposed homeostatic mechanisms for the regulation of heart rate, blood pressure, blood salinity, blood volume, blood glucose, and body thermoregulation. These mechanisms are highly simplified here. The same or similar homeostatic mechanisms are further proposed as operable for all forms of homeostasis in the body. In every case, it seems that neurotransmitters or hormones are involved with neuroreceptors or hormone receptors. The receptors are gates that, when binding to neurotransmitter or hormonal molecules, open to allow the passage of fluids or ions through an otherwise impermeable membrane that the receptors are embedded within. The neurotransmitters or hormones are generated and stored in the axonal terminals of special neurons. When the axons transmit action potentials, neurotransmitter or hormone is secreted into the surrounding fluid. It is proposed that in each case the number of receptors is controlled with the % Hebbian-LMS algorithm, and therein is the seat of homeostasis.

Regulation of any physiological variable is a very complex process. Regulation of blood pressure and regulation of blood salinity are two examples. The discussions of these and other cases are highly simplified, hopefully not oversimplified. The purpose is to explore the most basic mechanisms in order to gain an understanding of the role of adaptivity and homeostasis. Regarding the "forest and the trees," the forest will be studied, not the trees.

Chapter 11
Blood Salinity Regulation, the ADH System

Abstract Regulation of blood salinity (osmolality) is important for health and good functioning of the cells of the body. Increasing water in the blood reduces salinity, reducing water increases salinity. The kidneys play a major role in salinity regulation. Blood circulates through the kidneys for filtration. Toxins and waste in the blood are separated and excreted by urination. This process causes a loss of water that at some point must be replenished. Most of the water flow through the kidneys is retained and re-inserted into the bloodstream. The re-uptake of water from the urine is controlled by the effect of the hormone ADH acting on the kidneys. ADH is secreted by neurons in the supraoptic nucleus of the hypothalamus, presumed to be % Hebbian-LMS neurons. At rest, these neurons fire at their homeostatic rate and produce a steady stream of ADH, causing the kidneys to re-uptake and excrete water at a given rate. Water must be ingested to compensate for the water loss. Blood salinity is sensed by osmoreceptors neurons that fire at a rate proportional to salinity. Their signals are fed to the supraoptic neurons, closing an overall negative feedback loop. This loop mitigates transient extremes. The average salinity is not controlled by this loop, but is controlled by drinking water when thirsty.

Keywords Hypothalamus · Pituitary gland · ADH · Osmoreceptors · Kidneys, nephrons

11.1 Introduction

The bloodstream carries energy, oxygen, and signals to the brain and all the body organs. It carries away metabolic waste products, toxins, dead cells, etc. It plays a critical role in life support. Maintenance of blood salinity within narrow limits is important for the health and good functioning of all the cells of the body. The primary means for controlling blood salinity is control of blood dilution. Water is ingested when thirsty. Water is excreted in the urine. Drinking water and excreting more water or less water by the kidneys are parts of the control process, major parts.

B. Widrow, *Cybernetics 2.0*, Springer Series on Bio- and Neurosystems 14,
https://doi.org/10.1007/978-3-030-98140-2_11

11.2 Parts of the Blood Salinity Regulation System

Blood salinity regulation involves ADH, an anti-diuretic hormone, a neurotransmitter. Also involved are the hypothalamus, the pituitary, the supraoptic nucleus, osmoreceptors, the kidneys, their nephrons, kidney tubules and membranes, water channels or aquaporins, water, and blood. An excellent source of information about these subjects is three Khan Academy videos by Rishi Desai entitled "ADH secretion," "Aldosterone and ADH," and "Aldosterone raises blood pressure and lowers potassium." Additional sources can be obtained by Googling the subjects "ADH", "pituitary," and "osmoreceptor." Many more sources are available by Googling the relevant subjects. Wikipedia and Khan Academy have good articles. With this wealth of information, it is possible to "connect the dots" and extract a working model of blood salinity regulation and add to it the missing link, homeostasis.

This model should not be considered all-inclusive. Salinity control is a very complex process involving many organ systems.

11.3 Hypothalamus, Posterior Pituitary, and ADH

A diagram showing components of the hypothalamus and the pituitary is shown in Fig. 11.1. Blood osmolality is sensed by a group of neurons called osmoreceptors. Increase in salinity increases the osmolality and causes these neurons to fire faster. Thus, output signals from these neurons become stronger with increase in salinity. These signals are fed as excitatory inputs to a cluster of special neurons, about 300 in number, located in the supraoptic nucleus embedded in the hypothalamus. Stronger inputs to these super-optic neurons cause them to fire faster if these inputs are changing rapidly. If the changes are slow, the neurons will fire at close to their homeostatic rate, by and large unaffected by the slow inputs.

These supraoptic neurons manufacture a hormone, the neurotransmitter ADH. This neurotransmitter or hormone travels from the supraoptic neurons down their axons and pools in their axon terminals and buttons. The buttons have vesicles that secrete ADH whenever an activation pulse arrives. The axon terminals reside in the posterior pituitary. The higher the pulse rate of the supraoptic neurons, the more ADH is secreted into the posterior pituitary. Arterial and venous capillaries are there to absorb the ADH and transfer it to the bloodstream. ADH in the blood stimulates the kidneys and causes them to reabsorb water from the urine and deposit this water back into the bloodstream, thus reducing salinity. The rate of water excretion is reduced.

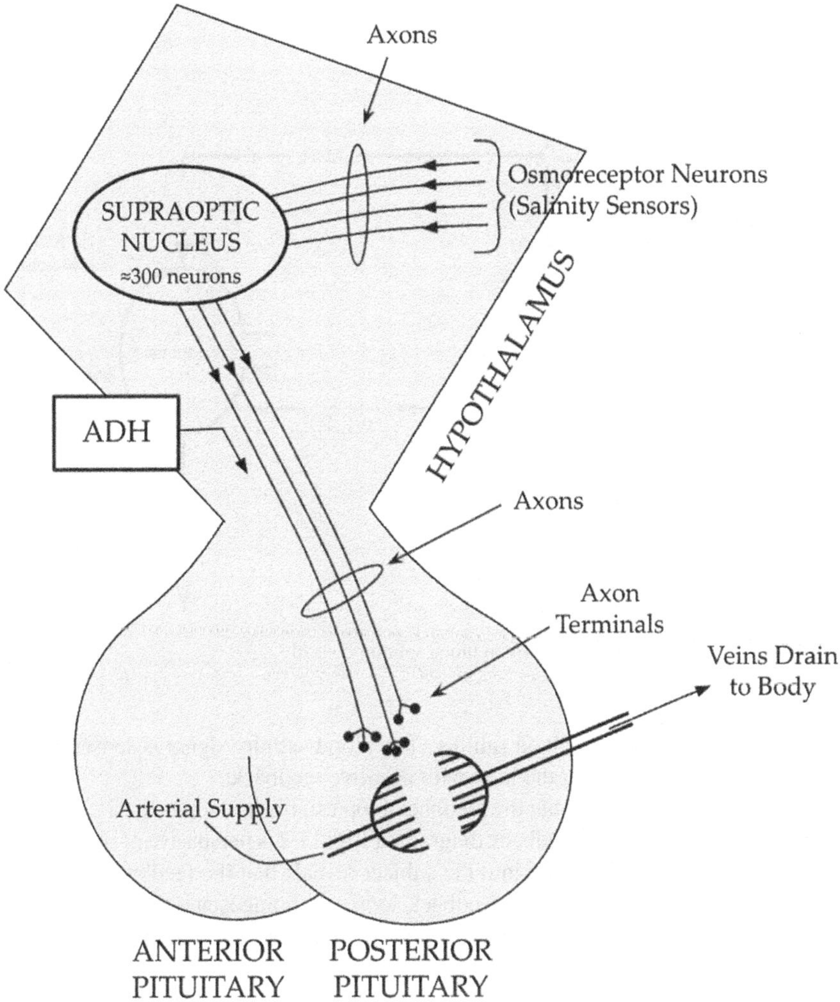

Fig. 11.1 Hypothalamus, pituitary, and ADH

11.4 The Feedback Loop

Figure 11.2 shows how the above described components are connected into a closed loop feedback control system. Starting with the osmoreceptor neurons that are connected to the supraoptic neurons, the hormonal outputs of the supraoptic neurons are carried by the blood in the form of ADH concentration. (The blood flows through the kidneys and is filtered producing urine.) ADH in the blood binds to ADH receptors in the kidneys opening gates that allow water to be passed from the urine back to the vascular system of the kidneys. Water re-absorption dilutes the blood, increases

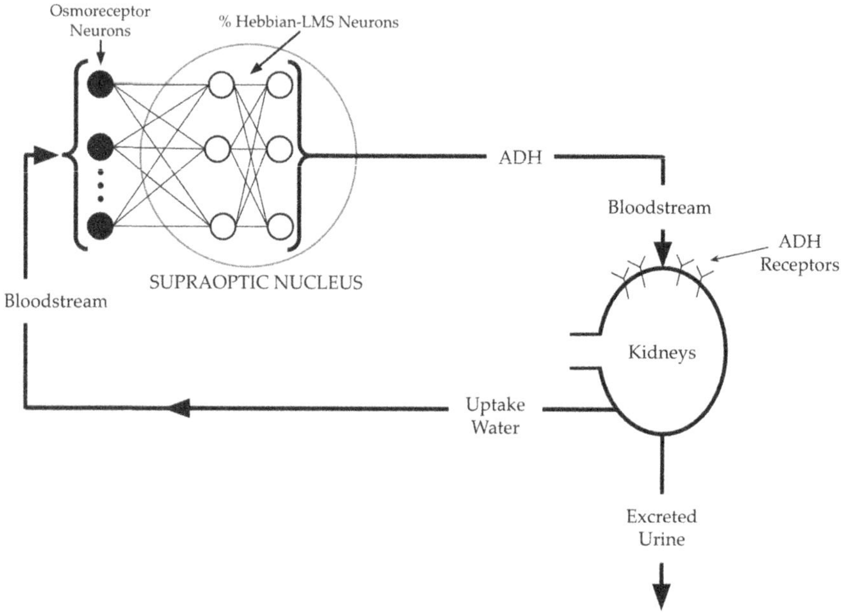

Fig. 11.2 The ADH system. Closed-loop blood salinity control

blood volume, and lowers blood salinity. The blood salinity signal is sensed by the osmoreceptors, thus closing the loop with negative feedback.

Many biologists believe that the feedback loop establishes homeostasis, thus regulating blood salinity. The feedback diagram of Fig. 11.2 is persuasive of this point of view. However, looking deeper into the subject reveals that the feedback loop does not implement homeostasis. The feedback loop and homeostasis are two separate mechanisms and both in their own ways contribute to blood salinity regulation.

11.5 A Brief Discussion of Classic Feedback Control Theory

A classic feedback control system is diagrammed in Fig. 11.3. The "plant" in control terminology is the system to be controlled, such as a chemical plant or a power plant. The purpose of the control system is to regulate the plant output in spite of disturbances. The set point is an input to the system that commands the plant to have an output at the value of the set point. An error signal, the difference between the set point and the plant output, is amplified by a coefficient k to provide a signal that drives the plant in a direction to reduce the error. Making k bigger makes the error smaller and this is good, but making k bigger could cause the system to go unstable and this is not good.

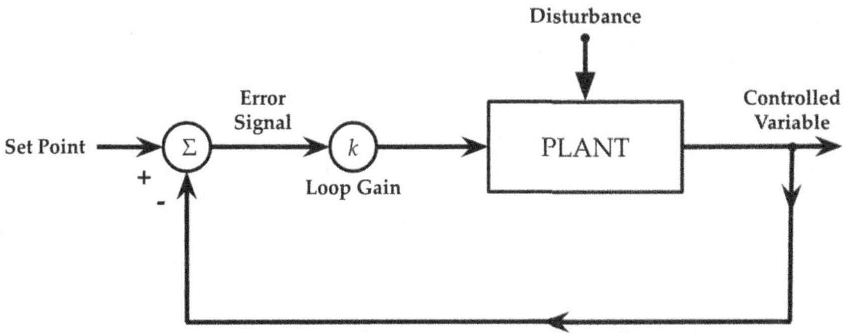

Fig. 11.3 A classical control system

An airplane and its autopilot is an example of a control system. The pilot inserts the desired altitude into the autopilot. This is the set point. The airplane flies steadily at that altitude until it flies into turbulent air with random updrafts and downdrafts. The airplane (the plant) and its autopilot are not fast enough to regulate altitude against the turbulent disturbances, so the airplane bobbles up and down randomly. What the autopilot actually does is regulate average altitude, averaging out the disturbance. The average altitude obeys the set point.

Plant disturbances tend to permeate the plant and cause undesirable responses that are apparent at the plant output. With negative feedback operating with a high value of the coefficient k, a high "loop gain," the error will be small and the amplitude of disturbance appearing at the plant output must therefore be small. The feedback mitigates the effects of plant disturbance keeping the response to it small. How well this works however depends on the dynamic response of the plant, the bandwidth of the disturbance, and the range of the loop gain which is subject to stability constraints. The autopilot example exemplifies the limitations of feedback mitigation of disturbance. Only the low frequency components of the disturbance can be mitigated, keeping the low frequency components of the airplane at or near the set point. The higher frequency components of the disturbance are felt by the passengers.

Every regulating control system must have a set point and an error signal. This is true for chemical systems, mechanical systems, and living systems. The closed-loop system of Fig. 11.3 illustrates this. In living systems, this is supposed to regulate against disturbances that could be caused by change in diet, change in water intake, change in exercise, etc.

With this in mind, an examination of Fig. 11.2 raises some good questions. What and where is the set point, and where is the error signal? What is the variable being controlled? What does homeostasis mean for this feedback system? We will propose answers to these questions and examine the consequences.

11.6 Blood Salinity Control: Homeostasis and Negative Feedback

The three hundred neurons of the hypothalamus, the neurons of the supraoptic nucleus, are hypothesized to have synaptic weights that adapt in accord with the % Hebbian-LMS algorithm. The synapses connect to the osmoreceptor neurons. The feedback loop of Fig. 11.2 is clearly different from the classic feedback loop of Fig. 11.3. The loop of Fig. 11.2 has an adaptive neural network within it. There is no precedence for anything like this in the literature of classic or modern control theory. Biological control systems are unique and in need of much more study. The goal of Part III is to begin with a few examples. Many more control systems in the body need to be looked at.

An engineer's block diagram of the blood salinity control system is seen in Fig. 11.4. The diagram shows ADH generated by the neurons of the supraoptic nucleus of the hypothalamus being fed to the kidneys. ADH causes water to be reabsorbed from the urine and reinserted back into the bloodstream. Otherwise, the water will be excreted and subtracted from the water in the blood. Ingested water adds and sweat subtracts from the water in the blood.

The amount of water in the blood relates negatively to blood salinity. The scale factor $-c_1$ converts blood water volume to blood salinity. Osmoreceptor neurons sense blood salinity and send their signals to the neurons of the supraoptic nucleus, thus closing the feedback loop. The question is, what is the set point for the loop? The answer seems to be, there is no set point for the loop! This seems likely, but there is no certainty.

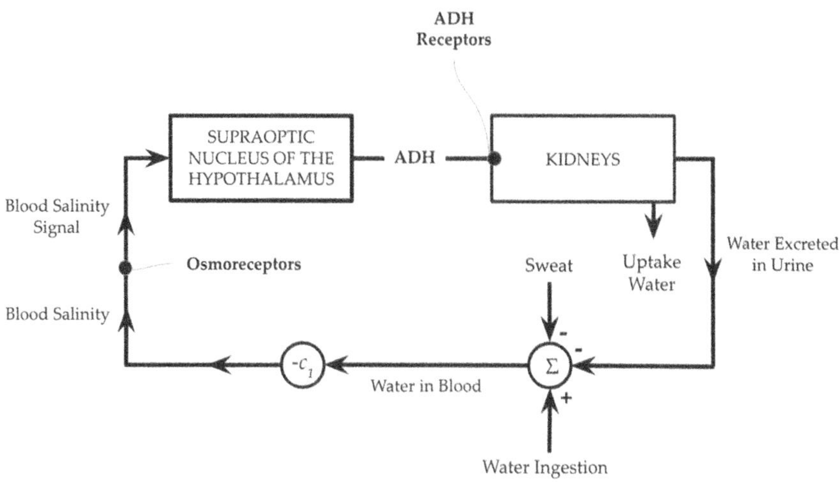

Fig. 11.4 The ADH system. Block diagram of blood salinity control system

Assume that everything is at rest. The three hundred or so neurons in the supraoptic nucleus are happy, firing at their homeostatic rate. The homeostatic rate does not depend on the osmoreceptor signals. At rest, these signals are steady and the supraoptic neurons adapt to this input and maintain firing at their homeostatic rate. In case of rapid change in the osmoreceptor signals, adaptation will not be fast enough and the firing rate can change rapidly away from the homeostatic rate. All this is in accord with the % Hebbian-LMS algorithm.

ADH is being produced at rest in proportion to the homeostatic firing rate. It is fed to the kidneys via the bloodstream. The ADH receptors in the kidneys are receiving ADH. These receptors are not incorporated within synapses. There are no neurons here.

Each kidney has millions of nephrons. The nephrons are the basic operational units of the kidneys. A portion of each nephron harbors ADH receptors embedded in a membrane. Each receptor is embedded in a membrane co-located with a gate that allows water to flow through the membrane from the urine into the blood when ADH binds to a receptor.

The water is electrically neutral and its passage has no effect on the membrane potential. The number of receptors will therefore not be controlled by a % Hebbian-LMS algorithm.

The literature is not clear. An assumption will be made here that the number of receptors is not controlled or varied. Accordingly, the rate of flow of water from the urine into the bloodstream is simply proportional to the concentration of ADH in the blood and proportional to the number of receptors that have captured ADH.

The concentration of ADH in the blood is proportional to the firing rate of the supraoptic neurons, a rate that is homeostatic under quiet resting conditions. Under these conditions, the firing rate depends on the neurons' set point and does not depend on the signals from the osmoreceptor neurons. % Hebbian-LMS adapts the synaptic weights so that the neurons fire at the homeostatic rate. The control of water uptake from the urine, the principal means of controlling blood salinity, is not affected by the actual blood salinity. The regulation of average blood salinity under resting conditions is not under feedback control. The steady-state average salinity is determined by the ADH concentration in the blood and the number of functioning ADH receptors in the kidneys. There is no precise feedback control of this. The resting blood salinity level is the average or (dc) level. Control of this level is "open-loop," no feedback.

So the question is, what is the purpose of the overall feedback loop pictured in Fig. 11.4? What is the purpose of the osmoreceptors? The purpose of the feedback loop is to mitigate the effects of disturbance. An example of disturbance is drinking a glass of water. This is not under system control. It is unanticipated. It may be a desirable and necessary act, but the system regards it as a disturbance. It would cause a drop in salinity and the feedback loop becomes activated and it works to stabilize the changing salinity level.

How this works can be explained. Refer to Fig. 11.4. Suppose a person has been at rest for some time and everything is peaceful and quiet. The neurons of the supraoptic nucleus are firing at their homeostatic rate in accord with % Hebbian-LMS. ADH is being produced, the kidneys are drawing water from the urine to re-inject into the bloodstream.

Everything has been peaceful and quiet. Now the person drinks a glass of water. This gradually but fairly rapidly increases water in the blood and this reduces salinity. The osmoreceptors sense this and gradually lower their firing rate, lowering the synaptic input signals to the supraoptic neurons. All this happens much faster than the adaptation rate of these neurons and their synapses. Therefore, the lowered synaptic input signals cause lowered membrane potentials and lowered firing rates of the neurons. This causes lowered ADH production, and this causes lowered re-absorption of water and an increase in urine excretion. This causes a lowered level of water in the blood which raises blood salinity, mitigating the effects of drinking the glass of water on the blood salinity level.

Mitigation of the response to disturbance is the function of the feedback loop. This is the main purpose of the feedback loop. Disturbance responses superposed on top of the average (dc) level is the (ac) component of the blood salinity level.

It is interesting to note that the (dc) is open-loop controlled. The negative feedback controls the (ac). There is no overall automatic control of average blood salinity.

There is overall feedback control, but it is not automatic. The person senses a dry mouth or feels thirst and drinks water. Also, the person sensibly knows that drinking water is necessary to replace the water lost in urine, sweated, and exhaled. In public buildings, one always notices drinking fountains located next to the rest rooms. Good advice: after you pee, take a drink of water.

The disturbance created by drinking a glass of water is one type of disturbance. An opposite type of disturbance is not drinking water for some time. This is a much more gradual disturbance, but is has an opposite effect on blood salinity regulation.

Assume that a person has been at rest for sometime. The supraoptic neurons are firing at their homeostatic rate, ADH is produced, and the kidneys are re-cycling water and at the same time are discharging water with steady production of urine. Water is also slowly lost by the sweat glands of the skin. To compensate for the water loss, water is sipped all day long. The blood salinity is maintained at a "normal" level, a level that is comfortable and healthy for all the cells of the body. Suddenly, the person decides to stop drinking water.

Refer to Fig. 11.4. Water in the blood gradually becomes depleted. The blood salinity rises and the osmoreceptor signals increase. Because the synaptic weights cannot adapt fast enough, the supraoptic neurons' firing rates begin to increase, ADH production increases, and uptake of water from the urine increases, causing reduction in urine discharge, adding to the blood water supply and counteracting some of the effects of lack of water intake, mitigating the effects of the disturbance.

This should not be done for long because the body is losing water and this must be replenished in order to maintain blood salinity level, blood volume, blood pressure, etc.

Now suppose that the intake of salt and water had been normal for some time when suddenly a bag of salted potato chips was consumed. Through the stomach and gut, an impulse of salt is injected into the blood. The osmoreceptors sense a sharp rise in blood salinity, causing a speed-up of the firing of the supraoptic neurons, causing an increase to the flow of ADH into the blood, causing increased re-absorption of water by the kidneys, causing dilution and reduction of blood salinity. The excess salt is gradually excreted and blood salinity returns to normal. The overall feedback loop had become active and limited the peak salinity response. This whole episode took place fairly rapidly, not giving the synaptic weights much chance to change and fight against the rapid firing rate of the supraoptic neurons. The mechanism of homeostasis did not significantly come into play in this case, not having been fast enough to respond. If it were fast enough in adapting, the effectiveness of the overall feedback loop would have been diminished, but this does not happen.

11.7 Abnormal Function: Diabetes Insipidus

If some of the neurons of the supraoptic nucleus die or have diminished capability for producing ADH, the result is a disease called diabetes insipidus. The reduction in ADH production, under resting steady-state conditions (the (dc)) and under disturbed transient conditions (the (ac)), causes reduction in water re-absorption. The kidneys do not compensate by increasing the number of ADH receptors. There is no mechanism to do this. The result is increased excretion of urine and an increased drinking of water. These effects could be excessive and hard to control. Treatment involves administration of a synthetic form of ADH.

Nephrogenic diabetes insipidus is a similar disease but caused by a reduction in the absorption of ADH by the kidneys. This could be caused by loss of ADH receptors or by reduced function of the kidneys' tubules. The result is reduced water re-absorption, increased water drinking, and increased excretion or urine.

Both types of diabetes insipidus are associated with intense thirst and heavy urination. These problems are not in any way moderated by homeostasis of the supraoptic neurons or by the overall feedback loop.

11.8 Summary

To summarize, homeostasis and overall negative feedback are separate items. The average rate of water uptake from the urine, the (dc), is open-loop controlled. Overall feedback mitigates, but does not totally eliminate, the effects of disturbance, the (ac). It should be recalled that the % Hebbian-LMS algorithm, accounting for homeostasis in the firing of the supraoptic neurons, does in fact involve negative feedback. This is local feedback within the neurons and their synapses. It is to be distinguished from the overall negative feedback. With all this feedback, there is no overall automatic feedback control of blood salinity level. Homeostasis and the overall feedback loop contribute to this, but final feedback regulation of the average blood salinity is accomplished by the person drinking water when experiencing a dry mouth or thirst. The body's thirst sensors are key to this.

11.9 Questions and Experiments

1. Figure 11.4 is a block diagram of the blood salinity control system. When referring to this diagram, describe what happens to blood salinity if the person has been at rest and then eats a bag of salted potato chips followed by drinking a glass of water.
2. The rate of uptake from the urine into the bloodstream is proportional to the concentration of ADH in the blood, and proportional to the number of ADH receptors in the nephrons. The number of receptors does not vary and is not controlled by % Hebbian-LMS. Search the literature to confirm that the rate of water uptake is proportional to ADH concentration in the blood and that the number of ADH receptors is not varied.
3. Explain why both forms of diabetes insipidus are not moderated by homeostasis of the supraoptic neurons or by the overall feedback loop.
4. The supraoptic neurons are subject to homeostasis. The ADH receptors in the kidneys are not. How could you verify and explain this?

Chapter 12
The Aldosterone System, Blood Volume Regulation, and Homeostasis

Abstract The purpose of the aldosterone system is regulation of blood volume. In the body, there are no sensors for blood volume. Baroreceptors located in the venous system serve as a substitute. The hormone aldosterone in the bloodstream acts on the kidneys to cause them to recover water and sodium from the urine stream and return this to the bloodstream, thus affecting blood volume. Aldosterone is generated by the adrenal cortices in response to the hormone ACTH in the blood. ACTH is generated by the anterior pituitary in response to the hormone CRH in the blood. CRH is generated by neurons in the paraventricular nucleus of the hypothalamus whose inputs come from the baroreceptor neurons and the hormone cortisol from the adrenal cortices. The entire system is connected together as a feedback loop. At rest, the production of these hormones is homeostatic with regulation at each organ independently determined by an algorithm, believed to be % Hebbian-LMS. Under transient conditions, the negative feedback of the loop mitigates swings in blood volume which is detected approximately by the baroreceptors.

Keywords Kidneys · Nephrons, principal cells · Aldosterone · ACTH · CRH · Cortisol

12.1 Introduction

Aldosterone is a hormone generated in the cortex of the adrenal gland. There are two adrenal glands, located above the kidneys. A primary function of aldosterone is to act on the kidneys to increase the flow of sodium and water from the urine into the blood, thus raising blood volume and blood pressure. Control of this flow, up and down, allows the aldosterone system to perform regulation of blood volume and pressure. Loss of volume through urination is limited by the action of aldosterone, and in that sense volume is increased. The ADH system was seen to perform regulation of blood salinity by increasing or decreasing the flow of water from the urine into the blood.

B. Widrow, *Cybernetics 2.0*, Springer Series on Bio- and Neurosystems 14,
https://doi.org/10.1007/978-3-030-98140-2_12

This increases or decreases volume. The aldosterone system regulates volume in spite of the variations caused by the ADH system and other sources of disturbance. Both systems work together to regulate both salinity and volume.

12.2 The Aldosterone System

The aldosterone system is diagrammed in Fig. 12.1. This is a closed-loop system consisting of four major components—the paraventricular nucleus of the hypothalamus, the anterior pituitary gland, the adrenal cortex, and the kidneys. The paraventricular nucleus contains a large number of neurons which are hypothesized to be operating with % Hebbian-LMS. The output is the hormone CRH (corticotropin-releasing hormone), secreted by these neurons. The anterior pituitary contains cells that are triggered by CRH to secrete ACTH (adrenocorticotropic hormone). The adrenal cortex contains cells that are triggered by ACTH to secrete aldosterone and cells that are triggered by ACTH to secrete the hormone cortisol. See the diagram of Fig. 12.1.

The kidneys filter blood and then return filtered blood to the bloodstream. While doing this, they extract urine containing water and waste matter. The kidneys have

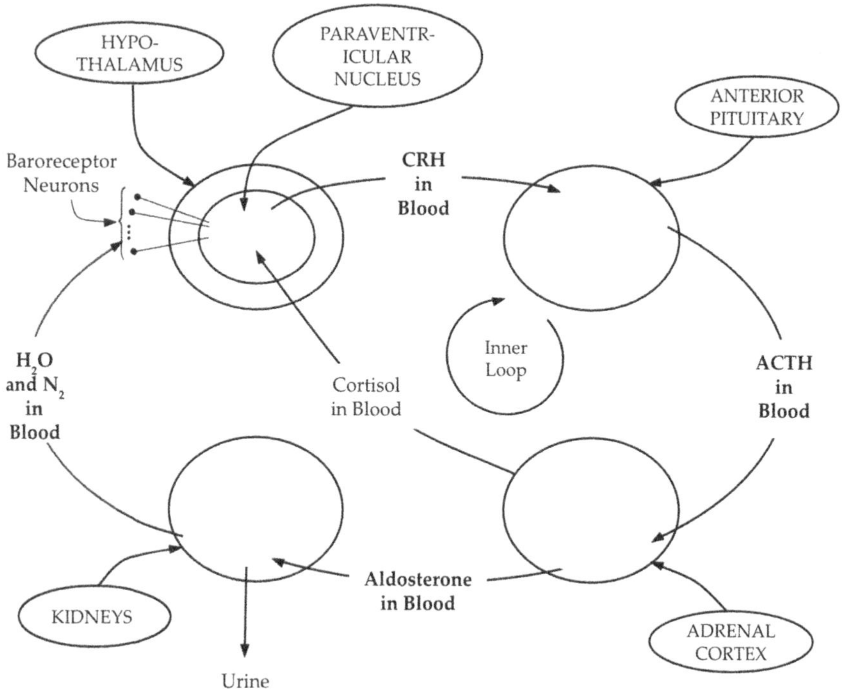

Fig. 12.1 The aldosterone system. The outer loop and inner loop

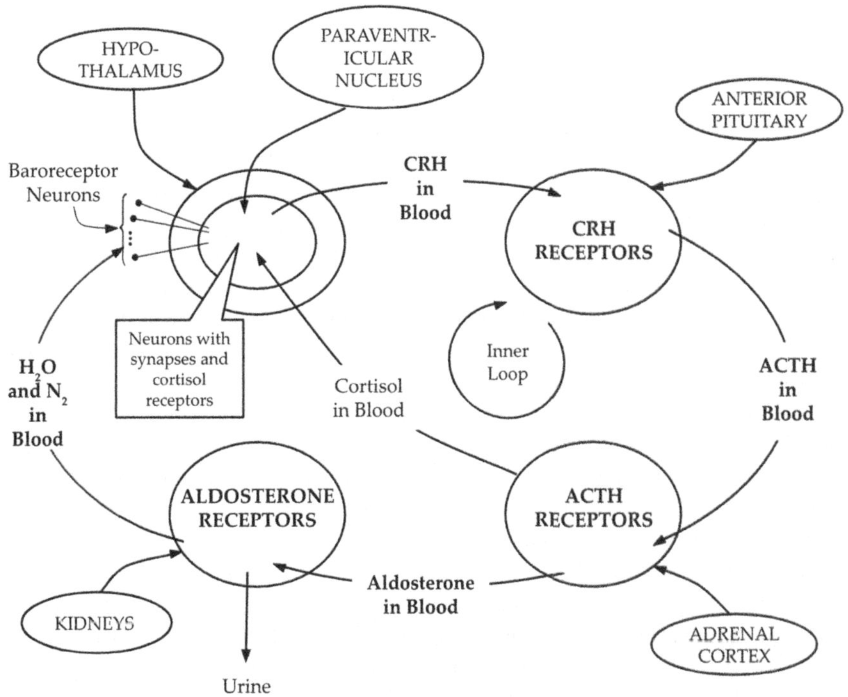

Fig. 12.2 The aldosterone system. Neurons and hormone receptors

a special structure that is triggered by aldosterone to extract water and sodium ions from the urine before it is excreted. The mix of water and sodium ions enters the bloodstream thereby increasing blood volume and blood pressure (this does not really increase volume; it reduces loss of volume due to urine excretion). Blood pressure is sensed by baroreceptor neurons located in major blood vessels of the venous system, among them the superior and inferior vena cava. Baroreceptor signals are sent to the paraventricular nucleus, thus closing the outer loop. An inner loop consisting of the adrenal cortex, the paraventricular nucleus and the anterior pituitary is closed by blood linkage with hormonal signals. Refer to Fig. 12.1 for clarification, and Fig. 12.2 for more details.

The linkages around the outer loop are all excitatory, stimulative, except for the baroreceptor inputs to the paraventricular nucleus which are negative, inhibitory. This is negative feedback. If the feedback were positive, the whole system would be unstable.

12.3 System Components

The components of the aldosterone system will be described in more detail. The first component is the paraventricular nucleus. This nucleus contains a large number of neuroendocrine neurons. These neurons project axons to the median eminence where they release CRH (Corticotropin-releasing hormone) into the small blood vessels, the portal vessels, that carry blood and CRH to the anterior pituitary gland.

The paraventricular neurons receive inputs from the baroreceptors and from the hormone (the neurotransmitter) cortisol generated by the adrenal cortex. The baroreceptor neurons project axons into the hypothalamus and synapse with the paraventricular neurons. The dendrites of the paraventricular neurons have conventional synapses and have in addition cortisol receptors that are not part of synapses. These receptors receive cortisol directly in the intracellular fluid of the neurons. Cortisol is an inhibitory hormone. The baroreceptor neurons signal synapses that are inhibitory, as an increase in blood pressure causes a stronger baroreceptor signal that causes a decrease in CRH production. These inhibitory connections create negative feedback around the outer loop and the inner loop.

The firing rates of the paraventricular neurons are determined by their inputs and their membrane potentials. The membrane potentials are determined by the number of inhibitory and excitatory receptors and the strength of the inputs to these receptors. The number of receptors is adjusted to maintain homeostasis, to regulate the firing rates. It is believed that the regulation is controlled by % Hebbian-LMS. The rate of production of CRH is proportional to the average firing rate. Slow variations in blood pressure and cortisol inputs are tolerated by the paraventricular neurons whose output of CRH is fairly steady because of homeostasis with % Hebbian-LMS. The weights adapt to maintain homeostasis. High frequency variations due to disturbance pass through these neurons. The weights cannot adapt rapidly.

The purpose of the aldosterone system is to regulate blood volume. Since there are no sensors that can measure blood volume directly, measurement of blood pressure by baroreceptors is the next best thing. So the system is designed to regulate blood volume by sensing venous blood pressure. To do this, blood pressure signals must enter the loop. The many neurons in the paraventricular nucleus receive signals from all over the body. It is most likely that baroreceptor signals enter the system there. The connection is not described in the literature. Intuitive reasoning requires such a connection, however.

The next system component to be described is the anterior pituitary. The cells of the anterior pituitary have CRH receptors and receive CRH in the blood from the portal veins. The CRH triggers production and release of ACTH (adrenocorticotropic hormone), which is sent to the adrenal cortex via the bloodstream. The cells in the anterior pituitary fire like neurons, and like neurons secrete a hormone (ACTH) with every firing. The production of ACTH is proportional to the average firing rate. Unlike neurons, these cells have no dendrites or axons. They receive hormonal inputs from blood and deliver their outputs into the bloodstream. Their firing rates are proportional to the membrane potentials of these cells. The membrane potentials are determined by the number of receptors on each cell and the concentration of CRH in the extracellular fluid. The mechanism that controls membrane potential by adjusting the number of receptors is believed to be the same as that of the neurons of the brain. As such the firing rate experiences homeostasis with % Hebbian-LMS. Slow variations in the CRH input concentration are tolerated and production of ACTH is well regulated by homeostasis, but high frequency variations are not filtered out by homeostasis. The synaptic weights cannot adapt quickly enough.

The next system component to be considered is the adrenal cortex. The adrenal cortex receives its input signal, the hormone ACTH, via the bloodstream. When triggered by ACTH, cells in the adrenal cortex are stimulated and release aldosterone. Other cells in the adrenal cortex are stimulated and release cortisol. Cortisol is an inhibitory neurotransmitter that enters the bloodstream. The bloodstream brings cortisol to the hypothalamus, to the neurons in the paraventricular nucleus that have cortisol receptors. The inhibitory effect provides negative feedback around the inner loop connecting the adrenal cortex to the paraventricular nucleus, then to the anterior pituitary, then back to the adrenal cortex. Both the outer loop and the inner loop provide negative feedback (such an arrangement of outer loop and inner loop is fairly common with mechanistic control systems). Aldosterone is released into the bloodstream and thereby stimulates the kidneys. The cells of the adrenal cortex do not fire. When continuously stimulated, they continuously secrete their neurotransmitters. The production of aldosterone is controlled by its cells' membrane potentials. When the average membrane potential is below the equilibrium point, more than average amount of aldosterone is produced. When the average membrane potential is higher than the equilibrium point, less than average amount of aldosterone is produced. The number of ACTH receptors increases or decreases to maintain the membrane potential at or near the equilibrium point, maintaining the production of aldosterone. Under normal resting conditions, the production of aldosterone is regulated by homeostasis in accord with % Hebbian-LMS.

The kidneys respond to aldosterone and recover water and sodium from the urine before it is excreted. Blood is filtered by the kidneys to remove waste products, toxins, etc. all of which are disposed in the excreted urine. To maintain proper filtration, the blood flow rate through the kidneys is high and the production of urine is high. If all of this were excreted, the loss of water would be excessive. Some of the water and sodium are recovered from the urine and re-inserted into the bloodstream. The recovery takes place in a part of the kidney's nephrons. Figure 12.3 is a diagram that is helpful for explaining how this works.

Each kidney contains millions of nephrons. The nephrons perform several functions, one of which is the extraction of sodium and water from the urine and inserting this back into the bloodstream. Channels for the flow of urine are lined with cells called principal cells. The outer surface of these cells is a membrane called the basolateral surface. Capillaries along side this surface exchange fluid and ions with it. Gates with aldosterone receptors penetrate the basolateral surface and the capillary wall to allow two way flow of sodium and potassium ions when stimulated by aldosterone in the blood. These gates are sodium potassium pumps that are powered by ATP (adenosine triphosphate) in the blood to cause sodium ions to flow from the

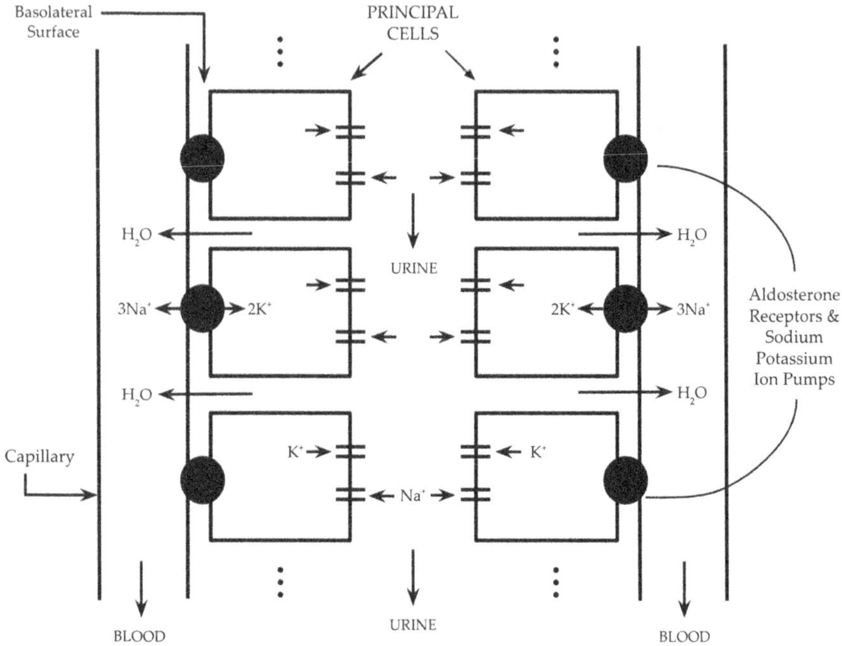

Fig. 12.3 In the nephrons. Extraction of sodium and water from the urine into the bloodstream. The residue of the urine is excreted

principal cells into the bloodstream and potassium ions to flow from the bloodstream into the principal cells. These flows are against concentration gradients and require pumping. There is a high concentration of sodium in the blood and a high concentration of potassium inside the principal cells. Leakage of sodium ions back into the principal cells limits the total flow of sodium ions into the blood. Three sodium ions are pumped for every two ions of potassium. The sodium ions bring positive charge to the blood. The potassium ions bring positive charge to the principal cells. The net result is a stronger positive charge in the blood relative to the charge in the principal cells. The resulting voltage difference is the membrane potential, the membrane being the basolateral surface and the capillary wall. Passive gates allow sodium ions to flow by osmosis from the urine channel into the principal cells and potassium ions to flow from the principal cells into the urine, to be excreted. The net effect is that sodium ions are transported from the urine stream into the bloodstream, and potassium ions are transported from the bloodstream into the urine. Also, water flows from the urine into the bloodstream accompanying the sodium ions in approximately the same ratio as already in the blood. The rate at which sodium and water enter the bloodstream is determined by the number of sodium potassium pumps and how hard they are pumping. The higher the concentration of aldosterone in the blood, the harder the pumps pump. The number of pumps can be increased or decreased.

The flow rate determines the membrane potential which in turn affects the flow rate by affecting the increase or decrease in the number of pumps.

The principal cells and their surrounding apparatus behave in some ways like neurons and their synapses in the brain. The input signal to the principal cells is the concentration of the hormone aldosterone. The weights are the sodium potassium pumps. The output signal is the membrane potential. Increasing the input signal causes the pumps to pump harder, contributing to a stronger output signal. The input signal is weighted by the many pumps in the principal cells, and the effect is a summed output, the membrane potential. This is analogous to signal flow in the synapses of a neuron adding together to create a sum, the neuron's membrane potential. A neuron has excitatory and inhibitory inputs, but the principal cells only have excitatory inputs from aldosterone.

The other aspect of behavior where there is analogy has to do with control, increase or decrease in the number of weights. In the brain, the rate of increase in the number of neuroreceptors in a synapse is proportional to the product of neurotransmitter concentration and the error signal. The error signal is once again a function of the membrane potential. In the nephron, the rate of increase in the number of neuroreceptor pumps is proportional to the product of aldosterone concentration and the error signal. The error signal is a function of the membrane potential. The error signal will be positive when the membrane potential is lower than an equilibrium point of the principal cell. The error signal reverses when the membrane potential exceeds the equilibrium point. The result is homeostasis of the membrane potential and homeostasis in the total pumping rate by adjusting the number of receptors. This seems to be the % Hebbian-LMS algorithm.

12.4 Homeostasis Everywhere

It is clear that water will be lost in the urine and in sweat. The water must be replenished to sustain proper hydration. Figure 12.4 shows how these factors enter into the aldosterone system. Figure 12.5 is a block diagram of the feedback system that is the aldosterone system.

The purpose of the aldosterone system diagrammed in Fig. 12.2 is to regulate blood volume. Blood volume is sensed indirectly by sensing blood pressure with baroreceptors. The overall system really regulates venous blood pressure. Each of the components of this system has its own homeostatic regulating system, believed to be nature's % Hebbian-LMS algorithm. Under normal resting circumstances, this system hums along with all of its components operating with their membrane potentials near their homeostatic equilibrium points. There is regulation on top of regulation. If there is then a slowly changing disturbance at some point in the system, receptors in the appropriate component will increase or decrease in number in order to restore homeostasis. Low frequency disturbances will accordingly be well tolerated by the system. Homeostasis is the ruling principle. But if the disturbance is of high frequency, the number of receptors will not be able to change fast enough to keep

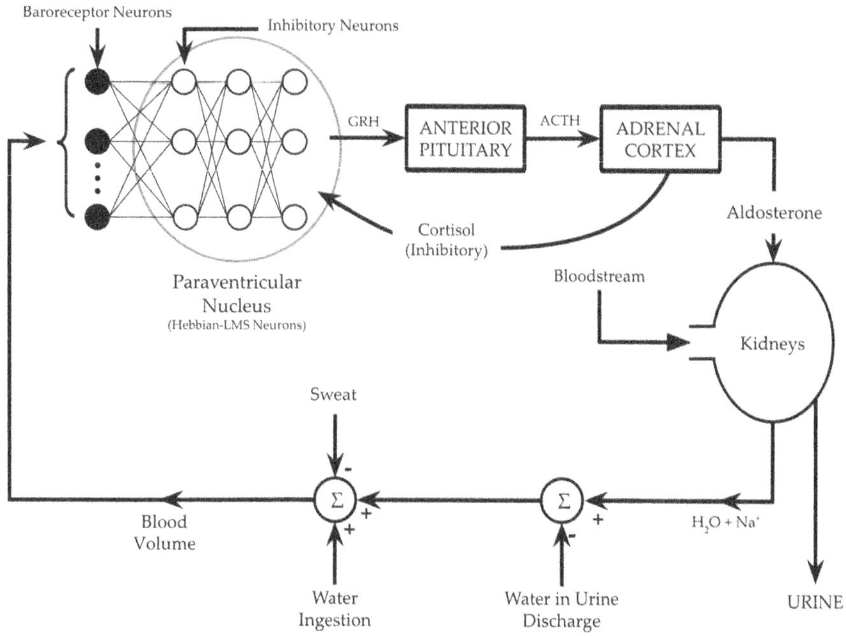

Fig. 12.4 The aldosterone system, showing external source of water and losses of water

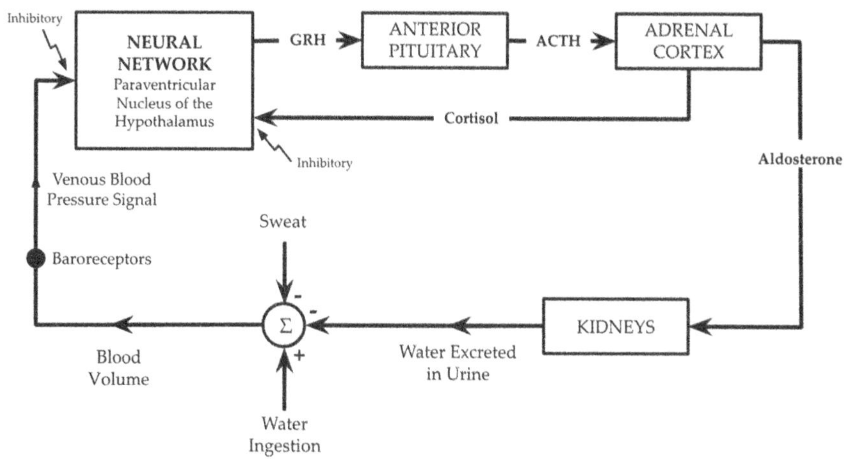

Fig. 12.5 A feedback block diagram of the aldosterone system

up. The disturbance will propagate instantaneously from component to component. The effects of the high frequency disturbance will be mitigated by negative feedback in both the outer and inner loops. In any event, homeostasis will be maintained on average.

Each element of the aldosterone system has its own unique means for establishing homeostasis, controlling its membrane potential. Each element has its own equilibrium point or set point. If one element is producing too much or too little output hormone, the following element will automatically adjust the number of its receptors to maintain its own homeostasis. Each component acting on its own maintains homeostasis to maintain its own healthy functioning. If the concentration of aldosterone in the bloodstream goes up, the number of pumps in the nephrons will decrease and maintain homeostasis. The rate of pumping on average will be constant and stay near the homeostatic set point, regardless of blood pressure and blood volume.

The purpose of the negative overall feedback is to assist the blood pressure regulation system under transient conditions. A sudden increase in blood pressure will cause a decrease in CRH production which will cause a decrease in ACTH production which will cause a decrease in aldosterone production which will cause a decrease of pumping in the nephrons which will result in a subsequent reduction of blood volume and a lowering of venous blood pressure. All this taking place at a speed that would not allow significant adaptation of the numbers of receptors. The overall feedback system regulates the "ac" component of the blood pressure, not its "dc" component. Other mechanisms are required to control average blood pressure.

In the nephrons, the average rate of pumping is determined by their set point. If some of the nephrons fail, the intact nephrons will not pick up the slack. They will continue to pump at the same equilibrium rate. This is bad news for the patient. Loss of nephrons leads to irreversible loss of kidney function.

12.5 Summary

In order to gain an overall picture of how the aldosterone system works, refer to Fig. 12.5. Assume that the person has been at rest long enough for all the weights in the system components to have converged. The number of receptors in each of the components has stabilized and homeostasis is everywhere. Homeostasis in the kidneys establishes a pumping rate that is dependent on the set point in the kidneys and independent of the homeostatic rates in the other system components. The pumping rate is independent of the system's objective. The variable that the system has available for control is the pumping rate. What has nature done here? The system's outer and inner loop feedback can only smooth out transient peaks. It does nothing for the steady state regulation of venous blood pressure or blood volume. This is out of control!

The answers to the dilemma is feedback control by the person drinking water. Blood volume is directly increased when one drinks water. Water is lost constantly, at varying rates, and it must be replenished. Water intake is not automatic. A person will drink water to overcome thirst or dry mouth. That is the feedback mechanism that really controls blood volume. Control of blood volume is not automatic.

At first, this seems shocking, that the body does not precisely control blood volume. But on reflection, it is clear that the average rate of water loss must be matched by the average rate of water intake. Control of intake is done by the person, but this is not automatic.

12.6 Questions and Experiments

1. The para-ventricular neurons have synaptic inputs and they also have cortisol receptors that are not parts of synapses. This is unusual. Do a literature search to find a description of these neurons, their dendrites, their attached synapses, and their cortisol receptors.
2. Sensing venous blood pressure is an indirect measure of blood volume. How is blood volume connected to venous blood pressure? Is this a reliable link?
3. It is most likely that baroreceptor signals enter the aldosterone system by synapsing with the neurons of the para-ventricular nucleus. Intuitive reasoning requires such a connection. Dig deeper and verify if this connection exists in the literature.
4. Under resting conditions, all the organs of the aldosterone system are functioning at their homeostatic rates. The baroreceptor signals are essentially ignored. The system is not really regulating blood volume. What is it regulating?

Chapter 13
The ADH System and the Aldosterone System Combined

Abstract Blood salinity and blood volume are determined essentially by the amount of water in the blood stream. Water is lost constantly by urination, sweating, respiration, etc. Water is gained by drinking it. The kidneys demand a high flow rate from the bloodstream in order to do proper filtration, resulting in high urine production, urine being mostly water. A large fraction of this water is recycled from the urine back into the bloodstream. Both ADH and aldosterone stimulate the kidneys to increase the fraction recycled, with the remainder excreted. This has a major effect on the amount of water in the bloodstream. The other major effect results from drinking water. Under steady-state resting conditions, the production of ADH and aldosterone are determined by homeostasis, not by sensed levels of salinity or blood volume. Transients in blood salinity are detected by osmoreceptors whose signals derive the feedback loop of the aldosterone system. Negative feedback from these loops mitigates transient effects. The steady-state average amount of water in the bloodstream is determined by ADH and aldosterone production, homeostasis, and by the volume of water ingested in satiating thirst.

Keywords Kidney function · ADH and aldosterone · The RAS system

13.1 The Aldosterone System

The purpose of the aldosterone system is to control the flow rate of water and salt from the urine back into the bloodstream. The flow rate is approximately proportional to the product of the aldosterone concentration in the blood and the number of pumps (the product of the input signal and the number of weights). The total flow is the sum of the flows of all the involved cells in the nephrons of the kidneys. The membrane potential is the average of all the membrane potentials of these cells. The membrane potential is proportional to the total flow rate. The adaptive algorithm adjusts the number of pumps to keep the long-term average membrane potential at the homeostatic equilibrium point. The adaptive algorithm is a feedback algorithm, believed to be % Hebbian-LMS.

B. Widrow, *Cybernetics 2.0*, Springer Series on Bio- and Neurosystems 14,
https://doi.org/10.1007/978-3-030-98140-2_13

The kidneys alone control the long-term homeostatic pumping rate. The negative feedback loops, the outer loop and the inner loop, have nothing to do with the long-term homeostatic pumping rate. These loops mitigate swings, up or down, of the effects of short-term disturbance in aldosterone concentration that could be caused by disturbances in ACTH or CRH concentrations. In contrast, long-term is days or longer. So, the kidneys control the long-term average pumping rate, the (dc), while the negative feedback loops help to reduce variations in pumping rate caused by system disturbances, the (ac).

13.2 The ADH System

The purpose of the ADH system is to control blood salinity by controlling the flow rate of water from urine back into the bloodstream. The flow rate is approximately proportional to the product of the ADH concentration in the blood and the number of ADH receptors in the nephrons. Since the flow is water, uncharged, the membrane potential does not depend on the flow rate. The number of receptors does not depend on flow rate and is not variable. The rate of flow of water is dependent on production of ADH in the hypothalamus. This production is proportional to the firing rate of the neurons in the supraoptic nucleus and is homeostatic in accord with the % Hebbian-LMS algorithm, and is not dependent on blood salinity. In steady state, the homeostatic flow rate of water uptake is fixed. An overall feedback loop works to moderate and smooth rapid variations in blood salinity if there were disturbances to the system, however.

13.3 ADH and Aldosterone Systems Together

Both the aldosterone system and the ADH system contribute water to the bloodstream. These systems work together and share some common components, yet have different purposes.

Figure 13.1 is a control engineer's block diagram of both of these systems functioning simultaneously. Both systems share the hypothalamus, the pituitary gland, and the kidneys. They occupy different portions of these organs, and this is indicated in the diagram of Fig. 13.1.

Assume that a person has been at rest long enough for all the weights to have adapted and converged. The ADH portion of the kidneys is pumping water from urine into the bloodstream at its homeostatic rate. The aldosterone portion of the kidneys is pumping water and sodium from urine into the bloodstream at its homeostatic rate. Water is being lost by excretion with the remainder of the urine. All components of both systems are operating at their homeostatic rates. The osmoreceptors are sensing

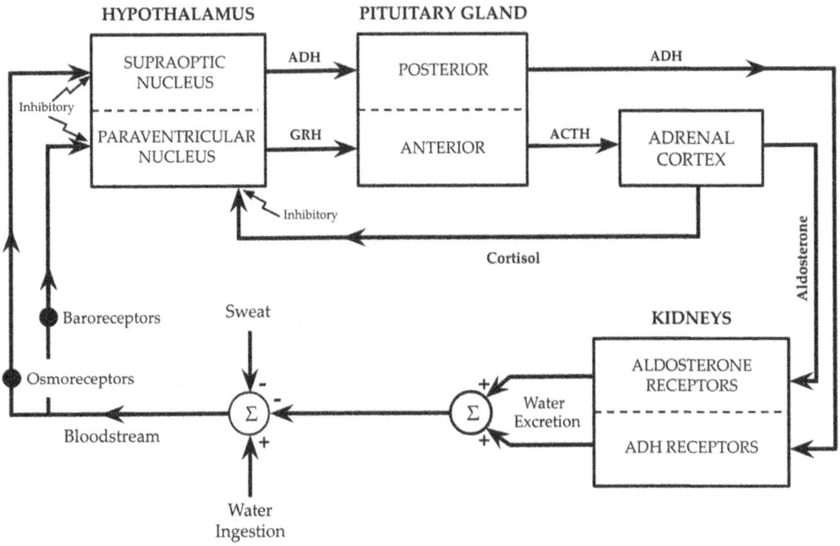

Fig. 13.1 The ADH system and the aldosterone system. Both ADH and aldosterone reduce rate of water excretion

blood salinity and the baroreceptors are sensing venous blood pressure (related to blood volume). The two systems interact, as they share a common bloodstream. Drinking water has an effect on both systems.

13.4 Homeostasis

In steady state, under the homeostatic conditions, both sides of the kidneys return some of the water from the urine to the bloodstream. This is advantageous, preserving water. But the blood salinity and venous blood pressure are not being brought to their proper levels by the two systems alone. The only way this might be done would be by the person drinking water. One degree of freedom, the rate of water intake, needs to be adjusted to satisfy two requirements, proper regulation of blood salinity and proper regulation of venous blood pressure. This is impossible.

The voluntary control of water intake, has an impossible task. What can the person do? The person ultimately drinks water to compensate for the water excreted in urine and the water lost in sweat. This maintains blood volume, maintains hydration, a must. But how is the hydration level sensed? How does a person know how much water to drink? The person drinks water to satisfy thirst or dry mouth. How that relates to blood salinity or blood volume is not clear. When a person drinks water to satisfy thirst, blood volume goes up, and blood salinity goes down. They are not going to be individually tightly controlled variables, unless other systems are involved. This may seem disappointing, but that is all that can be done with ADH and aldosterone systems.

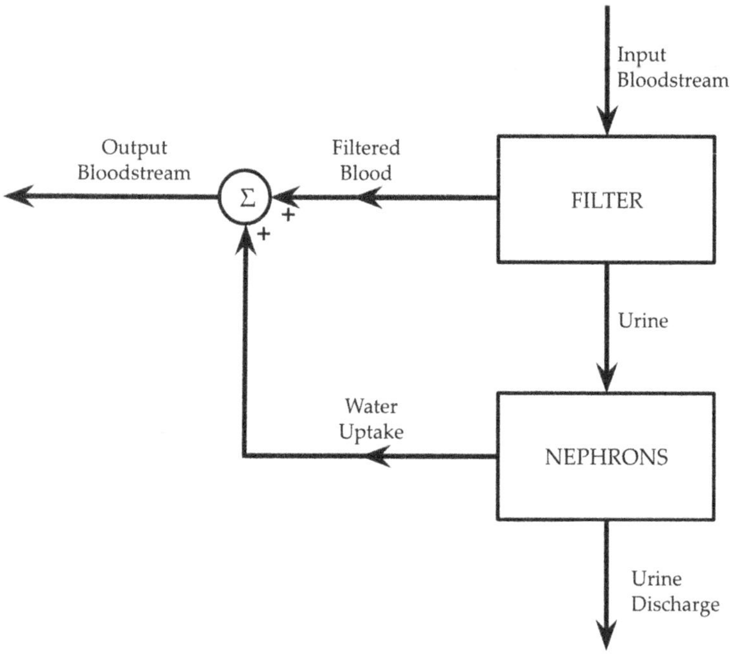

Fig. 13.2 Kidney function

13.5 Disturbance Mitigation

If a disturbance affects one system or the other, the disturbance will be detected by one or both sensors, the baroreceptors or the osmoreceptors, and the associated negative feedback loop will come into play to mitigate and smooth the effects of the disturbance. That appears to be the only function of the overall feedback loops. These loops do not control blood salinity and blood volume at their steady state homeostatic levels.

The block diagram of Fig. 13.1 is probably oversimplified. There are many other processes that effect blood volume and blood salinity. There is the RAAS system, where the enzyme renin may be released by the kidneys, having a significant effect on blood volume and salinity. The oversimplified picture gives insight into the effects of homeostasis and overall negative feedback on control and regulation of blood volume and salinity. Not only the control of these variables but the lack of absolute control over them.

13.6 Overall Kidney Function

Figure 13.2 is a block diagram illustrating kidney function, i.e. blood filtration and water recovery from the urine stream. The flow rate of the input bloodstream is proportional to blood pressure and is somewhat variable. When at rest, the flow rate of water uptake is regulated by homeostasis, while the flow rate of urine discharge is unregulated and dependant on the flow rate of the input bloodstream. The flow rate of urine discharge increases with increase of blood pressure. Excreted urine contains water that is lost. Water is also lost by sweating. The total water loss must be compensated for by drinking water.

The flow of water from urine into the bloodstream by the ADH system and the aldosterone system contributes to blood hydration, but does not totally control it. Preserving water by these systems reduces the volume of drinking water required and the volume of water excreted in urine. The ultimate control of hydration is done by the person feeling thirst or dry mouth. This is the ultimate feedback loop controlling hydration. It is not automatic. It is manually operated by the person feeling thirst.

13.7 Renin–Angiotensin–Aldosterone System (RAS System)

Renin is an enzyme that plays a major role in the secretion of aldosterone from the adrenal cortices. For simplicity of exposition, this role has not been discussed. The objective has been to describe control processes and how to diagram them, to illustrate the effects of negative feedback and homeostasis. The effects of renin are quite complex.

Renin is an element of the RAS system. This is a hormone system that regulates blood volume and vascular resistance, which have influence over cardiac output and arterial blood pressure. This system also has a strong effect on the production of aldosterone.

If renal blood flow were reduced, the juxtaglomerular cells in the kidneys convert prorenin in the blood to renin and discharges this into the bloodstream. The renin converts angiotensin from the liver to angiotensin-1. Angiotensin-1 is then converted to angiotensin-2 by the angiotensin-converting enzyme (ACE) located on the surface of vascular endothelial cells in the lungs and body. Angiotensin-2 stimulates the secretion of aldosterone. Angiotensin-2 is a powerful vasoconstrictor and has a strong effect on blood pressure.

Looking more deeply into the aldosterone system, the effects of the RAS system can be taken into account. The block diagrams and feedback loops can be modified. This however would greatly increase the complexity, and the simple explanations of negative feedback, homeostasis, and adaptivity would become much more obscure. Nevertheless, the influence of the RAS system is very important and could be accounted for.

13.8 Summary

If the water of the blood flowing into the kidneys is designated by 100%, then

$$100\% = (\%\text{ water uptake}) + (\%\text{ water excreted}).$$

At rest, in steady state, the % water uptake is determined homeostatically. The remainder is the % water excreted. The balance changes under transient conditions, however. That is when the overall negative feedback systems come into play to mitigate swings in salinity and blood volume.

A sudden change in blood volume will energize the feedback of the aldosterone system to take mitigative action. A sudden change in blood salinity will energize the feedback of the ADH system to take mitigative action. Their actions interact. Their actions together affect the % water uptake and the % water excreted. It is not clear which mitigation receives priority if there is a simultaneous sudden change in both blood volume and blood salinity. The remedy for both conditions is the same, i.e control water uptake. In any event, the average rate of water excretion must be compensated for by drinking water to satisfy thirst.

13.9 Questions and Experiments

1. Steady-state average (dc) contributions to blood salinity reduction and blood volume expansion are set homeostatically by control inputs to the kidneys. These control inputs determine the uptake of water and sodium ions from the urine back into the bloodstream. This affects blood salinity sand blood volume, but does not provide absolute regulation of these variables. Drinking water in response to thirst or dry mouth increases blood volume and decreases blood salinity. The average rate of ingestion of water contributes some regulation of these two variables but cannot provide them with individual control. The question is, what physical variables are responsible for the sensation of thirst? Would they be blood volume, or blood salinity, or some combination of both? In what way does drinking water affect control of these two variables?
2. Drinking a glass of water delivers a sudden step transient to both the ADH system and the aldosterone system. Refer to Fig. 13.1. The feedback loops of both systems will become engaged. Describe how they work individually and simultaneously to mitigate the transient effects.
3. The descriptions of the two systems in Chaps. 11, 12, and this chapter are very brief. Do further literature study to get a more complete understanding.
4. Trace back in history the evolution of the ADH system and the aldosterone system. In what historic animals did these systems begin, and how have they evolved.
5. If you were a design engineer, you probably would not develop a blood filtration process as complicated as the one that nature evolved. How would you design a

simpler system to do the same function? Would this resemble a modern dialysis machine? What more could you do with dialysis equipment to make its function closer to actual kidney function?

6. If you were able to "play god," using nature's parts and components, how would you configure a system simpler than the ADH and aldosterone systems to do equivalent work as nature's systems, but as simple as can be?

Chapter 14
Heart Rate and Blood Pressure Regulation

Abstract Blood pressure is proportional to the product of cardiac output and vascular resistance. Cardiac output is the product of stroke volume and pulse rate. The autonomic nervous system, originating in the brainstem, strongly influences pulse rate, stroke volume, and the vascular system. It has two components, the sympathetic nerves whose signals are generally excitatory and the parasympathetic nerves that are generally inhibitory. These nerves innervate the SA node, the heart's natural pacemaker, and can raise or lower its firing rate thus raising or lowering pulse rate. These signals also control vasodilation and vasoconstriction. A feedback loop connecting the heart, thc capillaries, and the cardiovascular center of the brainstem mitigates transient effects upon blood pressure and pulse rate. The steady-state resting heart rate and blood pressure are homeostatically controlled by an adaptive algorithm, believed to be % Hebbian-LMS. The autonomic system can pull heart rate and blood pressure away from homeostatic as might be needed for animal survival.

Keywords Autonomic nerves · Stroke volume · Pulse rate · Vasodilation, vasoconstriction

14.1 Introduction

Blood pressure in the body is primarily determined by cardiac output and blood vessel resistance. There are many factors that affect blood pressure but a major contributor is the cardiac output and resistance systems moderated by the sympathetic and parasympathetic nervous systems. The purpose of this chapter is to develop an explanation of the working of these systems and the role of adaptivity, particularly the role of % Hebbian-LMS, in their behaviour and performance. Homeostasis plays a major role under resting conditions. An overall feedback loop ameliorates the effects of disturbance.

B. Widrow, *Cybernetics 2.0*, Springer Series on Bio- and Neurosystems 14,
https://doi.org/10.1007/978-3-030-98140-2_14

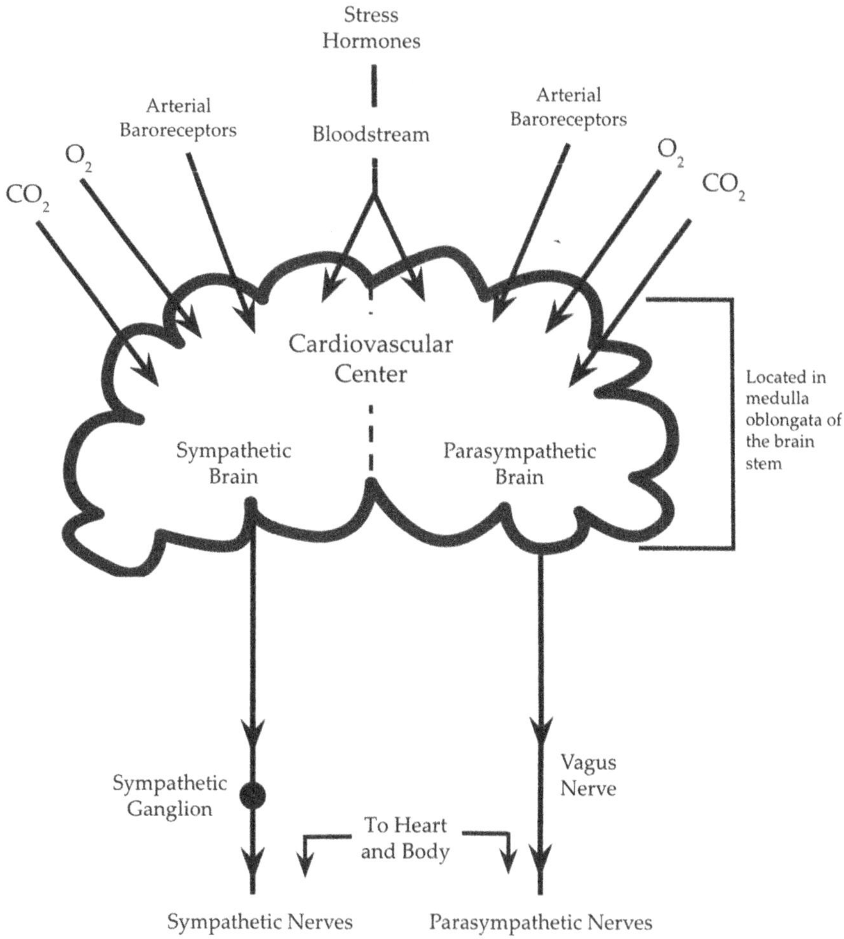

Fig. 14.1 Autonomic nervous system origin

14.2 Autonomic Nervous System

The autonomic nervous system consisting of the sympathetic nervous system and the parasympathetic nervous system has a primary signaling function, sending messages from the brain to organs of the body. These messages are control signals. The heart and the body's vascular system are innervated by the autonomic nervous system.

The autonomic signals originate in two portions of the brain located in the medulla oblongata of the brainstem. One portion generates the sympathetic nerve signals that are excitatory. The other portion generates the parasympathetic nerve signals that are primarily inhibitory. The two portions of brain comprise the cardiovascular center.

Figure 14.1 is a diagram of the autonomic nervous system at its origin in the brainstem. The cardiovascular center receives inputs from the baroreceptors located in the aortic arch and the two carotid sinuses. These arterial baroreceptors are specialized neurons that sense arterial blood pressure. Other sensors for carbon dioxide and oxygen feed their information to the cardiovascular center. The baroreceptors, oxygen sensors, and carbon dioxide sensors synapse their signals onto neurons in the cardiovascular center. Additional information arrives via the bloodstream relating to stress, fear, anxiety, etc. These are hormonal signals that directly affect the neurons of the cardiovascular center.

In response to these inputs, the neurons of the sympathetic and parasympathetic sides of the cardiovascular center act antagonistically. Under normal conditions, the sympathetic neurons' signals are essentially balanced with the signals of the parasympathetic system. An increase in blood pressure and/or oxygen will lower the firing rate of the sympathetic nerves and at the same time increase the firing rate of the parasympathetic nerves. An increase of stress hormones in the bloodstream and/or an increase in carbon dioxide levels in the blood will cause an increase in the sympathetic firing rate and a decrease in the parasympathetic firing rate. So, nature's "reasoning" seems to be that an increase in stress will cause an increase in heart rate since heart rate is increased by increased sympathetic nerve activity and decreased parasympathetic nerve activity. An increase in carbon dioxide causes an increased heart rate. Stronger circulation in the lungs enhances their ability to exhale carbon dioxide.

An increase in blood pressure has a negative effect on sympathetic activity. It seems that nature reduces the heart rate in order to reduce blood pressure, in order to mitigate the effects of elevated blood pressure. This negative effect is the basis of negative feedback in the control loop for blood pressure and heart rate.

Figure 14.2 is a diagram illustrating the heart, its connections, and some of its internal components. These components relate to heart rate and stroke volume. The total cardiac output is the product of the pulse rate and the stroke volume. Heart rate is affected by the sympathetic and parasympathetic nerves, and stroke volume is affected by the sympathetic nerves.

14.3 Autonomic Ganglia

The sympathetic and parasympathetic nerve signals propagate through stages of ganglia. The sympathetic ganglion, represented in a simplified manner in Fig. 14.2, can be thought of as a single layer neural network. The neurons work like the neurons in the brain and control of the number of input neuroreceptors, which are all excitatory, is believed to be regulated by % Hebbian-LMS. These neurons have a homeostatic firing rate corresponding to an intrinsic equilibrium point.

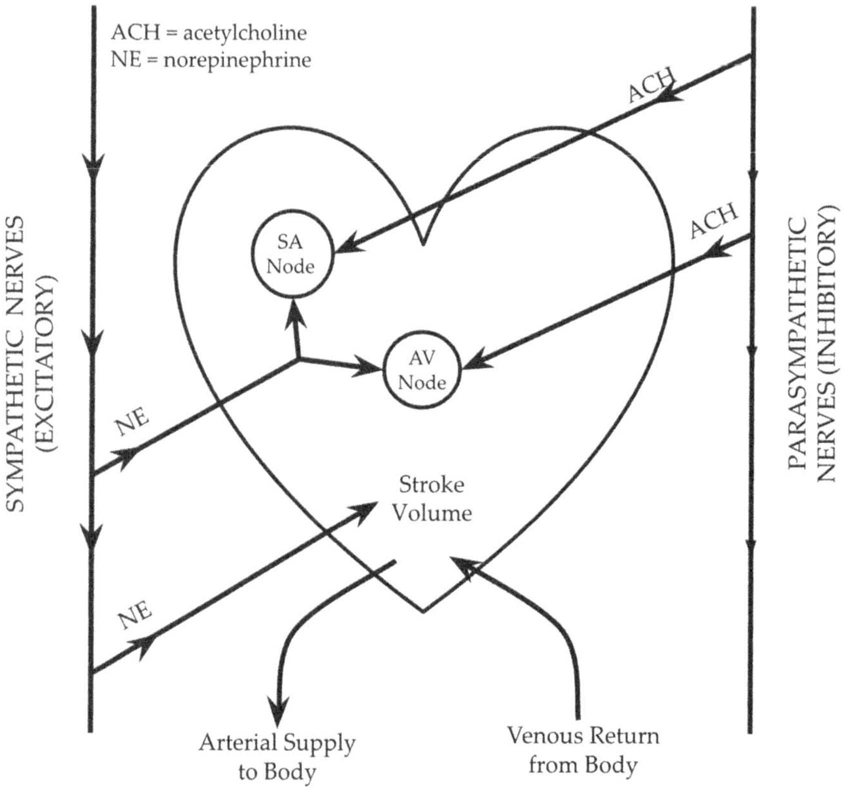

Fig. 14.2 Sympathetic and parasympathetic innervation to the heart

A sudden increase in the presynaptic firing rate input to the sympathetic ganglia will cause a sudden increase in the postsynaptic firing rate of the ganglia. The result will be a stronger sympathetic signal. If the elevated presynaptic firing rate persists, % Hebbian-LMS will adapt to this and will cause a reduction in the number of neuroreceptors and the subsequent gradual return of the firing rate back to the homeostatic equilibrium level.

The ganglia serve as relay nodes that allow rapid response but provide preliminary long-term homeostatic regulation of the sympathetic and parasympathetic inputs to the heart. The (ac) components of the ganglia inputs propagate straight through, but the (dc) components of the ganglia inputs are blocked and replaced by homeostatic firing at a rate determined by intrinsic set points.

14.4 The SA Node

The SA (sinoatrial) node is a cluster of myocytes (muscle cells) located in the upper right atrium. These cells are electrically active and in some sense behave like neurons, although they are not neurons. They have membranes with embedded neuroreceptors that receive innervation from both the sympathetic nerves and parasympathetic nerves. They fire spontaneously and continuously, generating action potentials that trigger the heart's electrical system enabling the beating of the heart. The SA node is the heart's natural pacemaker.

14.5 Innervation of the SA Node by the Autonomic Nerves

The cells of the SA node are different from the myocytes of the atria and ventricles. The latter need trigger inputs in order to fire and contract. The cells of the SA node will fire without inputs at a homeostatic rate of about 100 beats per minute. They can be speeded up by the sympathetic system or slowed down by the parasympathetic system. These two systems innervate by means of the neurotransmitters norepinephrine and acetylcholine. The SA node cells have excitatory receptors for norepinephrine and inhibitory receptors for acetylcholine. Excitation raises their membrane potentials while inhibition lowers their membrane potentials. Unlike neurons, the SA cells have no dendrites and synapses, but are innervated by axons whose terminals release neurotransmitter in the vicinity of the neuroreceptors.

Although the SA node cells are free running oscillators that fire spontaneously, the neurotransmitter inputs that can modulate their firing rates are needed in order that high-speed control of firing rates by the cardiovascular center and higher brain function becomes possible. The "fight or flight" response is enabled by action of the autonomic system and has effects on the heart and on other body organs and systems.

Under normal resting conditions the parasympathetic system emits stronger signals, higher firing rates, than the sympathetic system. This is apparently the homeostatic condition of the cardiovascular center. The seesaw balance between the sympathetic and parasympathetic systems favors the parasympathetic system when a person is at rest. The result is that the normal resting heartbeat is pulled down from the homeostatic level of 100 beats per minute to a resting level of 60–80 beats per minute.

Under stressful conditions, the seesaw balance shifts in favor of the sympathetic system and the heart rate increases from its resting rate. If the stressful conditions persist for a long enough time (for example, "trouble with your mother-in-law"), the SA node will adapt in order to bring the average heart rate down to a steady state level closer to the resting rate. In this case, the number of excitatory receptors will shrink and the number of inhibitory receptors will increase.

14.6 % Hebbian-LMS Control

Nature's homeostatic algorithm for controlling the number of SA node receptors, excitatory and inhibitory is believed to be % Hebbian-LMS. An explanation is best visualized by analogy to % Hebbian-LMS control of neurons in the brain. As with neurons, signals flow via neurotransmitters and neuroreceptors. The input signals to the SA node's cells come from axons of the sympathetic and parasympathetic nervous systems. When binding to receptors, neurotransmitters cause gates to open penetrating the cells' membranes and allowing ions to flow in or out of the cells, raising or lowering their membrane potentials, increasing or lowering their firing rates. Signal flow from input to output is very much like that of synapses and neurons.

> The resting firing rate is lower than the homeostatic rate. Inputs from the parasympathetic nerves make this possible. The % Hebbian-LMS algorithm, trying to maintain the homeostatic rate, is fighting against the parasympathetic system. Continual strong inputs from the parasympathetic nerves establish the resting firing rate.
> Maintaining the resting rate is not homeostatic, but this does seem to be something like homeostasis. Preserving the resting rate is regulating the (dc). Mitigating the effects of quick disturbances, controlling the (ac), is the function of an overall negative feedback loop which will be examined below.

It may seem strange that nature would choose a resting firing rate for the SA node of 60–80 pulses per minute, lower than the homeostatic rate of about 100 pulses per minute. There is probably a good reason for this bias. A "comfort zone" for the heart is approximately between 60 and 100 beats per minute. Holding the rate at the bottom end of the zone is probably a good strategy. A sudden change in situation or an emergency is likely to require a sudden increase in heart rate. With the resting rate at the bottom end of the comfort zone, there is plenty of headroom for a sudden increase.

While keeping the SA node's resting pulse rate below its set point, the number of excitatory receptors have increased and the number of inhibitory receptors have decreased in an attempt to bring the pulse rate up to the homeostatic set point in the face of seesaw pull favoring inhibition by the parasympathetic nervous system. As the adaptive algorithm requires an error signal in order to have a restoring force, the algorithmic correction of % Hebbian-LMS cannot totally correct the error. The resting pulse rate remains below the homeostatic set point. A considerable amount of experimental effort would be required to verify this with actual numbers, but the basic concepts conform to observed phenomena.

14.7 Vascular Resistance

Vasoconstriction and vasodilation in the vascular system are critical elements of the blood pressure control system. Blood pressure is the product of cardiac output and vascular resistance. Resistance increases with vasoconstriction and decreases with vasodilation, all controlled by the autonomic nervous system. An illustration of the vascular system with its connections to the rest of the body is shown in Fig. 14.3.

Blood vessels are innervated by neurotransmitters. Excitatory neurotransmitters released from the sympathetic nerves cause vasoconstriction. Inhibitory neurotransmitters released from the parasympathetic nerves are antagonistic to the sympathetic neurotransmitters and cause vasodilation.

The walls of blood vessels are muscular. With no innervation, the vessel walls would be totally relaxed and completely dilated. Lumens would be at maximum and resistance at minimum. With the body in a normal resting state, however, these muscles would be roughly half contracted, more strongly innervated by the sympathetic system than the parasympathetic system, maintaining muscle tone. Resistance would be moderate. The seesaw would favor the sympathetic system in this case. Note that if the sympathetic and parasympathetic effects were balanced, the effects would cancel and vessel walls would be fully dilated.

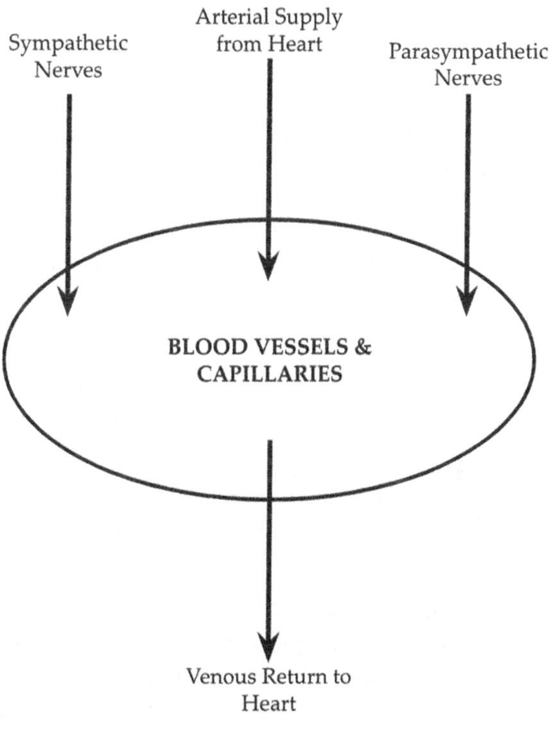

Fig. 14.3 Blood vessel system. Vasodilation and vasoconstriction control inputs from the parasympathetic and sympathetic nerves

If there were a need to increase blood pressure by vasoconstriction, the seesaw would tip further, favoring the sympathetic system. On the other hand if there were a need to lower blood pressure, vasodilation would be accomplished by tipping the seesaw in favor of the parasympathetic system. Under normal resting conditions, muscle tone would be maintained by tipping the seesaw in the sympathetic direction. The seesaw is biased and this is necessary because muscles can only relax or contract, they cannot expand. They can only pull not push.

The inner walls of blood vessels are lined with endothelial cells that comprise a membrane. This membrane has various embedded neuroreceptors. Some of these receptors bind to norepinephrine secreted from axon terminals from the sympathetic system. Other receptors bind to acetylcholine secreted from axon terminals from the parasympathetic system. Sympathetic activity causes vasoconstriction and parasympathetic activity causes the endothelial cells to discharge nitric oxide into the bloodstream. Nitric oxide is a vasodilator that can counteract the effect of norepinephrine. Once again there is an antagonistic seesaw, with the sympathetic nervous system causing vasoconstriction (muscle contraction) and the parasympathetic nervous system causing vasodilation (muscle relaxation).

The outer blood vessel layer consists of smooth muscle cells that comprise another membrane. Like endothelial cells, smooth muscle cells have receptors for the neurotransmitters embedded in their membrane. Neurotransmitters released from the sympathetic and parasympathetic axon terminals diffuse and interact with the vascular cells' receptors of both layers. The receptors are gates that penetrate the membranes and, when binding to neurotransmitters, allow the flow of ions, affecting their membrane potentials.

The two membranes have their own equilibrium set points. Homeostasis is achieved by raising or lowering the numbers of receptors of each layer in order to keep membrane potentials close to their set points. Although there are no neurons or synapses involved, the control algorithm regulating the numbers of receptors is believed to be % Hebbian-LMS.

Resting vasoresistance is determined by the cells' homeostatic set points (the (dc)). Rapid responses to the demands of the autonomic nervous system are accommodated, but extremes (the (ac)) are mitigated by an overall negative feedback control system to be described below.

14.8 Resting Conditions

The resting homeostatic conditions of heart rate and blood pressure can be examined in reference to the diagram of Fig. 14.4, a diagram that illustrates cardiovascular control by the autonomic nervous system.

The heart rate is controlled primarily by the firing of the SA node. The firing rate under resting conditions is held under the natural homeostatic rate of 100 beats per minute to a rate of about 60–80 beats per minute by the parasympathetic system supplying stronger inhibitory signal than the sympathetic system supplying excitatory

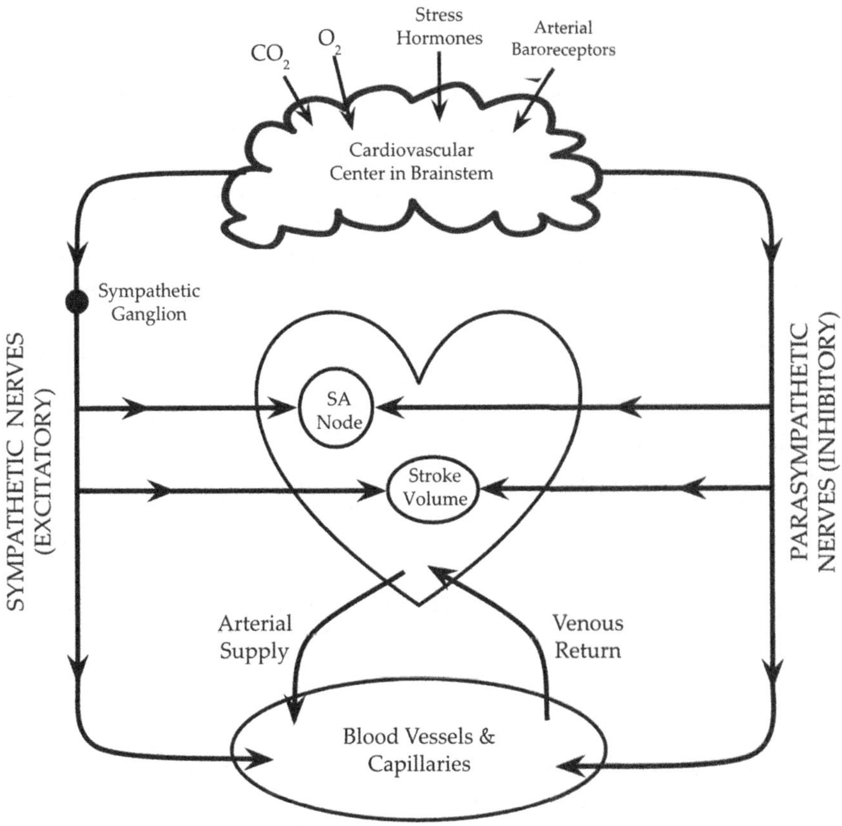

Fig. 14.4 Autonomic system contributes to regulation of heart function and blood pressure

signal. The autonomic excitatory and inhibitory signal levels are set homeostatically in the cardiovascular center.

The resting stroke volume is determined by excitatory and inhibitory signals from the autonomic system. These signals function at homeostatic rates set by the cardiovascular center.

When at rest, the capillary system operates with a balance between the influence of the parasympathetic nervous system that promotes vasodilation and the innervation of the sympathetic nervous system that promotes vasoconstriction. A seesaw balance is established homeostatically in the cardiovascular center that sets the dilation of the respective blood vessels at an in between level, leaving room for further dilation or further constriction. The capillary resistance will be approximately midrange.

At rest, the conditions of the cardiovascular system are established by autonomic homeostasis. These conditions are long term time averages, the (dc) components of every-day cardiovascular conditions.

14.9 Dynamic Conditions

The behavior of the cardiovascular system under dynamic conditions, with transient disturbances, can be studied with the help of the block diagram of Fig. 14.5. This diagram shows an overall loop with negative feedback.

The purpose of the overall feedback loop is mitigation of transient effects, control of the (ac). In the diagram of Fig. 14.5, blood pressure is the product of cardiac output and vascular resistance. The autonomic system controls cardiac output and also vascular resistance. Blood pressure is primarily controlled by the autonomic system, accordingly.

The autonomic signals are generated by the cardiovascular center of the brainstem. Inputs to this center are arterial blood pressure and blood oxygen level, which are inhibitory, and blood carbon dioxide level and hormonal inputs from other parts of the brain related to stress, which are excitatory.

The loop is closed by arterial baroreceptor signals which are inhibitory, thus giving the overall loop negative feedback. The other inputs to the cardiovascular center come from outside the loop and their variations amount to disturbances to the feedback system. Changes in blood pressure, carbon dioxide, oxygen, or stress level that take place faster than the system's weights adapt are disturbances that are mitigated by the

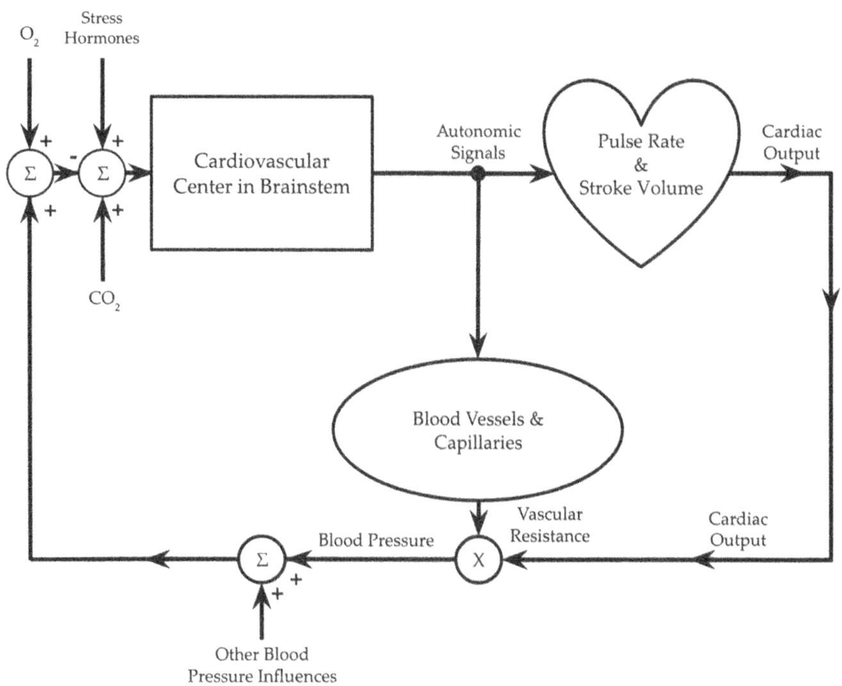

Fig. 14.5 Nonlinear feedback system controlling blood pressure

feedback loop. Negative feedback controls the (ac) components of the cardiovascular variables such as heart rate and blood pressure. The (dc) components are controlled by homeostasis.

An example of how this works is the following. Refer to the feedback diagram of Fig. 14.5. You are in a dark room, and out from the shadows another person shouts "boo!" You are startled and frightened. Stress hormones flood into the cardiovascular center causing the sympathetic nervous system to jump into action. Your heart starts racing and cardiac output increases rapidly. Blood pressure flies high and your cardiovascular system is overreacting. The baroreceptors sense the jump in blood pressure and send an inhibitory signal to the cardiovascular system, causing it to calm down. The negative feedback is mitigating the effects of sudden disturbance.

This example illustrates the feedback loop in action against a sudden transient. Homeostasis is not involved here in dealing with a transient disturbance.

14.10 The % Hebbian-LMS Algorithm Doing High-Pass Filtering?

% Hebbian-LMS neurons in the feedback loop of Fig. 14.5 and in the feedback loops of the previous chapters block the passage of slowly varying inputs and pass rapidly varying inputs through to their axonal outputs. Low frequency inputs to a given neuron are blocked by adaptivity. The synaptic weights adapt to maintain homeostasis, so the neuron's firing rate is unaffected by inputs to its synapses. A high frequency input signal does not cause the synapses to change. They cannot adapt fast enough, so the synaptic weights remain essentially constant and high frequency inputs propogate through to the output. In the field of digital signal processing, high pass filters block low frequency inputs. The question is, does the % Hebbian-LMS algorithm behave like a high pass filter?

Assume that the inputs to a given neuron are low-frequency. The synaptic weights will adapt and develop average values in order to sustain homeostasis. The low frequency inputs determine the average weight values. Now suppose high frequency signals are suddenly superposed onto the low frequency signals. The weights do not change and the high frequency signal components pass through. Figure 14.6 is a model of the neuron.

The block diagram of Fig. 14.6 shows the inputs applied to weights having average values. The weighted inputs are then applied to linear high pass filters whose outputs are summed. This model works at low frequencies that are in any event blocked. It works at high frequencies that propagate through. At intermediate frequencies, the model becomes problematic. The weights will vary somewhat and would not be constant. There is no way to model this neuron using conventional signal processing components. The % Hebbian-LMS neuron is unique and there is nothing else like it.

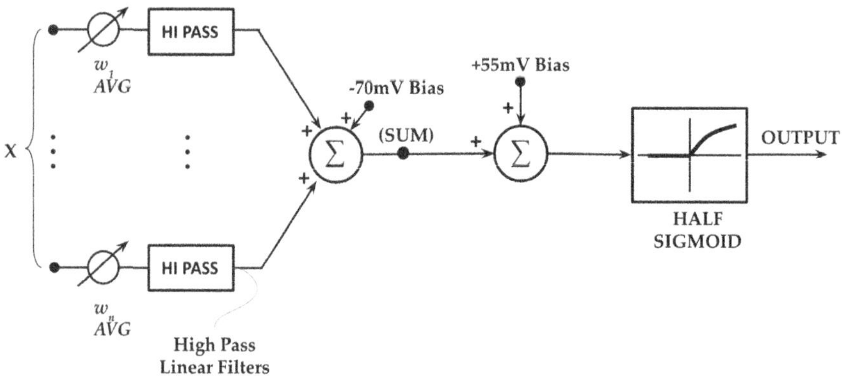

Fig. 14.6 A model of a %Hebbian-LMS neuron as a high-pass filter. This model works for low frequency and high frequency inputs and for a superposition of the two. It fails for intermediate frequency inputs

14.11 Summary

Figure 14.5 shows the major elements of the blood pressure control system and how they are interconnected. The blood pressure is the product of cardiac output and vascular resistance plus other factors in the body that influence blood pressure. Such influences are disturbance inputs. The autonomic system controls cardiac output by controlling pulse rate and stroke volume. The autonomic system also controls vascular resistance. Blood pressure is primarily controlled by the autonomic nervous system.

The autonomic signals emanate from the medulla oblongata of the cardiovascular center of the brainstem. The signals are generated by a neural network whose adaptive or learning algorithm is believed to be % Hebbian-LMS. Inputs to the neural network are signals from arterial baroreceptors and blood oxygen sensors that are both inhibitory. Inputs also from from blood CO_2 sensors and stress hormones carried by the bloodstream that are both excitatory. The O_2 signal, the CO_2 signal, and the stress hormonal signal are all external disturbance inputs. The baroreceptor signal is not external but is part of the overall feedback loop. The multiplier that forms the product of cardiac output and vascular resistance is part of the feedback loop and makes this loop nonlinear.

Figure 14.5 is useful for visualizing both the (dc) and the (ac) behavior of the blood pressure control system. In the resting state, homeostatic conditions prevail in the neural network of the brainstem. The signal strengths of the autonomic signals are homeostatic and are unaffected by the various inputs to the neural network. The cardiac output and vascular resistance are therefore under homeostatic control. The result is that blood pressure is homeostatically controlled. The pulse rate is homeostatically controlled.

The resting steady-state oxygen and carbon dioxide levels have no effect on the autonomic signals because the synaptic weights of the neural network adapt to preserve the homeostatic set point. In the resting state, blood pressure will be set homeostatically.

Under dynamic conditions, the weights cannot adapt quickly and so all elements of this system pass signals essentially instantaneously. The feedback system now functions and mitigates the effect of disturbances. The feedback loop has no effect on (dc) behaviour when at rest and comes into play to mitigate the effects of (ac) behavior when the system is subject to transient disturbances.

14.12 Questions and Experiments

1. The resting human heart rate is generally lower than the homeostatic rate. Can an experiment be done to verify this? Can this be done with humans? Is there an animal model for the phenomenon?
2. Referring to the diagram of Fig. 14.2, describe how homeostasis helps in establishment of resting heart rate and stroke volume. How does this work? What effect do the various ganglions have on this?
3. Referring to Figs. 14.4 and 14.5, explain how the negative overall feedback works how this mitigates transient effects yet has no influence over the resting heart rate and blood pressure.

Chapter 15
Regulation of Blood Glucose Level

Abstract Glucose in the blood is the energy source for all of the cells in the body. The membranes surrounding the cells prevent glucose from entering, except at specific gate locations. Only when gates are open can glucose be absorbed by the cells. Co-located at the gates are insulin receptors. When these receptors capture insulin molecules, the gates open and allow the cells to be fueled with glucose. The cells' consumption of glucose is homeostatically regulated. The blood glucose level is kept within 70–120 mg/l, and is not regulated homeostatically. It is regulated by food intake, satisfying hunger or satiation.

Keywords Glucose · Pancreas · Insulin, insulin receptors · Liver · Glycogen, glycogen receptors · Glucagon · Blood glucose electric charge

15.1 Glucose

Glucose is a simple form of sugar that enters the blood stream when carbohydrates are consumed. Various amounts of carbohydrate are present in foods. The concentration of glucose in the blood becomes highly elevated after a meal. Between meals, the concentration of glucose gradually declines.

It is important that the blood glucose level be maintained within a normal range of about 70 mg/dL to about 120 mg/dL. The purpose of blood glucose regulation is to keep blood glucose levels within this normal range, at least most of the time.

15.2 Insulin and Insulin Receptors

Insulin is a hormone that is generated by islet cells in the pancreas. This hormone binds to insulin receptors embedded in the membranes of all the body's cells.

In a given cell, the rate of flow of glucose into the cell is proportional to the blood glucose level and to the number of its insulin receptors multiplied by the blood insulin concentration. The number of insulin receptors is variable. An adaptive algorithm,

B. Widrow, *Cybernetics 2.0*, Springer Series on Bio- and Neurosystems 14,
https://doi.org/10.1007/978-3-030-98140-2_15

believed to be % Hebbian-LMS, adjusts the number of receptors slowly in accord with the average blood insulin level and blood glucose level.

Each cell has a set point for the rate of inflow of glucose. If the flow is greater than the set point, the number of receptors is reduced. If the flow is less than the set point, the number of receptors is increased. The number of receptors can change only very slowly. The number of receptors adapts to the average blood insulin level and blood glucose levels so that the average rate of glucose inflow matches the set point. This is the cell's homeostatic rate of glucose inflow.

Because of adaptivity, the cell can tolerate a significant range of average blood insulin levels and blood glucose levels. The blood glucose system only needs to keep the blood sugar level in the normal healthy range and is not concerned with the glucose inflow to the individual cells. The cells are able to adapt and take care of themselves by upregulating or downregulating the number of their insulin receptors.

15.3 The Pancreas and the Liver

The pancreas and the liver are major organs of the body. The pancreas has islet cells, beta cells, that release the hormone insulin, and it has other islet cells, alpha cells, that release the hormone glucagon. These two hormones play significant roles with regard to regulation of blood glucose levels.

The liver plays a major role in the regulation of blood sugar concentration. It is, among other things, a storage mechanism for glucose. When glucose enters the liver, it is converted to glycogen for storage. When the hormone glucagon is present in the blood, this induces the liver to reverse itself and to convert glycogen back to glucose, which is then released into the bloodstream.

The body's cells derive glucose from the bloodstream. They also derive insulin from the bloodstream to enable the cells to utilize the glucose. Glucose is the cell's power supply, vital for life and survival.

Figure 15.1, is a diagram that illustrates the connections between the pancreas, the liver, and the body's cells, to the bloodstream. All these organs utilize glucose to energize their cells. The liver in addition uses glucose to convert to glycogen for storage. The pancreas releases insulin and glucagon into the bloodstream. The liver sometimes stores glucose and sometimes releases it back into the bloodstream.

> The body's cells take up glucose at gates that are actuated by insulin. The liver takes up glucose also at gates that are actuated by insulin. The gates are all co-located with insulin receptors. Wherever there are receptors embedded in membranes co-located with gates, the average rate of flow through the gates will be controlled by choosing the number of receptors, subject to homeostasis. The "choosing algorithm" is most likely % Hebbian-LMS.

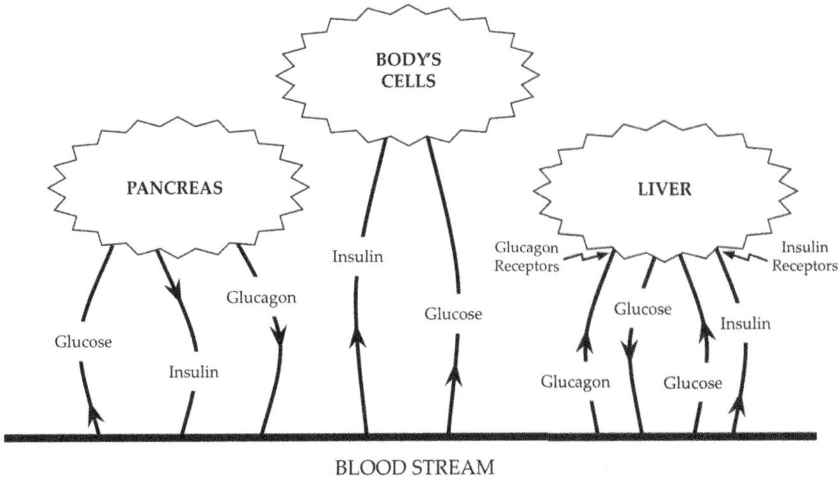

Fig. 15.1 Pancreas, liver, and body's cells connected to the bloodstream

15.4 The Resting State

Eating meals in the usual way causes blood sugar levels to vary widely up and down. Imagine eating small quantities of food continually, day and night. Now suppose that a person is at rest, only nibbling food and sipping water. The blood sugar level will stay essentially constant, day and night. This would be the resting state.

The question is, what determines the resting blood glucose level? This is a good question. An answer will be suggested.

At rest and nibbling food, the body's cells will consume glucose at their adaptive homeostatic rate. In spite of possible variations in glucose levels and variations in availability of insulin, adaptivity within the cell insures that glucose will be consumed at the homeostatic rate. Insulin will be produced by the pancreas at a relatively low-level, sufficient for maintaining proper functioning of the body's cells. The actual level of insulin production will not be critical. The number of insulin receptors will adjust homeostatically to match the homeostatic rate of consumption of glucose.

Another consumer of glucose is the body's muscles. At rest, muscular activity is relatively light and consumption of glucose for this activity will be relatively steady. Respiration and heart beating at rest are examples of such usage.

Another drain on the bloodstream's glucose, when in the resting homeostatic state, is caused by the liver storing glucose. The liver utilizes insulin, and will absorb glucose at its own slow steady homeostatic rate.

The cells' metabolism and muscular activity are the two main consumers of glucose. Storage in the liver creates additional glucose demand. The total glucose demand will be quite steady in the resting state.

Creation of glucose will be at a rate determined by the rate of nibbling. Energy balance occurs when the rate of glucose consumption equals the rate of glucose production. Slightly faster nibbling will cause the glucose level to go up. Slightly slower nibbling will cause the glucose level to go down. Hunger makes one nibble faster. Satiation makes one nibble slower. Blood glucose level is kept in the normal range by rate of food intake. It is a voluntary effort. It is not automatic.

15.5 Mitigation of Transient Effects by Overall Negative Feedback

In the resting state, the blood glucose level is controlled by the rate of food intake. The total glucose demand is partially homeostatic but essentially constant. Consuming enough carbohydrates to first satisfy this demand will stabilize the blood glucose at some level. Nibbling a little faster will make the glucose level begin to rise. Nibbling a little slower will make the glucose level begin to lower. Alternating between nibbling too fast and too slow could enable stabilization within the normal range and indeed at any desired level within that range.

In real life, the resting state does not often prevail. Eat a meal and a blood glucose level rises rapidly. Eat a candy bar and this level will spike upward. Vigorous exercise will cause this level to drop. The resting state corresponds to average conditions, the (dc). Conditions that deviate from the resting state are disturbances, the (ac). Transient effects caused by eating a meal as opposed to steady state nibbling is an example.

It is hard to think of eating a meal as a disturbance, but that is exactly what it is. It creates (ac) responses, deviations, on top of the (dc), the resting state.

Under transient conditions, if the pancreas were to suddenly release insulin at a higher rate, the consumption of glucose by the body's cells would suddenly increase, lowering the blood glucose level. The response is fast. Adaptivity, homeostasis, is not an issue here. The change in number of insulin receptors is too slow to keep up with transients. Transient mitigation is achieved by overall negative feedback.

If the blood glucose level were to go down to the low end of the normal range, the pancreas would sense this and start releasing glucagon. The liver would respond and begin converting stored glycogen to glucose which would be released into the bloodstream. Thus, the release of glucagon by the pancreas would cause blood glucose levels to go up.

The pancreas has the capability of sensing blood glucose level. At the low end of the normal range, the pancreas produces almost no insulin. At the high end, it produces insulin at its maximum rate. At the low end of the normal range, the pancreas produces glucagon. The two hormones have opposite effects on blood glucose level, insulin lowers it and glucagon raises it.

A feedback loop for control of blood sugar level under transient conditions, is shown in Fig. 15.2. The body's cells demand glucose and contribute to a lowering of glucose level. A sudden increase in blood glucose level causes a sudden increase in glucose demand. Figure 15.2 illustrates the connection of the body's cells to the bloodstream and their effect on blood glucose level.

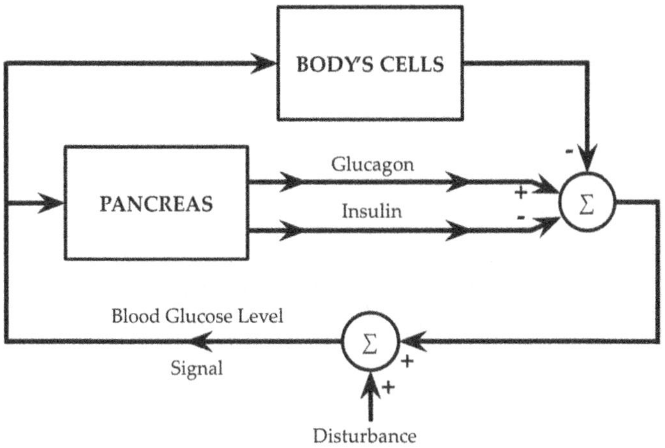

Fig. 15.2 Feedback control of blood glucose level

Insulin lowers the level and glucagon raises it in response to disturbance inputs. All this is too fast for adaptivity. A disturbance that would cause a rise in blood sugar level would excite the pancreas to produce insulin at a higher rate and this would in turn lower the blood sugar level. A disturbance that would lower the blood sugar level to the bottom of the normal range would excite the pancreas to produce glucagon and this would raise blood glucose level. It is clear that the loop provides negative feedback and this mitigates the extent of deviation in glucose level in response to rapid disturbances.

15.6 Summary

To summarize, in the resting state, the average consumption of glucose by the body's cells is regulated homeostatically. It is believed that the homeostatic algorithm is % Hebbian-LMS. The average rate of glucose uptake by the liver is controlled homeostatically. The average glucose demand by the body's muscles is not determined at rest by homeostatic means, but is quite steady. The total glucose demand is very steady at rest.

The average blood glucose level at rest is regulated by average food intake. This (dc) control is voluntary, not automatic. The driving forces behind this are hunger and satiation. The average blood glucose level is not homeostatically regulated.

Responding to transient disturbances, negative feedback is invoked to mitigate these responses. Positive feedback, on the other hand, results from secretion of glucagon, but that only occurs at very low blood glucose levels. Nature's objective is to minimize blood glucose level excursions about the average to keep the glucose level within the normal range as well as possible. Homeostasis does not play a role in disturbance mitigation. The (ac) is controlled by overall negative feedback. Eat a candy bar and the overall feedback kicks in.

15.7 Postscript: Diabetes Mellitus

Diabetes mellitus is a disease that is characterized by very high blood glucose levels, much higher than the normal range. This disease is widespread worldwide. It is a serious disease and often difficult to treat. There are two broad categories of diabetes:

1. Type 1: This is an autoimmune disease in which the immune system attacks the beta cells of the pancreas and causes them to stop producing insulin, or greatly reduces their production. This disease usually presents in young people and is often called juvenile diabetes. It is a rare disease. It could exist in the adult population as well.
   Lack of insulin results in greatly reduced glucose absorption by the body's cells. This causes high blood glucose levels. Treatment with insulin is necessary to bring glucose levels into the normal range.
2. Type 2: This disease results from defective performance of the insulin receptors. Even though insulin is available, the body's cells cannot utilize it very well. This is called insulin resistance. Without insulin, the body's cells cannot absorb glucose and blood glucose levels, become very high, much higher than the normal range. Treatment is essential in order to have a healthy life and avoid the complication of high blood sugar. Raising blood insulin levels is one form of treatment.
   A related disease is gestational diabetes. This is similar to Type 2 diabetes and is caused by pregnancy. It generally cures itself after birth of the baby.

15.8 Research Questions

When there are receptors of neurotransmitters or hormones embedded in the membrane surrounding a cell, one may expect that the increase or decrease in the number of receptors is controlled by a natural algorithm. This algorithm can be modeled by % Hebbian-LMS. Homeostasis is one of the properties of the % Hebbian-LMS algorithm.

Insulin and glucagon are hormones that bind to the respective receptors of the liver. The presence of insulin in the bloodstream is essential to opening gates that allow absorption of glucose into the body's cells. The presence of glucagon is necessary to opening gates that allow secretion of glucose from the liver back into the bloodstream.

Because of the arrangement of membrane, receptors, homeostasis is assumed to regulate the flow of glucose into the liver when glucose is stored. There is a problem with the assumption, however.

In most cases, gates are "designed" to allow the flow of charged particles, ions, through the membrane, changing the membrane potential in order to bring it closer to its homeostatic level. Recall the discussion of the ohmic current and the ionic current assumed taking place in the membrane giving the cell homeostasis. This makes sense when the particles flowing through the gates are charged. But glucose molecules are not charged. How could there be homeostasis? That is question number 1.

Research into this question begins with Google. The subject is "blood glucose electric charge." This leads to many websites regarding bio-batteries and glucose monitoring sensors. The articles about the bio-batteries are especially interesting. They describe present research into the development of devices like fuel cells that, instead of deriving energy from hydrogen, derive electric current from glucose in the presence of an enzyme. One could imagine running an electric car, filling up with a glucose solution instead of gasoline. One could imagine single molecules of blood glucose transiting the membrane with molecules of enzyme creating an electric current. Digging into the literature and imagining how this could work, question number 2 is this: How does homeostasis work with glucose?

About the bio-battery, Wikipedia has a simple article. Google gives many websites on this subject. The articles on glucose sensors are also relevant.

The hormone glucagon also has receptors on liver cells. Glucagon in the bloodstream causes glucose to flow from the liver into the bloodstream, probably at a homeostatic rate. An algorithm, probably % Hebbian-LMS, controls the number of receptors and thereby establishes homeostasis.

When glucagon binds to receptors, the co-located gates open and this stimulates the conversion of stored glycogen back to glucose for release into the bloodstream. Before leaving the liver, glucose molecules near an exit gate perhaps could, by bio-battery action enabled by the presence of blood enzymes, generate an ohmic electric current, this making homeostasis possible. Electric currents near the gates co-located with glucagon receptors would be independent of the currents near the gates co-located near the insulin receptors. This is highly speculative and raises questions.

Question number 3 is, does homeostasis exist for the flow of glucose from the liver back into the bloodstream? If so, how does it work?

Question number 4 is, how would you devise experiments to test these hypotheses, to support them or refute them?

Chapter 16
Thermoregulation—Control of Body Regulation/Bernie's Oscillation

Abstract Body temperature is regulated by the hypothalamus and the autonomic nervous system, alternately switching between a cooling system and a heating system. In the hypothalamus, there are cool sensing and warm sensing neurons whose firing rate determine which system is engaged. If body temperature were slowly cooling, the temperature at which the cool sensing neurons begin to fire is the body's temperature set point. Body temperature above the set point invokes cooling, below the set point heating. Switching affords precise control of body temperature. Negative feedback mitigates effects of thermal disturbances.

Keywords Preoptic area · Cool sensing neurons, warm sensing neurons · Thyroid · Capillary vasodilation, vasoconstriction · Sweat glands · Muscular shivering · Pyrogens

16.1 Introduction

In this chapter, regulation of core body temperature is discussed. The regulation system has the capability of separately controlling temperature in various parts of the body by local control of blood flow. For simplicity of explanation, this chapter will only analyze control of core temperature. Although highly simplified, this discussion should give one an overview of nature's methods for controlling temperature in the body.

In an internal combustion engine with pistons and crankshaft, a liquid coolant is circulated by a water pump to flow through the engine block and the radiator. A mechanical thermostat in the fluid stream regulates the flow rate, the cooling rate. From fuel combustion generating mechanical output, heat is generated that needs to be dissipated in a controlled manner. The engine runs best over a limited range of temperatures. Too hot or too cold is less than optimal. The purpose of the thermostat is to keep the engine temperature in the desired range.

B. Widrow, *Cybernetics 2.0*, Springer Series on Bio- and Neurosystems 14,
https://doi.org/10.1007/978-3-030-98140-2_16

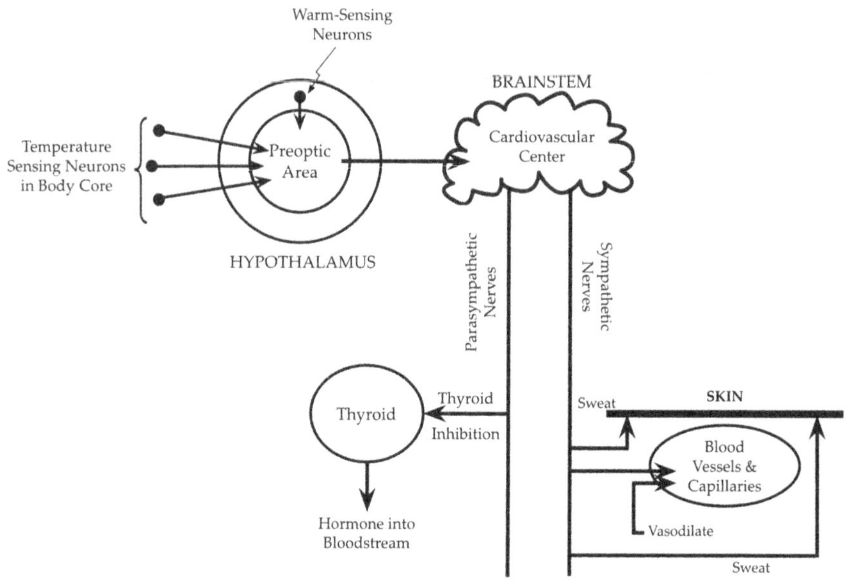

Fig. 16.1 The body's cooling system

Maintaining the temperature in the human body has some aspects and objectives in common with this. Maintaining body temperature is a much more complicated problem however. In the body, the liquid coolant is blood. The human system not only cools when the body is too warm but also heats when the body is too cold. It is more like an HVAC system, heating, ventilation, and air conditioning.

16.2 The Cooling System

The human cooling system is distinct from the human heating system, yet they both share some common components. They are controlled in different ways, however. The cooling system will be described first, then the heating system will be described.

Figure 16.1 shows the major elements of the cooling system. Warm sensing neurons in the preoptic area of the hypothalamus increase their firing rate as the body temperature warms. Signals from these neurons provide inputs to the cardiovascular center of the brainstem. The body's "radiator" is a system of capillaries located below the surface of the skin. Body heat carried by the blood circulating in the skin's capillaries is dissipated by radiation, conduction through contact with air or other surrounding media, convection with air flowing over the skin, or a combination thereof. The capillary system is innervated by the sympathetic nerves.

Another means of heat dissipation is sweating, resulting in evaporative cooling. Sweat glands under the surface of the skin deliver sweat to the surface. These glands are also innervated by the sympathetic nerves.

A third means of temperature reduction involves the thyroid gland. The idea is to reduce the production of internal heat by inhibiting the thyroid gland. This reduces the body's metabolism. The thyroid gland is innervated by the parasympathetic nerves. Thus, the autonomic nervous system plays a major role in thermoregulation, cooling and also heating.

16.3 Autonomic System

The source of signaling for the autonomic nervous system is the cardiovascular center of the brainstem. This center computes the sympathetic and parasympathetic signals based on inputs from thermal sensors located in the preoptic area of the hypothalamus. Temperature sensing neurons in the skin and all over the body also transmit their signals to the preoptic area, which integrates this information with signals from its own temperature sensors and sends a composite temperature signal to the cardiovascular center. Thus, the hypothalamus and the brainstem play major roles in thermoregulation.

The sympathetic and parasympathetic nerves are large nerve bundles with many signalling channels. In this chapter, references to these nerves include only channels having to do with thermoregulation.

The preoptic area of the hypothalamus is often called the body's thermostat. There is a general belief that a set point in the preoptic area's neurons determines and regulates the body's temperature. When the sensed body's temperature is above the set point, the cooling system is turned on to bring the temperature to the set point. If the sensed body's temperature is below the set point, the heating system is turned on to bring the temperature up to the set point. This may be correct, but the locus of the set point remains a mystery.

In contrast, all the neurons in the preoptic area have set points, equilibrium points. So do the neurons in the brainstem, in the thyroid, in the skin's capillary system, and the neurons controlling the skin's sweat glands. All of the neurons have synaptic inputs whose synapses have neuroreceptors responsive to neurotransmitters, and have systems that regulate the numbers of receptors in order to regulate their membrane potentials. Since all of these neurons have set points, the question is, whose set points determine the body's temperature?

To find an answer, focus on the three organs that directly impact body temperature, the skin capillaries and the sweat glands that provide cooling, and the thyroid that can be inhibited to lower the body's metabolism. The set points of their neurons determine their resting average homeostatic contribution to the cooling system, the (dc).

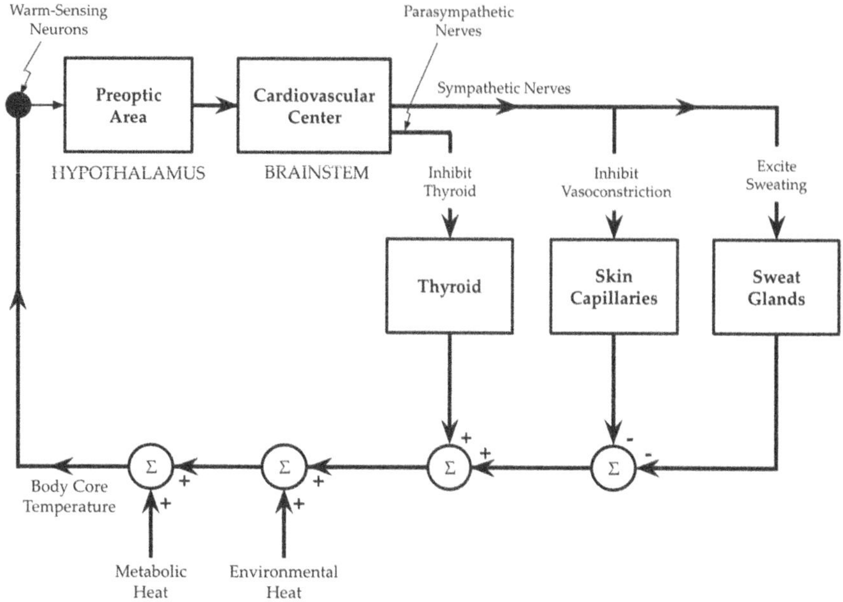

Fig. 16.2 Negative feedback for body cooling

16.4 Feedback in the Cooling System

Refer to Fig. 16.2 which shows a block diagram of a feedback system that plays a major role in controlling body temperature. Metabolic heat and environmental heat inputs increase body temperature. Cooling comes from sympathetic stimulation of sweat glands and inhibition of skin capillaries that add negatively to body temperature. Capillary inhibition causes vasodilation. Thyroid hormone in the bloodstream increases body temperature, but parasympathetic inhibiting of thyroid hormone production contributes to a reduction in body temperature.

Warm sensing neurons in the preoptic area of the hypothalamus increase their firing rate and their stimulative effect as body temperature warms. Body warming turns on the cooling system.

In the signal path around the feedback loop, the gain is negative. Negative feedback plays an important role in the cooling system. Its purpose is mitigation of transient effects (ac).

16.5 The Heating System

The essential elements of the body's heating system are shown in Fig. 16.3. Cool sensing neurons in the preoptic area of the hypothalamus increase their firing rate as body temperature cools. Signals from these neurons provide inputs to the cardiovas-

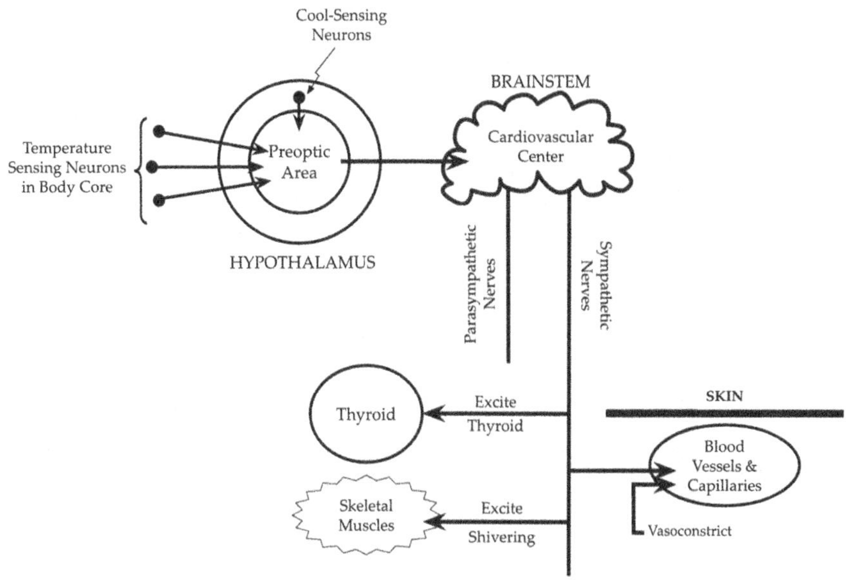

Fig. 16.3 The body's heating system

cular center of the brainstem which in turn activates the sympathetic nervous system. The thyroid gland is innervated by the sympathetic nerves such that increased sympathetic activity increases the production of thyroid hormone going into the bloodstream. Thyroid hormone enhances body metabolism, generating body heat.

The skeletal muscles are also innervated by the sympathetic nerves. Sympathetic nerve signals can induce a shivering reflex in these muscles, resulting in heat production. The skin's capillaries can be made to constrict by signals from the sympathetic nerves. Vasoconstriction reduces or cuts the bloodflow to these capillaries, reducing heat loss and thus conserving heat.

A sudden lowering of body temperature sensed in the hypothalamus will cause a large increase in body heat by means of these three effectors. An increase in heat by the three effectors will be sensed by the cool sensing neurons in the preoptic area causing a reduction of their firing rate. This will result in a reduction of excess heating. Thus, the loop is closed with negative feedback. The feedback loop moderates transient swings in heating. A feedback diagram of the heating system is shown in Fig. 16.4.

The heating system functions in a manner similar to the working of the cooling system, except in the opposite direction. Comparing Fig. 16.4 with Fig. 16.2, the warm sensing neurons in the preoptic area are replaced with cool sensing neurons. The three organs that produce cooling are replaced by the three effectors that produce heating. Otherwise, the two systems are alike. The response to input fluctuations and disturbances, the (ac), is mitigated by the negative feedback loops. The homeostatic operating levels of the thyroid, the skeletal muscles, and the skin capillaries control the average, the (dc), production of heat.

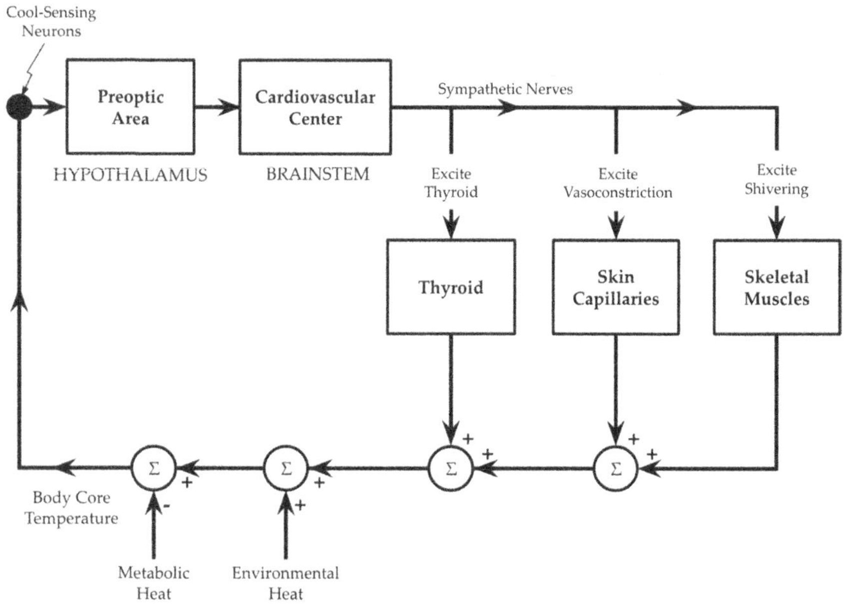

Fig. 16.4 Negative feedback for body heating

16.6 Heating or Cooling

How does the body know which system to use, the heating system or the cooling system? The following hypothesis suggests an answer to this question.

In the preoptic area of the hypothalamus, there are warm-sensing and cool-sensing neurons. The warm-sensing neurons fire at a rate that increases with temperature. It turns out that these neurons are wired to inhibit the firing of the cool-sensing neurons when the warm sensing neurons are firing rapidly. Now suppose that the body is gradually cooling to lower temperatures. The warm-sensing neurons' firing rate will become slower and slower until firing ceases. Then the cool-sensing neurons become uninhibited and start firing, faster and faster as the body temperature lowers further.

When the cool-sensing neurons are not firing, the body's heating system is inoperative. The cooling system is operating. When the cool-sensing neurons begin firing, the body's heating system becomes operative and the cooling system becomes inoperative.

When the cool-sensing neurons become uninhibited and they begin firing, a switch over of the two thermoregulation systems occurs. The signals from these neurons propagate to the cardiovascular center of the brainstem where the switch over is implemented by activation of the autonomic nervous system. Temperature sensing is done in the hypothalamus and the heating/cooling decision making and implementation is carried out by the brainstem.

It should be recalled that the sympathetic nerves and the parasympathetic nerves are nerve bundles, each having many nerve fibers, each fiber a separate channel.

Different control functions can be activated by the cardiovascular center by the signaling pattern that it uses to drive the autonomic nerves. Control of heart rate goes on independently.

16.7 Nature's Thermostat/Bernie's Oscillation

There is no thermostat in the hypothalamus, per se. Nothing has yet been found that you could identify with scalpel and microscope that would serve like the thermostat in a household heating and air conditioning system. There may be such a thermostat, if it exists, but its location and functioning remain undiscovered. But there is in effect a thermostat, and it regulates body temperature. It involves the warm-sensing and the cool-sensing neurons of the hypothalamus and in addition the brain which processes their signals. The brain function is performed by the cardiovascular center of the brainstem that decides at a given moment which system should be turned on, the cooling system or the heating system. This center implements the decision by activating the autonomic nervous system appropriately.

Thermoregulation of the body is effected by switching back and forth between the cooling system and the heating system. This switching occurs constantly, like a persistent oscillation, with a period of approximately half a minute. In the language of control engineering, switching is called "bang-bang control." This is used, for example, with the control of spacecraft whose rocket engines are either on or off, firing or not.

The body's thermoregulation system has two states, heating or cooling. Regulation is accomplished by determining the amount of time spent in each state. The oscillation is generally asymmetrical, not like a sine wave. Body cooling will be achieved by spending more time in the cooling state. Body heating will be achieved by spending more time in the heating state. More heating is needed in a cold environment, and more cooling is needed in a hot environment.

> The temperature oscillation described here is a physical phenomenon that should have a name. What would be a good name for it? Perhaps we might call it "Bernie's oscillation." If Alzheimer can have his disease and Parkinson can have his disease, then maybe Bernie could have his oscillation.
> When the warm-sensing neurons are firing, the cool-sensing neurons are inhibited. If the hypothalamus were gradually cooling, the temperature at which the cool-sensing neurons just begin to fire is believed to be the regulated body temperature. This is the switching temperature. This is the body's temperature set point. Body temperature above this point will turn on the cooling system. Body temperature below this point will turn on the heating system. Body temperature will oscillate above and below the set point. This goes on throughout life.

Switching backward and forward takes place at a rate much faster than could be followed by the adapting synapses. In the cooling system of Fig. 16.2, the synapses are part of the neurons that innervate the thyroid, the capillary system, and the sweat glands. These synapses over time will have adapted and converged to bring their associated neurons to their homeostatic firing rates. Likewise, in the heating system of Fig. 16.4, the synaptic weights will have converged over time to their homeostatic levels.

16.8 The Heating System, On and Off

Figure 16.4 shows the sympathetic nerves connected to excite the thyroid, the smooth muscle neurons of the capillaries, and skeletal muscles. Exciting the thyroid causes thyroid hormones to flow into the bloodstream causing an increase in body metabolism and an increase in body heat. Exciting the capillaries' neurons causes vasoconstriction and reduced blood flow and reduced heat loss from the capillaries below the skin. Exciting skeletal muscles causes shivering so that the increased muscle activity generates body heat.

These three forms of excitation take place only when the heating system is on. When the heating system is off, these excitatory input signals vanish. Production of thyroid hormone drops, the capillaries' neurons are no longer inhibited and the capillaries are free to relax, and the skeletal muscles no longer shiver.

16.9 The Cooling System, On and Off

Figure 16.2 shows the parasympathetic nerves connected to inhibit the thyroid, and the sympathetic nerves connected to inhibit the smooth muscle neurons of the capillaries and to excite the neurons of the skin's sweat glands. With the cooling system on, inhibitory signals inputted to the thyroid basically shut down thyroid production. With the cooling system on, inhibitory inputs to the smooth muscle neurons of the capillaries cause the capillaries of the skin to dilate and blood flow to increase, enhancing cooling. With the cooling system on, sweat glands' neurons are excited and sweat is produced inducing evaporative cooling.

With the cooling system off, all of these input signals vanish and the thyroid is no longer inhibited, the capillaries are free to constrict if so ordered, and the sweat glands no longer produce sweat.

16.10 Control of the Thyroid and Capillaries

When the cooling system is on, the heating system is off. When the heating system is on, the cooling system is off.

Control of the sweat glands is simple. With the cooling system on, sweat is produced, otherwise not. Control of the skeletal muscles is also simple. When the heating system is on, shivering actively generates body heat, otherwise not. Control of the thyroid and the capillary system is less simple as both of these systems receive command inputs when the cooling system is on and when the heating system is on. Both are inhibited when the cooling system is on, and both are excited when the heating system is on. When switching takes place, they go from excitation to inhibition, or vice versa, nothing half way.

Control of heating and cooling is implemented by the autonomic nervous system. The difference between heating and cooling is dramatic. This should cause strong fluctuations in body temperature except for the effects of feedback. The feedback loops enable the switching oscillation. By rapid oscillation, temperature extremes are avoided. They are averaged out. As with examples in previous chapters, feedback mitigates large amplitude fluctuations in variables being controlled.

16.11 The Synaptic Weights Adapt

The neurons that control the thyroid, capillaries, skeletal muscles, and sweat glands are all synaptically connected to the autonomic nervous system. The synapses all change and adapt in accord with the % Hebbian learning rules, it is believed.

When the body cooling system is on, the sweat-gland synapses, which are excitatory, slowly adapt to bring the sweat gland neurons' firing rate down toward their homeostatic firing rate. Sweating slows to the homeostatic level. Progress toward the equilibrium point is gradual over a long time with ultimate convergence to the homeostatic rate. When the cooling system is off, the inputs to these synapses is zero and no adaptation will take place. In steady-state operation, when the cooling system is on, the neurons fire at the homeostatic rate and sweat is produced. When the cooling system is off, no sweat is produced.

Similar adaptation takes place with the synapses of the skeletal muscles. The result is that when the heating system is on for some time, the skeletal-muscle neurons with adaptation ultimately fire at their homeostatic rate and shivering takes place at the homeostatic level. When the heating system is off, there will be no shivering.

The neurons of both the thyroid and the capillaries are excited when the heating system is on and inhibited when the heating system is off. Adaptation of these neurons' synapses is similar. These neurons have both excitatory and inhibitory synapses, and the adaptive process is therefore somewhat more complicated.

When the body's heating system is on, the inputs to the excitatory synapses of both the thyroid neurons and the capillary neurons are finite and strong. The inputs to the inhibitory synapses are zero. The excitatory synapses will adapt to cause the firing rates of the thyroid neurons and the capillary neurons to progress toward their homeostatic rates. The inhibitory synapses will not change.

When the heating system is off, the cooling system is on. The inputs to the inhibitory synapses of the thyroid and capillaries are finite and strong. The inputs to the excitatory synapses of the sweat glands are zero. The excitatory synapses do not adapt, but the inhibitory synapses do adapt. In accord with the % Hebbian learning rules, the inhibitory synaptic weights will increase. Adapting to increase the inhibitory weights will continue whenever the heating system is off and the cooling system is on. The weight increasing will continue since, with the neurons inhibited and not firing, there is no homeostatic goal. The weights will continue increasing until they saturate.

The result of this adaptation is that, in steady-state, the thyroid neurons and the capillary neurons will all fire at their homeostatic rates when the heating system is on. The thyroid gland will produce thyroid hormone and the capillaries will experience vasoconstriction. When the cooling system is on, the thyroid gland will be shut down and the capillaries will be fully dilated.

It should be noted that other inputs to the thyroid and to the capillaries could possibly override the commands from the cooling system or the heating system.

16.12 Temperature Extremes

Thermoregulation of core body temperature is set at its normal level by the "thermostat" in the hypothalamus. This is accomplished by switching between the heating system and the cooling system which is very effective at room temperature and in a finite range of temperatures above and below room temperature. Body temperature can however be pushed above or below normal by environmental heat or cold. Visualizing how this works is aided by reference to the feedback diagrams of Figs. 16.2 and 16.4.

Starting at room temperature and gradually reducing the room's temperature, a higher percentage of time will be spent with the heating system on, cooling system off. When the temperature is low enough that the heating system is on full force 100% of the time, the low end of the environmental range is reached. Further lowering of the temperature will require wearing heavier clothing. Otherwise, core body temperature will drop below normal. The result would be hypothermia. The thermoregulation system is saturated and no longer can regulate body temperature.

Refer to Fig. 16.4. With hypothermia, the thyroid is excited and producing thyroid hormone at a maximum rate. The capillaries are constricted and stay that way. Strong shivering will take place with a very high firing rate of the skeletal muscles' neurons. This firing rate will gradually slow somewhat as the synaptic weights adapt and shrink to re-establish homeostasis. The heating system will be making its best effort but this would not be enough and the core body temperature will drop below the normal. Hypothermia can be life threatening. The only way to prevent this in a cold climate is to dress with heavy clothing. That will bring body temperature into the range where the thermoregulation system can function. With sufficient insulation, a person can safely live in a very cold climate, even above the Arctic circle.

In the other direction, with a sufficiently warm environment, the cooling system can be overwhelmed. Hyperthermia is a result when the core body temperature exceeds normal. The thermoregulation system no longer controls body temperature. Severe hyperthermia is heat stroke and this can be life threatening.

Refer to Fig. 16.2. In a very hot climate, with the cooling system at maximum capacity, the thyroid will be inhibited, the capillaries will be fully dilated, and the sweat glands will be producing at a maximum rate, higher than homeostatic. Production of sweat will gradually subside as the synaptic weights adapt to re-create homeostasis with the sweat glands. The cooling system will be unable to cope with the environmental heat. One would wear the lightest of clothing or none at all, wash with water or swim, find shelter and shade. If even these measures cannot bring core body temperature into the normal range, a serious problem is at hand.

In the normal temperature range with body temperature under the control of the thermoregulation system, vasodilation is one of the means for cooling. If, on the other hand, air temperature is above body temperature, vasodilation will contribute to overheating. A portion of the cooling system becomes self defeating.

As global warming worsens, it may not be possible to live in the Tropics. If this happens, mass migration will be necessary. Overheating is intolerable. People can be insulated to live with cold, but they cannot live with excessive heat.

16.13 Three Scenarios

16.13.1 Going from Indoors to Hot Outdoors

Imagine a person having been indoors at room temperature for several hours, not being engaged in significant physical activity. Basically, this is a body at rest. Metabolic and environmental heat inputs are constant. Oscillation is taking place between the heating system and the cooling system. This could be called "resting conditions."

Assume that this person suddenly went outdoors on a hot day. The question is, what effect does this action have on body temperature?

A theoretical plot of skin temperature as a function of time is shown in Fig. 16.5. Starting from resting conditions, skin temperature rises rapidly upon going outdoors. Temperature sensors in the skin then turn on the cooling system full force, 100% of the time. Environmental heat input jumps up. Refer Fig. 16.2. The feedback system is energized, having turned on the cooling system. The thyroid is inhibited and the capillaries are fully dilated. Sweating is profuse. The sweat-gland neurons are firing substantially faster than their homeostatic rate. The synapses of these neurons

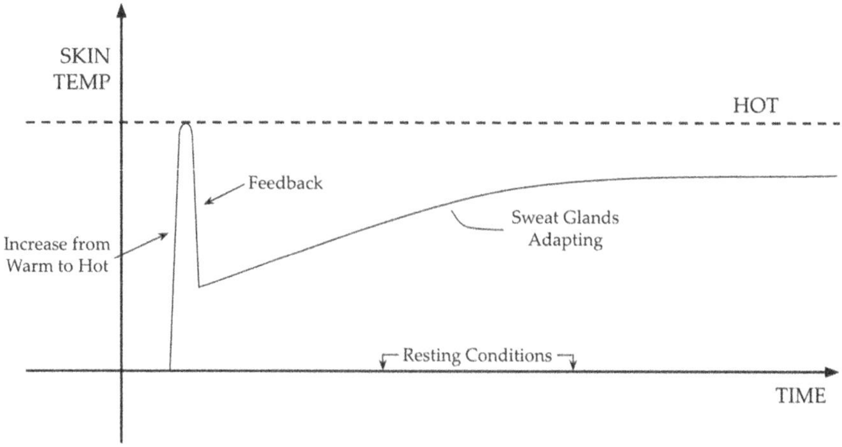

Fig. 16.5 Going from warm indoors to hot outdoors

gradually adapt to restore homeostasis. Some of the cooling effect is lost by this, and the plot of Fig. 16.5 shows this.

A change in the environment had an immediate effect on skin temperature. When going from room temperature to the hot outside temperature, the cooling system is rapidly engaged. The feedback loop is activated. As a result, a major element of the cooling system, the skin's capillaries, are activated by the sympathetic nervous system and they dilate. Sweat glands are activated. This causes rapid cooling at the skin. The outside temperature is oppressive, but lowering the skin temperature provides some relief from the heat that is quite noticeable. The feedback mitigates the effect of the sudden environmental temperature change.

The feedback loop has really controlled skin temperature, not core body temperature. Body temperature follows skin temperature, but lags behind it. Skin temperature influences body temperature through circulation of the bloodstream and by heat conduction. The body has large thermal inertia, and the heat conduction paths have resistance. In electrical engineering terminology, body temperature follows skin temperature through low pass filtering. The skin temperature is the input and the body temperature is the output of a one-pole low pass filter. Body temperature is a smoothed, lagged version of skin temperature.

The sweat-gland weights, when adapting, would partially undo the mitigating effects of the feedback loop. The weights would be adapting in directions to restore the operation of these organs to their homeostatic states. This would amount to a reduction in the cooling effect back toward what it was before going outdoors into the heat. Skin temperature and body temperature would gradually rise. Refer to Fig. 16.5.

16.13.2 Going from Hot Outdoors to Indoors

An individual has been outdoors on a very hot day, sitting and resting for several hours. Suddenly, this person goes indoors to ambient room-temperature and sits at rest. What happens to skin temperature and body temperature?

Figure 16.6 is a theoretical plot of skin temperature versus time. An explanation for the plot of Fig. 16.6 is the following. Refer to the feedback diagrams of Figs. 16.2 and 16.4. Initially, the skin temperature is high, given the heat of the outdoor atmosphere. Going indoors, the skin temperature drops quickly to the new ambient temperature. Initially, in the heat, the cooling system had been on full strength at 100% of the time. Going indoors, the feedback system of Fig. 16.2 becomes operative. A sudden change in ambient temperature is equivalent to a large negative environmental input step. This will turn on the heating system.

Now that the feedback system of Fig. 16.4 is operative, the heating system will be on full force. The skin temperature rises, as shown in Fig. 16.6. The thyroid is fully excited, the capillaries are fully constricted, and the skeletal muscles are shivering. The corresponding neurons are firing at rates higher than their equilibrium rates. Gradually their synapses adapt to return these systems to their homeostatic rates and the skin temperature returns to that of "resting conditions."

Coming inside from a hot outside, one would initially feel a cooling effect, then a minute later the heat returns, and the heat will gradually fade away over an hour or so so that everything ultimately returns to "resting conditions." One can try this as an experiment. It is a very interesting and strong effect.

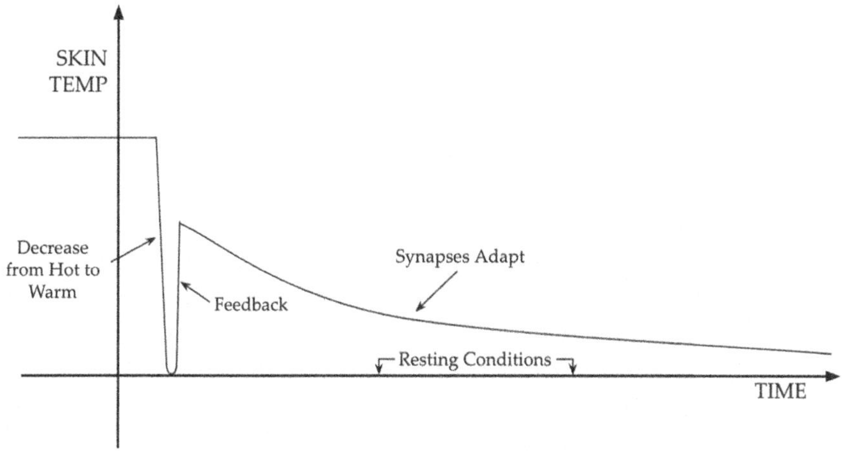

Fig. 16.6 Going from hot outdoors to warm indoors

16.13.3 Going from Warm Indoors to Cold Outdoors

An individual has been sitting for several hours at rest at room temperature under "resting conditions." The person suddenly goes outside on a very cold day. Figure 16.7 is a theoretical plot of skin temperature versus time. An explanation follows.

Initially, the skin temperature and body temperature are normal under initial "resting conditions." Going outside, the skin temperature drops fairly rapidly and the heating system is tuned on at full force. The thyroid is fully excited, the capillaries are fully constricted, and the skeletal muscles are shivering. The environmental heat input of Fig. 16.4 has just experienced a large negative step. The skin temperature had a large drop, but in a minute or so, this temperature rises rapidly due to the action of the heating system.

The neurons of the thyroid, capillaries, and skeletal muscles are all firing well above their homeostatic rates. Gradually, their synapses adapt to bring the firing rates to their homeostatic levels. The heating effect drops to the homeostatic heating level.

The experience one feels going outdoors on a cold day is to sense an immediate chill. Then after a minute or so, one acclimates to the cold. The chill is now not so onerous. Gradually, over an hour or two, the synaptic weights adapt and bring heat production down to the homeostatic level, and the chill returns.

A similar experience can be had when going to bed on a cold night. Crawling into a cold bed, one feels a chill. Resting flat on the back with arms aside, the chill subsides and one acclimates. The chill returns however if one stays awake for an hour or so. Falling asleep, one could be awakened by the chill.

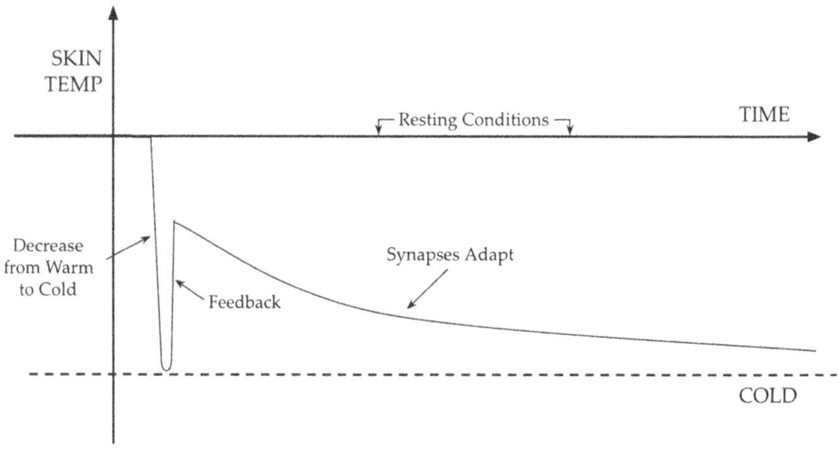

Fig. 16.7 Going from warm indoors to cold outdoors

16.14 Summary

Core body temperature is a physical variable that is very tightly controlled. A cooling system and a heating system are involved. Temperature regulation is accomplished by switching between cooling and heating, continually alternating between them. (This is Bernie's oscillation). The thermostat consists of warm sensing neurons and cool sensing neurons.

When the warm sensing neurons are firing, the cool sensing neurons are inhibited. If the body were not cooling, the temperature at which the cool sensing neurons just begin to fire is the set point of the thermoregulation system. The cool sensing neurons turn on the heating system, and the warm sensing neurons turn on the cooling system. Thermostatic signals from these neurons are transmitted by the autonomic nerves to the body's organs.

The receptors of the organs are most likely regulated by the % Hebbian-LMS algorithm. The cooling system inhibits the thyroid, thus reducing body metabolism, inhibits capillary vasoconstriction, thus increasing vasodilation, and excites the body's sweat glands. The heating system excites the thyroid, excites capillary vasoconstriction, and excites the body's muscles inducing shivering. During the switching cycle, if more time is spent with the heating system on, body temperature will rise, and vice versa. The heating and cooling systems have separate feedback loops that are engaged when each of these systems is turned on one at a time. The switching occurs day and night, throughout life to keep tight control over body temperature.

16.15 Questions and Experiments

1. There are good reasons for hot walking a racehorse for an hour or so after a race. Which of the above three scenarios most closely resembles the racehorse situation? What would the horse's body temperature as a function of time look like before, during, and after a race?
2. This is an experiment to observe Bernie's oscillation. To do this visually, it is necessary to have a subject with white skin, the lighter the better. Rest the hand, palm up, on something soft on a table top. Illuminate with incandescent light or LED with 3,000 K or lower. The light should have components from the red end of the visible spectrum. Try different lightning angles. Carefully examine an area at the center of the palm. Stare at this area. Try not to blink. Notice the color change from white to slightly red back to white back to red, etc. A full period of this oscillation is about one half minute. This is a subtle but real effect and requires some care to observe it. What is happening is that the body's thermoregulation system is switching back and forth between heating and cooling. This constricts the capillaries giving white skin or dilates the capillaries giving ruddy skin.

3. Extend the study of Bernie's oscillation to observe the effect of variations in room temperature on the percent times spent with white skin and with ruddy skin.
4. With an electronic oral temperature sensor and recording apparatus, observe and record Bernie's oscillation. With the sensor in place, sit in a room whose temperature can be varied above and below normal room temperature. Normal room temperature is comfortable. At normal room temperature, Bernie's oscillation should be apparent. Now gradually lower the room temperature and confirm that the test subject is comfortable and Bernie's oscillation remains apparent. When the temperature of the room is lowered so that comfort just vanishes, Bernie's oscillation should correspondingly just vanish. Now reverse the process and test on the warmer-than-room-temperature side of the comfort range. The presence of Bernie's oscillation indicates that body temperature is being controlled by the thermoregulation system. Verify that when going outside the comfort range, one enters an extreme temperature range and Bernie's oscillation will not be present.
5. Perform experiments with situations like those of the above scenarios and confirm the predictions of Figs. 16.5, 16.6 and 16.7.
6. Many of the assertions and statements of this chapter correspond and agree with the relevant literature. Many of the statements cannot be found in the literature however. By searching the literature, point out the statements in this chapter that correspond to existing general knowledge and those statements that are new, appearing here for the first time. What do you agree/disagree with?
7. In a stall shower, take a hot shower for about 15 min or so, allowing the body temperature to rise. Now exit the shower stall. Feel the chill. The room temperature is much cooler than in the shower stall. Evaporative cooling from the wet skin also contributes to the chill. Now towel dry. The chill rapidly goes away and the body feels warm. What happened? Why? What happens next to skin temperature and body temperature? Make a plot of skin temperature vs. time and body temperature vs. time, starting from in the shower. Which of the scenarios is this sequence closest to?
8. Weigh yourself with a digital scale at the same time of day and record a log of your weight over a period of two months. Do you observe a periodic variation in your weight of the order of $\pm 1\frac{1}{2}$ pound or 0.68 kgm, with a period of a month or more? Study the body's weight regulating system and see if there would be a basis for oscillation. Would the oscillation be periodic or of irregular periods? Where does homeostasis come into the regulation of body weight?
9. Go outside on a cold day, lightly clothed. You are shivering indicating that the heating system is on. After a few minutes of shivering what is the color of the palm of your hand? Describe the reverse scenario for a hot day. You need to be light complexioned for this.
10. Fever is the body's natural reaction to infection. High temperature helps fight the invader. You are shivering, have goosebumps and shaking. The hypothalamus has turned on the heating system and inhibited the cooling system. The core temperature rises. Because of the shivering, you feel cold, but your body is very hot. Fever can give you the chills. How does infection fighting turn on

the heating system and inhibit the cooling system? Ever heard of pyrogens? Do some research.

If the fever suddenly breaks, its effect on the hypothalamus stops. The high core temperature now turns on the cooling system and you sweat profusely. Describe what is happening within the hypothalamus. Do some research.

Part III
Reflections On

In part III of this book, several examples of biological control systems have been featured. There are many many more such systems that could be analyzed. There is need for many more books! It is interesting to know that most physical variables are controlled and subject to homeostasis, but are not tightly controlled. Blood pressure varies fairly widely over the course of a day, so does heart rate, for example. One of the variables controlled very tightly is core body temperature. On the Fahrenheit scale, a temperature of 98.6° is considered normal. A temperature of 100° is considered almost feverish.

To study the immune system would be fascinating. This system is trainable with vaccines, and in many cases a single dose is sufficient to train, and the training could last for years if not for a lifetime. It is interesting to note that the immune system has endorphin receptors. It will be challenging to discover how the immune system learns and adapts. Where does the immune system reside, in the bloodstream?

Another interesting system to study would be that of control of body weight. Steady-state average (dc) is under homeostatic control. Slight ups and downs (ac) occur due to variations of diet, exercise and hydration. But body weight average over many months is remarkably stable.

A very nice discussion of body weight is given on pp. 179–182 of the book by the late Dr. Robert Ornstein and the late Dr. Richard F. Thompson, Emeritus Professors of Psychology and Neuroscience at Stanford, and entitled "The amazing brain." Houghton Miffin Co., 1984. This Book is a fine introduction to brain anatomy and brain function. It is a simple and well illustrated tutorial.

Given that control of average weight is homeostatic, there must be neurons and synapses and neural receptors in the part of the brain that is responsible for weight control. The hypothalamus seems to be a part of the system. Assuming that the number of neuroceptors is controlled as part of the weight control process, it would be logical that the number of receptors would be determined by % Hebbian-LMS. In light of this idea, it is interesting to note what Ornstein and Thompson say about experience people have with weight reducing diets.

They don't work. One is fighting against the set point of homeostasis. When first starting the diet by reducing caloric intake, weight goes down rapidly and significantly. But after a period of a few weeks or so, the weight starts coming back. Why? % Hebbian-LMS seems to hold the answer.

Initially, before dieting, the body weight was at the set point. The neurons involved are firing at the homeostatic rate. The sudden drop in calorie intake causes a sudden change in the neurons firing rate and a sudden onset of a high rate of weight loss. The synaptic weights start adapting in an attempt to maintain the homeostatic firing rate. Gradually, over weeks, the body weight converges with the firing rate back closer to the set point. The synaptic weights start adapting in an attempt to maintain the homeostatic firing rate. Gradually, over weeks, the weights converge to bring the firing rate back closer to the set point. % Hebbian-LMS will not restore homeostasis perfectly. There needs to be some error to have a restoring force. Body weight goes back up, but settles at a level slightly lower than the set point.

Staying on the diet, some weight will be lost but nowhere near what one would expect based on the initial "great results".

Discontinuing the weight loss diet and returning to normal caloric intake, one can predict what will happen with body weight. Assuming that the weight loss diet was followed for weeks, the synaptic weights will have adapted to return body weight close to the homeostatic level. Going off the diet, the synaptic weights now adapted to increase body weight, will do so in the absence of the dietary force pulling body weight down. The results will be that the body weight will overshoot and go to a level above homeostatic. Gradually over time, the synaptic weights will adapt to bring body weight back down to the homeostatic weight level.

The study of body weight control needs to be done in a great detail, and then another book chapter would result. The same could be said about the study of the immune system. The same could be said about an almost infinite number of systems of the body.

Part IV
Virus, Cancer, and Chronic Inflammation

Introduction to Part IV

Beginning with the % Hebbian-LMS algorithm as a model for synaptic plasticity in the brain, it was natural to use this model to explain mood disorders and addiction. The application of this model to the control systems that regulate the activity of the body's organs was not so obvious, but the model seems to work. The area of application of this model is expanded here to a study of viral infection, cancer, and chronic inflammation.

What does synaptic plasticity have to do with viral infection? It turns out that viruses infect by binding to receptors on the surfaces of normal cells, enter the cells, take over, and explosively multiply releasing copies of themselves to infect more normal cells.

A particular virus infects by masquerading as the natural ligand to the receptors of a particular normal cell. It is a matter of lock and key. Once binding takes place, the virus has easy entry into the captured cell.

The more receptors that exist on the surface of the normal cell, the higher the probability of infection and the more rapid will be spread of infection. What determines the number of a cell's receptors? Could the receptors be downregulated by % Hebbian-LMS to reduce the probability and severity of infection?

What does synaptic plasticity have to with cancer? Normal progenitor cells have receptors and their cancerous counterparts have many times more of these receptors. Could this disparity be used to develop treatment that would kill the cancerous cells with minimal damage to normal cells?

An important method of treatment works by reducing the ligand concentration. Breast cancer cells with estrogen receptors are attacked by reducing or eliminating the ligand estrogen. This reduces the cells' membrane potentials, killing the cells. Taking an opposite approach, could better results be obtained by substantially increasing the concentration of ligand, raising the membrane potentials of the cancerous cells? With the large number of receptors on the cancerous cells, membrane potentials will be pulled high above homeostatic, killing them with little damage to normal cells. Would this work?

What does synaptic plasticity have in common with chronic inflammation? The answer is interesting forms of homeostasis. Myeloid cells of the body's immune system seem, with ageing, to defy the common notions of homeostasis. They can go rogue and develop into sources of inflammation which is destructive to the brain and all other organs. Myeloid cells produces the hormone PGE2, increasingly with age. These cells have receptors, EP2, whose population also increases with age. Their ligand is PGE2. When the concentration of PGE2 increases with age, the population of EP2 receptors should downregulate to maintain homeostasis. But this does not happen. When PGE2 binds to EP2, the cell, instead of dying, lives and assumes a new "personality" as a source of inflammation. Ageing causes inflammation and inflammation causes ageing. Blocking the EP2 receptors restores the myeloid cells to a youthful state. Testing a blocking agent with mice restored the cognitive capability of old mice to that of young adult mice. Homeostasis was denied.

These are the subjects of Part IV.

Chapter 17
Viral Infection

Abstract In the midst of a pandemic, the interest in viruses and how they infect is very strong. Viruses infect by binding with receptors on normal cells. Upon binding, the virus has an easy path to the cell interior. The ultimate solution for prevention of infection is vaccination. The idea is to train the immune system to attack the virus. Proposed here is another approach, focusing not on the specific virus but instead on the receptors of the cells that may be subject to infection. By delivering a synthetic version of the receptors' natural ligand, the excess ligand will cause downregulation of the receptors, thus making it difficult for the virus to find a receptor, thus reducing the probability of infection. The concept, which should work with any virus, is to train the cell and its receptors as an alternative to training the immune system. It would be possible to use both methods simultaneously. These ideas need to be tested.

17.1 Introduction

In the years 2019, 2020, and beyond, the world was enveloped with the COVID-19 pandemic. The disease is caused by a virus designated by the World Health Organization as SARS-COV-2. This virus is extremely contagious and potentially lethal. Research is being done worldwide to develop vaccines and to take all measures to prevent the spread of this virus. A novel approach to this problem is presented in this chapter that may possibly be useful to counteract many forms of viral infection.

In the Science Times section of the New York Times, Tuesday, March 17, 2020, p. D2, there was an article about the SARS-COV-2 virus and how it infects cells of the body. Reading this article, and viewing the illustrations, an idea of a new method to stop or slow this virus was born. The article explains that the COVID-19 virus and all the coronaviruses such as SARS and MERS and the common cold infect by binding to the ACE2 receptors on epithelial cells of the airway. The virus binds to these receptors, enters the cell, then replicates.

The cells of the airway, in the nose, throat, trachea, and the lungs, are exposed to droplets and airborne coronaviruses and thereby contract SARS-COV-2 infection. The world wide pandemic is the result.

B. Widrow, *Cybernetics 2.0*, Springer Series on Bio- and Neurosystems 14,
https://doi.org/10.1007/978-3-030-98140-2_17

Most approaches to the COVID-19 problem focus on the virus itself. Understanding the properties is essential for the development of a vaccine and other mitigating techniques to help stop the spread of the virus.

17.2 Reduction in Number of ACE2 Receptors

The approach proposed herein focuses not on the virus itself but on the ACE2 receptors of the epithelial cells. The idea is to reduce the number of receptors, without introducing unwanted side effects. The epithelial cells and their receptors can be trained to reduce the number of receptors, to downregulate them. With fewer receptors to bind to, the viral particles and their replications have a lower probability of finding an ACE2 receptor. This reduces the incidence and severity of COVID-19 infection. Slowing the spread and propagation of the virus gives an immune system more time to attack and fight it.

ACE2 receptors on the surface of the epithelial cell membrane are designed to bind to the ACE2 ligand. This ligand is normally found in the extracellular fluid surrounding the epithelial cell in a low-level concentration, a level that maintains cellular homeostasis. At this concentration, the membrane potential of the epithelial cell is at its equilibrium point. If the concentration of the ligand were raised, the membrane potential would deviate from the homeostatic level and % Hebbian-LMS would kick in to reduce the number of ACE2 receptors in order to restore homeostasis.

The ACE2 ligand can be manufactured as soluble, human recombinant ACE2 protein. ACE2 could be administered to raise the level of ACE2 ligand in the extracellular fluid. This would cause a reduction in the number of ACE2 receptors. Training the epithelial cells to downregulate their ACE2 receptors, to reduce their number, can be accomplished by controlling the concentration of soluble ACE2 ligand as a function of time.

17.3 Training the Epithelial Cells and Their Receptors

The ACE2 input signal to an epithelial cell is proportional to the product of the ACE2 concentration in the extracellular fluid and the number of ACE2 receptors. Homeostatic control of this product is essential. If the concentration of ACE2 in the extracellular fluid is increased, adaptive control will be effected automatically to reduce the number of receptors.

The concentration of ACE2 in the extracellular fluid can be increased through the bloodstream by injection or by ingestion. Increasing the concentration of ACE2 will be sensed by every epithelial cell in the body, causing receptor reduction. This effect will be needed in the airway and wherever else epithelial cells are that could be infected by the SARS-Cov-2 virus, such as the GI tract. Elsewhere, epithelial cells would not want a higher concentration of ACE2 and indeed this raises concern about side effects.

A sudden increase in ACE2 concentration in the epithelial cells' extracellular fluid would very likely cause side effects. If this occurred impulse like, as from an injected bolus, the ACE2 presence at the epithelial cells would suddenly jump high above normal. This would happen before the adaptive algorithm could start reducing the number of receptors. The ACE2 input signal would jump well above homeostatic, and this could cause side effects. A sudden increase in ACE2 concentration in the bloodstream as a therapeutic technique might not be a good idea.

A better way to administer soluble ACE2 would be to start at a very low dose level and gradually ramp up the dosage level to a maintenance level. The maintenance dosage needs to be sustained during the epidemic. It should not be stopped abruptly. Once the epidemic subsides, the dosage can be slowly ramped down to zero. This is a way to train the receptors.

Training the epithelial cells and their receptors initially by slowly ramping up the concentration of soluble ACE2 has a simple justification. Start with a very small dose. After a certain time increment, the dose is increased by a small amount. After another time increment, the dose is increased by another increment, and so forth. After the first dose, the adaptive algorithm adapts and converges causing a small reduction in the number of receptors. After the second dose, the adaptive algorithm adapts and converges with an additional reduction in the number of receptors, and so forth. Ramping up slowly gives the adaptive algorithm, % Hebbian-LMS, time to converge and to continue to keep the epithelial cell essentially homeostatic all during the ramp-up. Reaching the maintenance level, the percent reduction in the number of receptors is established and the adaptive algorithm is converged.

Once at the maintenance level, if the dose of soluble ACE2 were suddenly stopped, the ACE2 concentration would rapidly decline and the number of receptors would start to slowly increase toward normal. The adaptive process would be slow, since the ACE2 input signal would be small. Given the large reduction in the number of receptors, a sudden drop in ACE2 concentration would cause the ACE2 input signal to drop well below homeostatic. The epithelial cells all over the body will be deprived of ACE2 input, and side effects could result.

Lowering the dosage from the maintenance level with a gradual down ramp allows the adaptive algorithm to converge at each step and to keep the epithelial cells at homeostasis all the time.

The concentration of soluble ACE2 in the extracellular fluid can be changed up or down, but if done slowly so that the % Hebbian-LMS adaptive algorithm remains essentially converged at all times, this will keep the epithelial cells "happy" and side effects will be obviated.

Treatment against the virus requires that the concentration of soluble ACE2 be ramped up as fast as possible. How fast is a good question. The answer depends on the speed of convergence of % Hebbian-LMS. How fast can the ramp-up be done safely without side effects would need to be determined clinically. So, acclimate on the way up, reacclimate on the way down.

17.4 Receptor Blocking

The idea of blocking the epithelial cells and their receptors originated early in 2020. An effort using soluble ACE2 as a therapeutic against COVID-19 had been underway at Northwestern University. A research team under the direction of Dr. Daniel Battle was exploring the possibility of using soluble ACE2 to block the receptors of epithelial cells to prevent viral binding. Blocking would be done by ACE2 molecules binding to the receptors. In vitro evidence is given in the brief paper: "Soluble angiotensin—converting enzyme 2: a potential approach for Coronavirus infection therapy?" by Daniel Battle, Jan Wysocki, and Kalar Satchell, Clinical Science, pp. 534–545, Portland Press, March 13 2020.

With the blocking approach, it would once again be necessary to ramp up dosage slowly to prevent side-effects. A sudden high dose of ACE2 would cause a very large ACE2 input to the epithelial cells and this would drive their membrane potentials away from homeostatic.

The same remedy is proposed for receptor blocking as is proposed here for receptor down-regulating, reducing the number of receptors. If both methods were successful, the number of receptors would be reduced and the remaining receptors would be blocked from viral binding. The combination could be very effective against the virus.

17.5 Nasal and Oral ACE2 Inhalers

It is possible that introduction of soluble recombinant ACE2 into the bloodstream would cause side-effect reactions in the body. Another way of utilizing ACE2 as a therapeutic agent is the following. Nasal and oral inhalers exist for Asthma and COPD, spray inhalers, or inhalers of powder. Recombinant ACE2 could be administered by spray or powder into the nose and separately into the throat. This would bring ACE2 to the epithelial cells of the airway, without engaging the epithelial cells or any other cells in the rest of the body. It would not be necessary initially to gradually ramp up the dose. It would be possible to hit hard with a strong initial dose. This would cause a rapid reduction in the number of receptors. The epithelial cells in the airway would initially be driven away from their homeostatic state, but this would only affect airway cells without disturbing any other cells in the body. This would give rapid initial protection and could be followed with maintenance dosage to sustain the receptor reduction and allow for blockage of the remaining airway receptors.

17.6 Other Viruses

The idea of fighting against viral infection of host cells by introducing manufactured molecules like the molecules of the natural ligand of the host cell receptors to cause a reduction in the number of host cells receptors is an idea that can be extended to apply to many other infectious viruses. For the family of coronavirus such as SARS, MERS, and the common cold, the ACE2 receptors would be the target.

Viral pneumonia is a complex disease that has a number of causes. An article by researchers at University of Pennsylvania identifies the IL27 receptor on cells of the airway as being involved in the disease. Their article can be found by Googling "Viral pneumonia receptors." The article is entitled "Missing Link for fighting viral pneumonia identified" Science Daily, February 3, 2017, Pearlman school of medicine at the University of Pennsylvania. The following is a quote from their conclusions:

"More animal studies are needed to determine whether the administration of IL27 can successfully treat severe illness from these viral infections. In principle, IL27 could be delivered via an intra nasal spray, so the effect would be isolated to the airway, and not affect other parts of the body."

They are not sure how the ligand IL27 works to mitigate the effects of pneumonia infection. They demonstrated in rats that survival is enhanced and severity of infection is reduced. They think that IL27 administration keeps T-cells from causing great damage due to inflammation resulting from fighting the virus.

The effectiveness of treatment with the ligand IL27 is very likely a result of downregulation of The IL27 receptors. There may be other benefits, but a reduction in the number of IL27 receptors will diminish viral infection.

Human influenza is another respiratory disease caused by viruses. The virus H5N1 for example infects epithelial cells of the airway by binding to alpha 2, 6 galactose receptors of the lower airway. Other influenza viruses infect by binding to alpha 2, 3 galactose receptors of epithelial cells of the upper airway. Some animals, for example, pigs and chickens, also have these receptors and when infected serve as reservoirs for the human viruses. Sialic acids are receptors for influenza viruses and are usually bound to galactose in an alpha 2, 3 or alpha 2, 6 configuration. The natural ligands for these receptors have sialic acid as one of their components.

Googling "viral influenza receptors," an article was found entitled "Influenza virus receptor specificity and cell tropism in mouse and human airway epithelial cells," published in the Journal of Virology by Aida Ibricevic et al. in 2006. An interesting comment was made "... Finally, targeting a specific cell type for gene or protein therapy could be augmented in mice by the use of a ligand for the alpha 2, 3—linked SA receptor." This comment seems to echo the thesis of this chapter.

Measles is another viral disease that can affect people of all ages. The name of the virus is measles virus (MV). It infects the epithelial cells of the airway, and can also infect cells in other parts of the body. In the airway, the receptor on the epithelial cells is CD46.

CD46 is a ubiquitous human cell surface receptor. It is a receptor to a variety of infectious viruses and bacteria. It seems to be part of the immune system yet it is a receptor for so many viruses and bacteria that it is called "pathogens magnet." Its functions and properties are still being discovered. CD46 has two ligands, C3B and Jagged 1.

An interesting paper about CD46 was obtained by Googling "Ligand of CD46 receptor." The title of the paper is "ligand binding determines whether CD46 is internalised by clathrin-coated pits or macropinocytosis." It was published in J. Biol. Chem., Nov 21, 2003 by B. Crimeen-Irwin et al. The article states that CD46 is the receptor for measles virus and human herpes virus-6. Quoting from the article "ligand binding to CD46 affects (1) protection of antologous cells from complement attack by breakdown of complement components, (2) intracellular signals that affect the regulation of immune cell function, (3) antigen presentation, and (4) downregulation of cell surface CD46. Recent evidence indicates that CD46 signaling can link innate and acquired immune function." Of greatest interest here is affect (4) which is: ligand binding to CD46 effects downregulation of cell surface CD46.

When a molecule of ligand binds to a CD46 receptor on an epithelial cell, a signal is sent thereby to the membrane of the cell. This signal has an effect on the cell's membrane potential.

If a therapeutic dose of ligand were administered to the epithelial cells of an airway, the cells' membrane potentials' would be altered. Homeostatic feedback would come into play, to restore the membrane potentials. Membrane input signals from the receptors would be too strong, and the way to correct this would be downregulation, reduction in the number of receptors on each cell.

What kind of input signals would be delivered to the cell membranes by the receptors, and how would these signals affect the membrane potentials? It turns out that the ligands carry electric charges. Their attachment to the receptors would create an electric current that would immediately affect the membrane potentials. Homeostasis would be restored by downregulation which would increase resistance to infection.

There are many other infectious diseases that are caused by viruses that bind to cell receptors. Examples are mumps, rubella (German measles), chicken pox, poliovirus, hepatitis A, B, C, meningitis etc. They are all different from each other, but except for possible side effects, all might be treated by therapeutic doses of ligand. The idea is to cause receptor downregulation by training the cell and its receptors. The method should work against all infectious viruses and their variants.

17.7 Post Script

Since completing this chapter, a new reference has surfaced entitled "Inhibition of SARS- COV-2 infections in engineered human tissues using clinical—grade soluble human ACE2" by Vanessa Montiel et al. Cell Press, 2020. Their rationale for therapeutic use of soluble ACE2 is blocking the ACE2 receptors of the epithelial cells of the human airway. The 18 authors of this paper are each involved in specific aspects of the study. They came from 11 institutes and universities, from Karolinska Institute in Stockholm to the University of British Columbia. The group is led by Professor Josef M. Penninger of UBC. They have demonstrated the effectiveness of soluble ACE2 in vitro and, progressing with unusual but appropriate speed, have brought this to simultaneous Phase 1 and Phase 2 human trials. Soluble ACE2 has been used to treat human heart and lung problems, so it is safe. Based on their evidence so far, these trials will most probably prove to be successful. If so, they will have in their hands a means to mitigate the COVID-19 world crisis.

For the USA, the big problem will be to bring this to the attention of the FDA and to seek their approval. The world will be watching. If everything is successful, deployment could be very rapid.

17.8 Summary

Trials all over the world are underway testing the use of soluble ACE2 against the SARS-CoV-2 virus. Success or failure will certainly be determined.

Some of these trials are based on the idea of blocking the ACE2 receptors. Other trials are based on a different idea, that high concentration of soluble ACE2 molecules in the extracellular fluid will act as decoys and trick a virus into binding to them instead of binding to the ACE2 receptors of the epithelial cells. A third idea proposed in this chapter involves downregulation of ACE2 receptors.

The three ideas are rationales for the same soluble ACE2 therapy. Blocking, decoying, and down-regulation are hypotheses, unproven. They are mutually compatible and could all be correct. If they were all working at the same time, this could turn out to be very effective therapy.

As the COVID-19 pandemic propagates worldwide, more and more is being learned about the SARS-CoV-2 virus and how it is affecting humanity. It is surprising that when sickened by this virus, the severity of illness is much less in children than in adults and much less in women than in men. Among men under 65, there is considerable variation in severity. Some men can be infected and be unaware of this. Other men can become very sick, and others yet could die from the infection. Likewise, there is great disparity in severity of illness among women, and a similar disparity in illness among children. Averaged over the population, men have worse cases than women, and women have worse cases than children. Comparing infected men with infected women, the men's death rate is almost four times that of women. The question is, why?

Soluble ACE2 in the blood might be the answer. Blood serum level of ACE2 might be different from one individual to the next, and that may relate to severity of infection. To gain an understanding of the difference in severity between men and women, the following search topic was presented to Google: "soluble ACE2 level in women vs men." Titles of a large number of publications was found by Google.

One paper was particularly helpful. Its title is "COVID-19 infection and circulating ACE2 levels: protective role in women and children," by Elina Ciaglia et al., Frontiers in Pediatrics, 23 April, 2020. Children have the highest levels of ACE2, same for both boys and girls. Adult women have a high level of ACE2, but not as high as children. Men have the lowest level of circulating ACE2. Children have the lowest level of expression of ACE2 receptors, women higher, and men the highest level of ACE2 receptor expression. It seems that high level of ACE2 receptor expression is associated with severe infection. It also seems that high level of ACE2 receptor expression is associated with low levels of circulating soluble ACE2. Putting all this together, one might conclude that high level of soluble ACE2 is protective. That is the theme of the cited paper.

Ciaglia et al. suggest using human recombinant soluble ACE2 therapeutically. Although all people seem to be equally susceptible to infection with the SARS-CoV-2 virus, Ciaglia et al. suggests that a test could be devised to measure the level of soluble ACE2 to predict who, if infected, would have a mild case and who would have a severe case. If the test worked, it would be very valuable.

17.9 Questions and Research

1. For various viruses,
   (a) What are the receptors?
   (b) What are the ligands?
   (c) Are there many cases where the chemicals that make up the ligands are the same as the chemicals that make up the receptors?
2. Devise experiments to determine the speed of adaptation during up-regulation, and down-regulation, of animal receptors and of human receptors. How fast can one safely ramp up or ramp down dosage?
3. What percentage reduction in number of receptors would be required to stop an epidemic or substantially slow its progress?
4. If trials of recombinant human soluble ACE2 as a therapeutic against SARS-CoV-2 turn out to be successful, success could be attributed to blocking, decoying, or downregulation. How could one determine which of these three are the means of success?

References

Keywords

(a) SARS-Cov-2
   i. Soluble ACE2 level in women vs in men
   ii. Soluble ACE2 hormone
   iii. ACE2 receptor ligand
   iv. ACE2 ligand
   v. ACE2 receptor purpose of

(b) Other viruses

    i. Viral pneumonia receptors
    ii. Viral influenza receptors
    iii. Ligand of CD46 receptor

Chapter 18
Cancer

Abstract Hormones and hormone receptors play a major role in the development of cancer. With most human breast cancers, for example, the hormone estrogen is present and responsible when cancer develops and propagates. Present treatment methods utilize drugs to block estrogen receptors and to minimize the body's production of estrogen. There are almost about 10x as many estrogen receptors on breast cancer cells as on their normal progenitor cells. Could this difference be used to advantage against breast cancer? A novel approach is proposed here to treat in the opposite direction, to increase estrogen levels to well above normal. Cancerous cells with large numbers of receptors would experience large increases in their membrane potentials, pulled high above the homeostatic level, killing them. Normal progenitor cells would be spared since their departure from homeostatic would be much smaller, thus minimizing side effects. The same method might be useful for fighting other forms of cancer.

Keywords Melanoma ligands · Melanoma receptors · Prostate cancer receptors · Breast cancer receptors · Progesterone inhibitory receptors · Normal breast receptors · Normal breast progesterone and estrogen receptors · Breast cancer · Lung cancer receptors · Wikipedia-epidermal growth factor

18.1 Introduction

Cancer is a disease of cells that have gone wrong. These cells originate within tissues. They grow and divide without the normal checks and balances of non-cancerous cells. They proliferate at a very high rate while crowding out normal cells and stealing their resources and sources of nourishment. They originated as mutated versions of normal cells. They become cancerous after many mutations occur to the various genes that control cell proliferation and cell death. These mutations occur during cell division and lead to unchecked cell growth. As a mass of cancerous cells grows, it becomes a tumor in situ. Over time, metastasis is possible where tumor cells break out, enter the bloodstream or the lymphatic system, and travel to other parts of the body establishing additional cancers.

B. Widrow, *Cybernetics 2.0*, Springer Series on Bio- and Neurosystems 14,
https://doi.org/10.1007/978-3-030-98140-2_18

Normal cells and cancer cells derive their energy from the glucose of the bloodstream. Normal cells and cancer cells have enveloping membranes that support surface receptors. Some of these cells have both surface receptors and receptors in their interiors. The receptors are like antennas that receive signals from other cells. These signals control machinery in their membranes and in their interiors. The receptor inputs are crucial in maintaining the cells' membrane potentials and homeostasis.

The receptors relay signals from the outside world to the membranes and cell interiors. It is a curious fact that all cancer cells express far more of the critical receptors than their progenitor normal cells. EGFR (Epidermal Growth Factor) receptors exist in normal lung tissue and in cancerous lung tissue. In many types of tumor including lung, prostate, ovary, gastrointestinal tract, and brain, the EGFR receptor is expressed approximately 100-times more than the normal number of EGFR receptors found on the surface of normal cells. EGFR, its family members HER2, ErbB, HER4, and their ligands are involved in over 70% of all cancers. Many other receptors are also overexpressed in cancerous cells.

The range of receptor overexpression in a wide variety of tumor cells seems to be roughly 25–100 times. In cases where the tumor cells have a proliferation of a specific receptor and the progenitor normal cells do not have that type of receptor at all, the range of overexpression is infinite.

The question of why some cells become cancerous is understood in some cases but is generally not understood and is a subject of research in other cases. The question of why cancerous cells overexpress certain receptors is not understood at all. This subject is hardly even discussed in the literature.

Receptor overexpression seems to indicate a desire of the cancerous cell to have an increase in receptor signal, no doubt to maintain membrane potential at its homeostatic level. One can only speculate about this. Why the need for more signal? One answer might be that the homeostatic membrane potential of the cancer cell is much higher than that of its normal progenitor cell. Another answer might be that the mechanism (refer to Chap. 5, Inside the membrane, Figs. 5.1 and 5.2) that controls and implements upregulation and downregulation is defective in the cancerous cell and simply needs more signal in order to function. Either one of these conditions could be existing or both could be existing at the same time. All of these conditions are possibilities.

Whatever the reason, the receptor overexpression of the cancerous cells compared to the normal expression of their progenitor cells is tantalizing when one thinks about cancer therapy. Can this disparity be used to some advantage? The answer is yes, or maybe.

18.2 Melanoma

Melanoma is a form of skin cancer that originates in melanocytes, cells that give skin its color. If caught early, it can be removed surgically. If it becomes invasive and metastasizes to the bloodstream, it is deadly. Metastatic melanoma is currently almost impossible to cure. A great deal of research seeking a cure is underway worldwide.

A group of researchers in the Netherlands have been studying human melanoma transplanted into mice. They inoculated a number of mice with human A375M melanoma cells and this started tumor growth. After two weeks, their research procedure was initiated. Half the mice were controls. All the mice were inoculated with A375M. A 14-day treatment schedule was established. Each day the controls were injected with water and the treated mice were injected with a water solution of rimcazole.

The results of their study were reported in a paper entitled, "In vivo responses of human A375M melanoma to σ ligand: F-FDG PET Imaging," by Anna A. Rybczynska et. al., The Journal of Nuclear Medicine, August 12th, 2013.

Melanoma possesses moderate to high levels of σ receptors. There are two types, σ-1 and σ-2 receptors. The population of these receptors is upregulated with the onset of cancer. They are over expressed by a factor of 25–50 times compared to the normal progenitors. Rimcazole is a σ ligand.

After 14 days of treatment, the treated mice had a 4.8x reduction in tumor volume compared with the controls. This is a 77% reduction, caused by tumor cell death. Treatment with rimcazole was successful. The paper made an interesting observation, that the expression of the receptors was preserved. This is surprising since one might expect σ-receptor downregulation given the treatment with excess σ ligand.

The question is, how can one explain these results? The method seems counterintuitive. The normal way to slow or stop cancer would be, aside from surgery, to treat with drugs that block the relevant receptors and that lower the level or eliminate their circulating ligand. The reported method seems to do just the opposite. Although it does not affect the receptors, it does increase the level of ligand. And it works. Why?

One can only speculate. There is no short answer. We return to the homeostatic theory. Cells can be killed if they are forced to function far away from their homeostatic conditions.

Refer back to Chap. 7, Synaptic plasticity, Figs. 7.1, 7.2 and 7.3. The input, The X-vector, is proportional to the ligand level. Bringing the ligand level to zero would cause the membrane potential, proportional to the (SUM), to be much lower than its homeostatic value. This could kill the cell. The large number of receptors on the cancerous cell would not affect the result if the ligand level were essentially brought to zero, or if the receptors were blocked. This is the usual approach to cancer treatment.

The reported method takes a different approach. Before treatment, the cancerous cells were homeostatic. Given the level of circulating σ-ligand and given the large number of receptors, homeostasis was established at a much higher membrane potential than normal. With treatment, increasing the level of σ-ligand acting on an overexpression of receptors forces the membrane potential to go much higher than homeostatic. With strong treatment, the membrane potential will go very much higher than homeostatic, resulting in cancer cell death. The normal progenitor cells, having a normal number of receptors, are largely unaffected. If the extent of cancerous receptor overexpression is 50 times, then the effect on the cancerous cells will be 50 times stronger on the cancerous cells than on the normal ones.

The number of receptors on the treated cancer cells remain static. There may have been downregulation, but this did not happen. The treated mice were injected once per day. At the time of injection, the level of rimcazole must have been very high, and gradually reduced to a low-level thereafter. When the level of the rimcazole was very high, some lethal damage was done to the cancerous cells. Thereafter, the adaptive algorithm corrects for changes in the number of receptors and re-establishes homeostasis. This is merely an hypothesis, not yet proven.

The presence of circulating ligand is considered essential for the growth and survival of tumor cells. This is correct, but the energy for the cancer cell comes from glucose from the bloodstream. What the ligand binding to the receptors does is establish the membrane potential at the equilibrium level, the homeostatic level, which is essential for the health of the cell gone bad, making it possible for it to divide and proliferate without check. Since the ligand is essential for the cancerous cell, treating with more ligand is indeed counterintuitive. But it works.

18.3 Prostate Cancer

Abnormal cells growing in the tissues of the prostate gland is a characteristic of prostate cancer. This is a disease that becomes more common as men age. About 20% of all men will be diagnosed with prostate cancer during their lifetimes. Hormone therapy is the standard treatment for advanced prostate cancer. Androgens (male sex hormones, including testosterone) are essential for normal cellular function, but can cause prostate cancer to grow. Drugs, surgery, or hormones are the means of treatment. Drugs are used to reduce androgens and to block androgen receptors. These receptors are overexpressed on cancer cells. Androgen deprivation kills cancerous cells and shrinks the prostate gland while inhibiting proliferation of cancerous prostate epithelial cells.

Androgen deprivation therapy decreases circulating testosterone to very low levels. This creates a condition called chemical castration. This method is quite effective for about 80% of cases, and usually keeps the tumour at bay for approximately 12–18 months.

Some androgen is derived from the adrenal gland. Abiterone is a drug that further reduces androgens. Flutomide and Enzalutamide are drugs that block the androgen receptors that help even further to reduce the effects of androgens. In spite of all this effort, the cancer relapses.

The cancer appears to take a new form called hormone therapy resistant prostate cancer (HRPC), also called castration-resistant prostate cancer (CRPC). Mutations in the androgen-receptor genes cause mutations to the androgen receptors. They no longer bind to androgens. Curiously, they bind to progesterone and estrogen. These androgen-resistant cells may have been present when the tumor first formed, but they were overshadowed by the androgen-sensitive cells. When the androgen-sensitive cells were killed off by androgen deprivation, the androgen-resistant cells took over, presenting to the oncologist a different tumor to control.

It is proposed that an alternative treatment for prostate cancer be tried. Instead of androgen deprivation, one could employ the opposite approach. Inject the patient with androgen. The dosage would need to be adjusted to be harmless to normal cells but lethal to cancerous cells with overexpressed androgen receptors.

To kill the androgen resistant cancer cells, inject just enough progesterone and estrogen to do this, but hopefully not enough to create significant side effects in male patients. If other types of cancerous cells arise, find the ligands and treat accordingly.

Androgens are thought to promote prostate cancer. The energy for the cancer cells come from glucose. The androgens provide control signal input for the cancer cells necessary to maintain homeostasis. They are not so much promoters as they are enablers. Treating prostate cancer with the androgen ligands is counterintuitive and maybe dangerous, but would be worth trying. The dosage and schedule (once a day, once a week?) would need to be determined.

18.4 Breast Cancer

Breast cancer forms in the cells of the breast. It occurs most commonly in women, but could also occur in men. Cancer cells divide and proliferate without the normal checks. They can metastasize through the lymphatic system and the bloodstream. Estrogen seems to play a major role in the development and subsequent propagation of this cancer.

Breast cancer is very complicated. There are many identifiable conditions. There are dozens of treatment options, the scope and complexity of which is beyond this discussion. Only the most elemental form of treatment will be considered here.

Normal breast tissue has both estrogen and progesterone receptors. Estrogen is excitatory. Progesterone is inhibitory. Under normal conditions, considering the circulating levels of these two hormones, the number of receptors of both kinds is adjusted by an adaptive algorithm to maintain homeostasis.

Breast cancer cells most often have estrogen receptors and in about 50% of cases also have progesterone receptors. The number of receptors is adaptively adjusted to maintain homeostasis, albeit at an abnormal level. There are about 10x as many estrogen receptors on cancerous cells as on their normal progenitor cells.

Treatment of breast cancer usually involves surgery, radiation therapy, and chemotherapy. In addition, hormone therapy may be applied long-term to prevent recurrence.

Estrogen is generally associated with breast cancer. Progesterone not. Hormone therapy targets estrogen, either lowering estrogen levels or stopping estrogen from acting on breast cancer cells. Tamoxifen is a popular drug that blocks estrogen receptors. In premenopausal women, a drug such as Zoladex stops the ovaries from making estrogen. In postmenopausal women, aromatase inhibitors can be used to stop further estrogen production after the ovaries are shut down. Both reducing estrogen and blocking estrogen are commonly practiced methods.

Breast cancers that express estrogen receptors are called (ER) positive. Otherwise they are called (ER) negative. If they express progesterone receptors, they are called (PR) positive. Otherwise, they are called (PR) negative. Most breast cancers are (ER) positive and two out of three are also (PR) positive. They will be able to benefit from hormone therapy.

In breast cancer patients, reducing or eliminating estrogen lowers the membrane potential of cancer cells driving them away from homeostasis. Removing an excitatory input will lower the membrane potential. Leaving the realm of homeostasis, the cancer cells die. This form of treatment has a much stronger effect on the cancerous cells because of overexpression of estrogen receptors. This treatment will be felt by normal cells and various side effects will occur all over the body. But having cancer is a lot worse.

Sometimes standard hormone therapy is augmented in (PR) tumors by the use of the progesterone-like drug Megace. Since progesterone is inhibitory, this drug will further reduce the membrane potential, speeding the death of the cancerous cells. If caught early, these methods can cure breast cancer. But this does not always work and new methods are constantly being sought.

An alternative approach is proposed that would raise the membrane potential above the homeostatic level. This exact opposite method would treat the tumor by giving it an excess of estrogen. Because of receptor overexpression, the effect on cancer cells would be much stronger than on normal cells, although some side effects could occur in other parts of the body. High doses of estrogen should kill the cancer cells and leave normal cells unharmed. Adjusting the dosage and determining an injection schedule would be necessary to shrink the tumor without creating serious side effects.

Once again, the proposed method is counter-intuitive. Hit the tumor cells with therapeutic ligand. It is not yet known if this would help or make matters worse. It is hoped that this would work.

18.5 Lung Cancer

There are many kinds of lung cancer. It is a multifaceted disease. There are dozens of remedies. The lung cancer patient is not aware of having the disease until symptoms appear. At that point the cancer is advanced and difficult to treat. New treatments are constantly being devised with mixed results.

A common form of lung cancer involves epithelial cells with EGFR receptors (Epidermal Growth Factor). Some types of melanoma also involve EGFR, hence the name. Under normal conditions epithelial cells and their EGFR receptors and their ligands are at equilibrium, in the homeostatic state. Including EGF, there are seven major ligands of EGFR. They are similar to each other, in the same family, but have somewhat different functions in the cells. The major purpose of EGF is promotion of cell growth and wound healing. In case of a serious wound, EGF promotes the growth of cells to replace lost tissue. When the wound is filled, creation of additional cells miraculously stops. Sometimes injection of recombinant human EGF at the wound site is done to speed healing.

The healing mechanism is subverted when cells go bad and engage in uncontrolled cell division and propagation. Cancerous epithelial cells will start to express more EGFR receptors. Their homeostatic set points will change. A new homeostatic state is established accompanied by an approximately 100-fold overexpression of EGFR receptors. Cell proliferation takes place. EGF is the ligand. A tumor forms that over time could metastasize and spread over the body.

Treatment generally begins with surgery and chemotherapy. This is followed by techniques that deny ligand to the cancerous cells. Drugs exist to lower the concentration of EGF in the body. Other drugs exist, such as the antibody Zalutumumab, that inhibit and block EGFR receptors. Denying ligand pulls cancerous cells away from homeostatic equilibrium and kills them. If other receptors in addition to EGFR are involved, treatment to deny their ligands would need to be included. Treatment would also affect normal cells, reducing or cutting off their EGF inputs. Side effects are possible.

Another approach would pull away from homeostasis in the opposite direction. Because of the overexpression of EGFR, it might be possible to pull further away from the homeostatic level with little effect on normal cells. The idea is to increase the concentration of EGF. With the 100-fold overexpression of EGFR, not a great deal of therapeutic EGF will be needed. Perhaps increase the level of EGF by about 50% so that the cancerous cells would be hit hard and the normal cells not so much. This could be dangerous and could cause greatly increased proliferation of cancer cells, but it might work and kill them with minimal side effects. If this works, it could be a major new weapon in the fight against cancer.

18.6 Summary

Cancerous cells generally have substantial overexpression of certain critical receptors compared with normal expression of these receptors on the progenitor cells. It is proposed that a general method for attacking cancer would be to use a ligand of the receptors as a therapeutic establishing a concentration of ligand of perhaps 50% above normal or perhaps several times normal. This could disturb homeostasis and disproportionately affect cancer cells as compared with their progenitors. The idea is to kill cancerous cells with only minor upset to normal cells. The large number of receptors on the cancerous cells binding with ligand would be associated with large disparity in their membrane potentials. Small changes in membrane potentials of normal cells would not be lethal and would only cause minor side effects. This idea is unproven and needs to be tested. It is contrary to existing practice.

18.7 Questions and Experiments

1. For cancers whose involved receptors and their ligands are known, how would you devise experiments to prove or disprove that therapy with ligand would work?
2. Use experimental method on several different kinds of cancer in animal models. How would you determine dosage and schedule?
3. If some cancers are successfully treated with ligand therapy and other cancers not, why would this be?

Chapter 19
Chronic Inflammation

Abstract With aging, the immune system gradually becomes impaired and less effective. Chronic inflammation in the myeloid cells of the immune system is a major cause. These cells produce the hormone PGE2, and production greatly increases with age. These cells have many kinds of receptors on their surfaces, among them is the receptor EP2. With aging, the population of EP2 receptors is greatly increased. When PGE2 binds to EP2, a cell undergoes remarkable changes. Glucose metabolism drops and the cell becomes a source of inflammation. The increase in EP2 receptor population is unexpected as increase in the hormone PGE2 should have caused downregulation of EP2 to maintain homeostasis. The PGE2-EP2 binding caused a great departure from homeostasis. Drugs that block the EP2 receptors prevent binding and this allows the restoration of homeostasis. The cells return to normal and inflammation stops. Use of a blocking drug with old mice restored youthful metabolism. Their recall and navigational skills became like that of young adult mice. Similar blocking drugs for humans are under development.

19.1 Introduction

Chronic inflammation is a disease state that is often associated with aging. It can affect the brain and body, causing disease and could ultimately cause death. There are many forms of inflammation and much research is being done to combat this. The goal is to gain understanding of its causes, and to develop possible cures. Chronic inflammation is believed to contribute to Alzheimer's disease, cancer, and frailty. Reducing inflammation could slow the aging process and delay or obviate the gradual loss of mental acuity that can result from aging.

A group at Stanford University led by Dr. Katrin Andreasson, Professor of neurology and neurological sciences, is studying inflammation of the immune system and its effect on brain function. They have made some important discoveries and have formulated possible cures that could be of great benefit to human health. This chapter summarizes their work and provides an interpretation of their results in terms of homeostasis theory.

B. Widrow, *Cybernetics 2.0*, Springer Series on Bio- and Neurosystems 14,
https://doi.org/10.1007/978-3-030-98140-2_19

19.2 Inflammation in the Immune System

Why certain cells of the immune system grow rogue and cause inflammation is not generally understood, but the Stanford group are shedding new light on this. Their work was described in Stanford science of December 28, 2021 [1] and they published a paper on this in January 21 of Nature [2].

Myeloid cells in the brain, in the circulatory system, and in the bodys peripheral tissues are at issue. These cells are normally fighting infectious intruders, cleaning up dead cells and other debris, and providing nutrients to yet other cells. These are vital functions of the immune system.

With aging some myeloid cells could begin neglecting their normal functions and instead go rogue and begin inflicting damage to the body's tissues. The Stanford group found that blocking a particular receptor on the myeloid cells restored youthful metabolism and temperament in older mice. Their recall and navigational skills were restored to be like that of young mice. Similar restoration was observed with human myeloid cells in vitro.

19.3 Aging Myeloid Cells

Myeloid cells are the body's main source of the hormone PGE2 which is of the prostaglandin family. PGE2 does many useful things in the body, but when binding to the EP2 receptors on the very same myeloid cells, inflammatory activity develops within myeloid cells. This can happen with aging. It is interesting to note that myeloid cells generate a ligand that can bind to some of its own receptors. The PGE2-EP2 binding accelerates inflammational development in the myeloid cells.

Older mice and older humans have much higher production of PGE2 by myeloid cells than that of younger mice and younger humans. The blood and brains of old mice were found to have much higher levels of PGE2 than that of young mice.

The population of EP2 receptors on the surfaces of myeloid cells was formed to be very much greater with older cells than with younger cells.

With aging, the production of PGE2 and of EP2 greatly increase. This causes a great multiplicative increase of PGE2-EP2 binding. Dr. Andreasson calls this a "double whammy."

19.4 Glucose and Myeloid Cell Metabolism

Glucose entering the myeloid cells is fuel to support metabolism. However, as a result of PGE2-EP2 binding, less of the glucose is consumed for metabolism and the remainder is stored as glycogen. Storing glucose as glycogen in aging myeloid cells is analogous to converting glucose to glycogen in the liver. In the liver, this storage process is a feature of normal operation. This is not normal in the myeloid cells. Glycogen can be stored but not consumed as a source of energy. The consequence of this is aging with myeloid cells being glucose deprived.

The glucose starved myeloid cells react by going into an inflammatory mode. Aging causes this and this causes aging.

19.5 Rejuvenation

Restoration of youthful vigor in mice has been demonstrated. The Stanford group gave old mice a drug that blocks the EP2 receptors. Old myeloid cells began to metabolize glucose like young myeloid cells, reversing inflammation in the old cells.

The drug reversed the mice's age-related cognitive decline. They could perform recall and spatial navigation like young adult mice.

These results have inspired several pharmaceutical companies to develop drugs to block the EP2 receptor that would be suitable for human use. If successful, this could be the "fountain of youth" for adult brains. However, no drug could restore neurons and synapses destroyed by Alzheimer's, but the drug might slow or halt the progression of the disease.

19.6 Homeostasis and Aging of Myeloid Cells

The process of inflammation in myeloid cells can be described from the point of view of homeostasis theory. This gives a different perspective, another point of view.

With aging, the production of PGE2 greatly increased. This should have caused a reduction, a downregulation of the EP2 receptors in order to maintain cell homeostasis. This did not happen. The opposite happened. The population of the EP2 receptors greatly increased.

The membrane potentials of the myeloid cells have, no doubt, deviated far from homeostasis with PGE2-EP2 binding. This should have killed the cells, but it didn't. The cells' metabolism became abnormal and the cells became glucose starved. This too should have killed the cells, but it didn't.

Being both glucose starved and deviated far from homeostatic, this is another kind of "double whammy." Instead of dying, the cells go rogue and produce chronic inflammation.

19.7 Summary

Blocking the EP2 receptors allows the membrane potentials to be restored to their homeostatic levels, under the influence of the cells' adaptive algorithm. Normal metabolism results and inflammation stops. The aging myeloid cells become restored to a youthful state. Finding the right blocking drug is key. The Stanford team is looking for it. If this works, it will be great for humanity. Rejuvenating the immune system may be possible.

References

1. Study reveals immune driver of brain aging | news center | stanford medicine. https://med.stanford.edu/news/all-news/2021/01/study-reveals-immune-driver-of-brain-aging.html. (Accessed on 06/09/2021)
2. Minhas, P.S., Latif-Hernandez, A., McReynolds, M.R., Durairaj, A.S., Wang, Q., Rubin, A., Joshi, A.U., He, J.Q., Gauba, E., Liu, L., et al.: Restoring metabolism of myeloid cells reverses cognitive decline in ageing. Nature **590**(7844), 122–128 (2021)

Part V
Computer Simulations

Chapter 20
Hebbian-LMS Neural Networks, Clustering

Abstract A network of % Hebbian-LMS neurons can perform clustering, i.e, pattern association. If input data or patterns have natural clusters, the clustering network will find them. A trained clustering network can also perform "deja vu". When given an input pattern without designation, if it belongs to one of the clusters, it was "seen before." A sum of the neurons' errors will be small, and this is a "hit." If the unknown input pattern belongs to none of the clusters, it has not been seen before and the sum of errors will be large, and there is no hit. The discrimination between hit or no hit is important for data or pattern retrieval with human long-term memory. Clustering and pattern association is important for thinking and reasoning.

Keywords Clustering algorithms · Unsupervised learning · Deja vu

20.1 Introduction

Clustering is the process of association of input patterns that share some form of commonality. Pattern association seems to be a major part of thinking and deductive reasoning. "Connecting the dots" is an example of clustering and reasoning.

At the macroscopic level, the basis of association is learned by experience. This is valuable for survival and productive life. Young people are taught by older people the technique of making associations.

At the microscopic level, at the level of the synapse and the neuron, making associations cannot be taught. The ability to make associations develops spontaneously. In this chapter, it will be shown how neurons trained by the % Hebbian-LMS algorithm without supervision can learn to make simple associations.

This chapter has two main parts. The first part is about clustering i.e pattern association. The second part is about "deja vu." An input pattern satisfies deja vu if it has been seen before or if an associated pattern has been seen before. Questions of deja vu can be resolved by application of the % Hebbian-LMS algorithm. Deja vu is an important part of the retrieval process in a content addressable memory, an auto-associative memory system.

B. Widrow, *Cybernetics 2.0*, Springer Series on Bio- and Neurosystems 14,
https://doi.org/10.1007/978-3-030-98140-2_20

20.2 Clustering

It has been demonstrated in Chap. 5, Fig. 5.3, that the % Hebbian-LMS neuron is capable of a simple form of clustering. In this chapter, a network of % Hebbian-LMS neurons is used to perform more elaborate kinds of clustering. This however is nowhere near the sophistication of clustering that no doubt occurs during the thinking process.

Here is the problem. Can thousands of unlabeled input patterns be separated into clusters, spontaneously and without supervision? Suppose that there is natural clustering in the input data. Can the clustering algorithm find the clusters? Could a neural network train itself to form clusters and be able to place new unidentified input patterns into the same clusters?

20.3 Input Data Synthesized

Experiments were done to find some answers. Not knowing what input patterns would look like in a living system, a set of artificial training patterns was synthesized.

The 26 letters of the English alphabet were chosen with a 16×16 format. The 256 pixels of each letter were black or white. The black pixels were designated as 0's. The white pixels were designated as 1's. Each letter of the alphabet was thus represented as a binary pattern. When presented to the clustering system, these patterns were merely abstract binary patterns. The association of these patterns with letters of the alphabet was of no significance to the system. These associations were meaningful only to human observers who were doing performance evaluation (Fig. 20.1).

The binary patterns of the letters of the alphabet became vectors in a 256-dimensional space. A given binary pattern was "vectorized" by taking the value of each pixel, 1 or 0, and making that value be one vector component. The pixels were assigned to specific vector coordinates. However this was done, the same assignment would be kept throughout an experiment.

Input data was created to have clusters like in Fig. 20.2. The vector corresponding to each letter of the alphabet was made the centroid of a cluster. Relatively small random noise vectors were added to the centroid vector to form the cluster. One thousand noise vectors were generated for each of the 26 letters. These vectors,

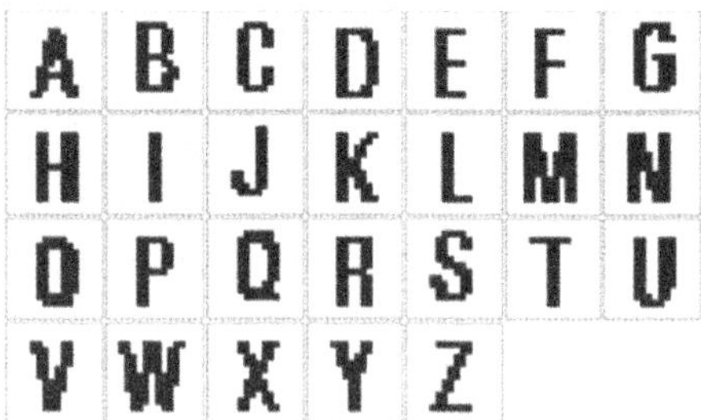

Fig. 20.1 English Alphabet as 16×16 letters

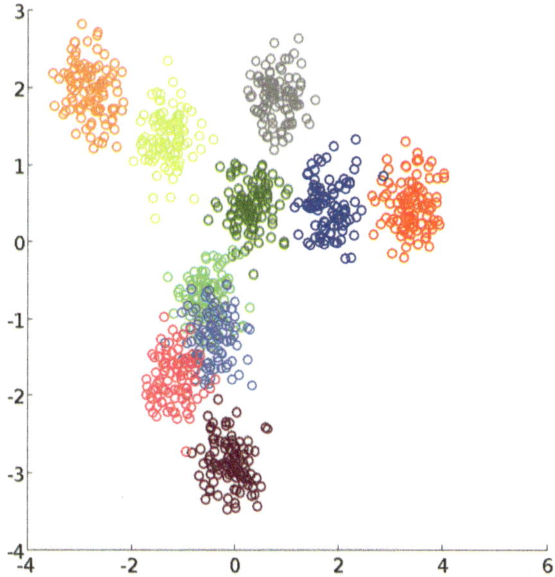

Fig. 20.2 Clusters of points seen in two dimensions. Reproduced from [1] with permission from IEEE @ IEEE 2021

26,000 of them, were used for training the clustering system. They were all presented to this system without any identification. The question was, could the system make sense of this?

The additive random noise vectors were generated as follows. For every input vector component, a noise number was chosen independently from a uniform distribution ranging from 0 to $+0.15$. The noise number was subtracted if the input vector component was 1 and added if 0. The reasoning was that the synaptic inputs generally came from activation function output, and with half sigmoids, the maximum value was 1 and minimum was 0. The pixels of the training patterns were no longer binary, but a grayscale. The vectors within a cluster were associated as they were similar to each other. In two dimensions, an example of clusters is shown in Fig. 20.2. Each cluster point or dot is the tip of a vector coming from the origin of the space.

20.4 A Clustering Experiment

The results of a clustering experiment with % Hebbian-LMS are demonstrated in Fig. 5.3 of Chap. 5. In a large-dimensional space, points were dispersed randomly. Each point was the tip of a pattern vector. The neuron's weights were randomly initialized. Accordingly, when the patterns were presented to the neuron, the (SUM) values were randomly distributed. Some values were positive, some negative.

Adaptation commenced to minimize mean square error. The learning curve shows the error becoming smaller and smaller until convergence. Now the values of (SUM) were clustered at or near the equilibrium points. Most values of (SUM) that were positive at the outset trained toward the positive equilibrium point, and some switched sides and trained toward the negative equilibrium point. Likewise, most that were initially negative trained to the negative equilibrium point, and some trained to the positive equilibrium point. All input patterns were decisively classified as either positive or negative. At the output of the half sigmoid, the positive were designated as 1's and the negative were designated as 0's.

20.5 A Clustering Network

The same method has been used to form a clustering network of many neurons. Input patterns were presented in parallel to an array of neurons. The weights of each neuron were randomly independently set initially. Each neuron was then trained independently with % Hebbian-LMS [1].

Upon convergence, the array outputted a multi-bit binary "word" with the presentation of each input pattern. A diagram of the neuron array can be seen in Fig. 20.3.

Imagine next a high-dimensional space with pattern vectors randomly dispersed in it. There are no natural clusters. The tips of the vectors are randomly located in the space. Let these vectors be inputted to the network of Fig. 20.3.

With independent random initial weights, train the neurons with % Hebbian-LMS. Upon convergence, the outputs corresponding to the inputs will be binary. Presenting any of the training vectors to the network, the set of outputs will comprise a binary word. Input vectors that are close to each other in the space will output similar binary words. If very close, the output binary words will be identical bit by bit. If the input vectors are far apart, the corresponding binary words will be very different.

With many points in the space, points in a neighborhood in the space will correspond to output words that are very similar. Similarity of output words implies association, although no association was intended in this case.

Consider now the clustering example, the 26 letters of the alphabet. Each letter was digitized with a 16×16 pixel format. All 26 digitized letters are shown in Fig. 20.1.

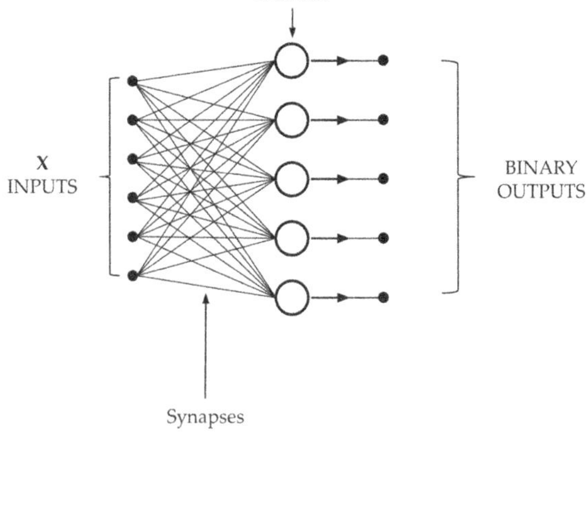

Fig. 20.3 A clustering network

For each letter, there is a vector in a 256-dimensional space. By adding noise to each of the 26 vectors, clusters can be created. Examples of some of the clusters projected onto two dimensions is shown in Fig. 20.2. In this case, the points within a cluster are associated. A human observer would understand the association. The system sees only a collection of input vectors, without any "understanding." Each neuron of the array had 256 weights.

One thousand vectors were generated for each cluster. The total number of training vectors was 26,000. None of these vectors has designations, they were all unidentified. The question is, could the network make sense of these vectors? Could clusters be detected? The answer is yes. After convergence, all 26 clusters were identified. All thousand vectors of each cluster yielded identical output words, and completely different output words from cluster to cluster. The system made the correct associations. It connected the dots.

The amount of additive noise with cluster creation determines the amount of spread of the cluster points. With a large enough spread, there will be cluster overlap. It would be impossible to determine if pattern vectors at the overlap belonged to one cluster or the other. Then the system will make errors. The standard deviation of the spread of the clusters should be compared to the standard deviation of the spread of the centroid points. The ratio of the standard deviation of the centroids to the standard deviation of the clusters is of interest. With a ratio of 5:1, there were no errors. The 26,000 input vectors were perfectly associated with their clusters. The system develops an individual binary-word representation for each cluster. With a 3:1 ratio, some errors occurred. Some of the 26,000 input vectors were not perfectly associated with their clusters. Even with errors, the number of output bits that were misclassified was typically very small. When the associations are perfect, a human observer would see 26,000 noisy letters being correctly associated with their centroids, the original noise-free letters. These experiments were done with a network of 500 neurons. When results are perfect, all 500 bits of the output words are perfect. Examples of noisy letters when the ratio of standard deviations is 5:1 and 3:1 are shown in Fig. 20.4.

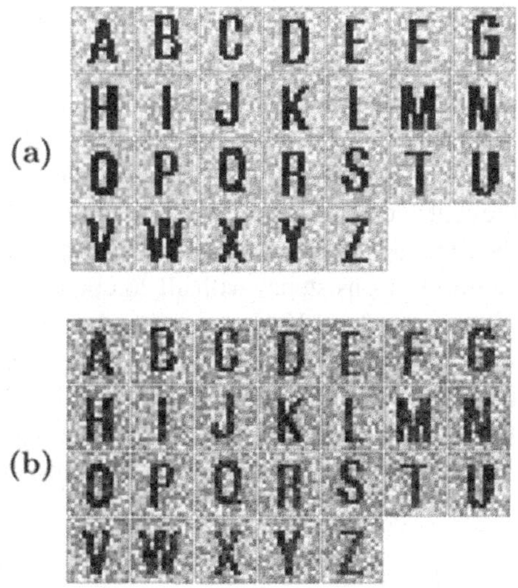

Fig. 20.4 Noisy letters with a ratio of standard deviation of centroids to clusters of **a** 5:1 **b** 3:1

20.6 Experiments Done by Abhipray Sahoo

A set of experiments were performed by Abhipray Sahoo to test the performance of the clustering network. In his network, there were 500 neurons. Each output word therefore consisted of 500 bits. All input patterns were 256-dimensional. Accordingly, each neuron had 256 weights. There were two versions of input vectors that were used for both training and subsequent testing. Version A generated input vectors from random vectors in the 256-dimensional input space. Version B generated input vectors from the letters of the English alphabet as in Fig. 20.1. Each cluster originated with a binary vector, the centroid. The clusters were formed by adding noise to the centroids. The centroid vectors were binary, but the vectors of the associated cluster were no longer binary because of the additive random noise.

20.6.1 Version A: Random Cluster Centroids

With Version A, 26 random binary vectors were generated. Their components were random choices of 1 or 0. These became the centroids. Noise was added to the centroid vectors to create clusters about them. For each cluster, 1000 vectors were generated. Two datasets of 26,000 vectors each were generated, one for training and one for testing.

The network was first trained with only the cluster centroids. This is training with noise free data and the results were perfect. When inputted to the trained network, the outputs for each centroid was a distinct 500-bit binary word. The network assigned the binary words, each a unique representation of the centroid. The assignment resulted partially on the initial random choice of the weights, the initial conditions, and partially on the adaptation.

The network was retrained with noisy data. Starting with random initial weights, with all 1,000 vectors of the 26 clusters of the training dataset. Training with 26,000 noisy vectors, the process converged to MSE of 9.6e-4 with 100 passes of the data. The number of clusters found in the training data was 25. Evidently, two of the clusters were so close to each other that the network gave both the same output representation, i.e, made them into a single cluster.

Test vectors were generated from each of the clusters with random noise added to the centroids. The number of test vectors generated was also 26,000. Inputting them to the trained network, their output binary words matched those of their corresponding centroids. Consistency with all 26,000 test vectors was perfect. However, just as with the training data, 25 clusters were detected.

The experiment was repeated many times with different initial conditions each time. Most of the cases resulted in finding 26 clusters. Each time, the clusters had different unique output binary representations, but consistency was essentially perfect. The actual assignment of cluster representation is not important. What is impor-

Table 20.1 Training details

Number of layers	1
Number of neurons	500
Bias	0
Excitatory ratio	Top half of input vector treated as excitatory and bottom half as inhibitory
Gamma, γ	0.5
Learning rate, μ	0.03
Initial weights	Uniform distribution in [0.0, 0.5]

tant is that all members of the cluster have a consistent representation, that they are associated with each other and unassociated with the other clusters.

Further details on training are in Table 20.1.

It should be noted that for these experiments, half of the components of the input vectors were excitatory and the other half were treated as inhibitory. The 50-50 split is not critical. Which input components are excitatory and which are inhibitory does not matter, only once an assignment is made, it must be sustained for the experiment.

20.6.2 Version B: English Letters

Version B, with centroid vectors formed from the letters of the alphabet, was a similar example. The digitized letters are as in Fig. 20.1.

There were 26 clusters. When pictured in two dimensions, these clusters were randomly dispersed like the clusters shown in Fig. 20.5. They represent vectors in a 256-dimensional input space. The experiments were the same as in Version A.

The 26 centroid vectors were used as a noise free dataset. Training was perfect. 26 clusters were found with no inconsistency.

Training with 26,000 noisy vectors from the clusters and testing with another different set of 26,000 noisy vectors, the network performed perfectly as well.

These experiments demonstrate that if the trained clustering network were presented with thousands of unidentified points in a high dimensional space, the network would discover associations between the points if an association exists.

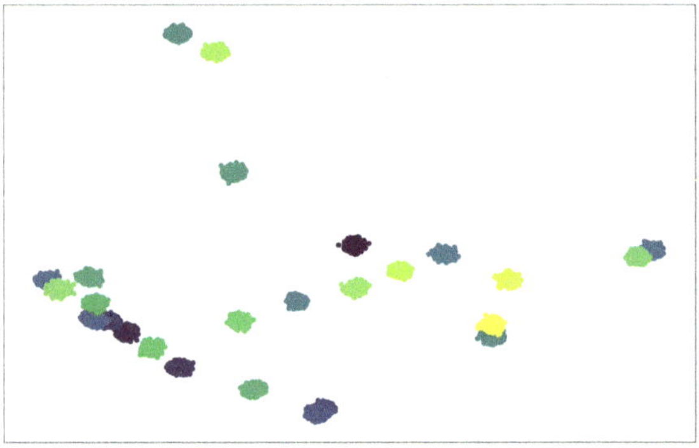

Fig. 20.5 2D visualization of the clusters formed by adding random noise to the English letter vectors in Fig. 20.1. Each point is plotted along the top two principal components of the noisy letters

20.7 Deja Vu?

Deja vu means already seen. Deja vu? means have I seen this before?

Given a trained clustering network operating in a high-dimensional space with many clusters, when a new unidentified input vector is presented, could the network be used to determine if this new input has been seen before? Does this new vector fit into any of the existing clusters? Deja vu?

When the network is being trained with vectors taken from a set of clusters, the initial mean square error is very high. Upon convergence, the MSE is very low. If an unidentified input vector is presented that was not previously trained in, if the vector is taken from any one of the clusters, the sum of the square of the errors of all the neurons will be very low. On the other hand, if the unidentified vector was a random vector not from any of the clusters, the sum of the squares of the errors generally will be very high.

Small sum of the squares of the errors means that the unidentified input vector is associated with one of the clusters. There is deja vu, a "hit." If the sum of the squares of the errors is large, there is no deja vu, no hit. A threshold could be set to distinguish hit from no hit.

In a living system, the errors are available for each neuron. But squaring the error by nature is probably not possible. Instead, a sum of the absolute value of errors could be created.

A deja vu experiment was done with the English letters. The network was trained with 26,000 vectors taken from the clusters. New unidentified vectors were taken at random from the clusters, thousands of them. Probability distributions of absolute value of error are plotted in Fig. 20.6. The blue distribution shows that for vectors taken from the trained clusters, we see a high probability of low values for sum of absolute errors. Similarly the orange distribution shows that for randomly chosen vectors, the network produces high sum of absolute errors. The two distributions have little overlap. A good threshold for sum of absolute errors in this experiment is 49. This experiment shows that is easy to distinguish "hit" from "no hit."

Deja vu discrimination is very useful for information retrieval with content-addressable memories. That is the subject of the next chapter.

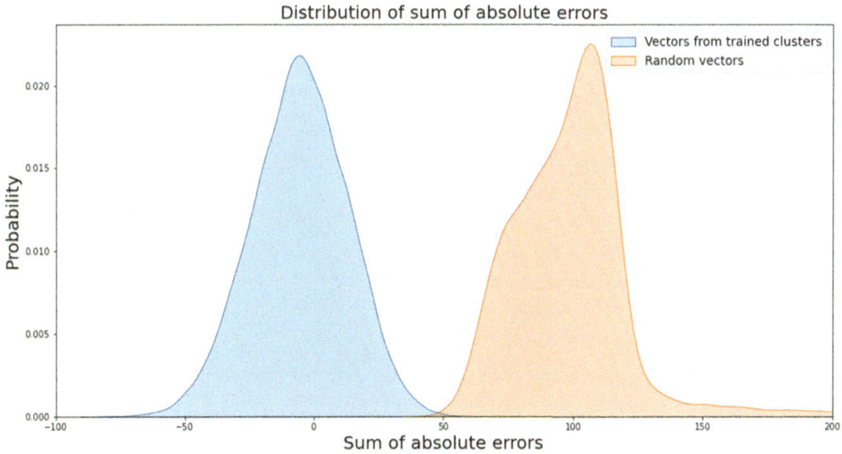

Fig. 20.6 Deja vu with sum of absolute errors. Blue distribution shows low error for points taken from trained clusters (Hit). Orange distribution shows errors for random input vectors (No Hit)

20.8 Summary

Neural networks trained with the % Hebbian-LMS algorithm can identify relationships between unknown input patterns. Unsupervised learning with % Hebbian-LMS can do clustering, finding associations in undesignated input data and patterns.

Inputs to a neural network are vectors that either belong to a cluster or not. A network trained with vectors taken from clusters will identify an unknown input vector if it belongs to one of the clusters. Deja vu is the question. Does this unknown pattern belong to any one of the clusters or not? If it belongs, it was "seen before," deja vu.

Reference

1. Widrow, B., Kim, Y., Park, D.: The Hebbian-LMS learning algorithm. IEEE Comput. Intell. Mag. **10**(4), 37–53 (2015)

Chapter 21
Cognitive Memory

Abstract Cognitive memory is long-term human memory. Thinking about nature's design leads one to design a machine memory that would in some respects behave like human memory. The machine version is useful in the engineering world. It could also serve as a model of human memory. Cognitive memory is content-addressable. Retrieval of data and patterns is responsive to associative pattern matching between a prompt input and data or patterns previously stored in the memory. When a prompt input is presented, a parallel search is made of the entire memory hoping to find a storage folder having a pattern or patterns that are associated with the prompt pattern. A successful hit would cause the folder contents to be delivered to the memory output bus. Simultaneously stored sensory patterns, such as visual, auditory, tactile, olfactory, etc., would be outputting in response to a prompt hit.

Keywords Cognitive memory · Content addressable memory · Pattern recognition and memory

21.1 Introduction

Cognitive memory is long-term human memory. The terminology of cognitive memory could also be applied to machine memory that imitates human memory. Such memories are content-addressable.

In a conventional computer, data, patterns, etc. are stored in registers that are numbered. The program, the computer code, must know the "register" number of every specific piece of data. When a piece of data is needed, the computer approaches the memory and asks for the contents at a specific address. The computer program controls the storage of data and the retrieval of data. A computer's RAM (Random Access Memory) is designed for retrieval by address number.

A content-addressable memory operates on a different principle. It allows data or patterns to be stored in folders wherever there is an empty folder. The folders are not numbered. The computer has "no idea" where a specific piece of data is stored. Data retrieval is accomplished by the computer approaching the cognitive memory and asking for data having a certain content or a certain characteristic. It is believed that

B. Widrow, *Cybernetics 2.0*, Springer Series on Bio- and Neurosystems 14,
https://doi.org/10.1007/978-3-030-98140-2_21

human memory is content addressable. Data and patterns are retrieved in response to input prompting patterns. If the prompting pattern is associated with a pattern or patterns stored somewhere in the memory, the memory would divulge the contents of the folder containing the associated pattern. Making the association is functionally deja vu.

In the human brain, sensory inputs such as visual, auditory, tactile, vestibular, olfactory, etc. are most likely stored in the same folder if they were initially synchronized in time. The evidence for this is for example that a visual cue could make one remember the smell of mother's cooking. These sensory patterns could not be retrieved together unless there were recorded together. The consequence of this is that visual patterns are not stored in the visual cortex, auditory patterns are not stored in the auditory cortex, etc. They are stored together, somewhere else.

Where in the brain is memory stored? This is not understood. Parts of the cortex destroyed or diminished by Alzheimer's disease might be the locus, based on loss of memory. But the loss in memory is most likely a loss in retrieval ability, not a loss of the original stored data. On a bad day, an Alzheimer's patient might not remember a certain piece of data, but on a subsequent good day, memory of the certain data comes pouring out. Permanent data storage is probably stored digitally in DNA. Properly encoded, a great deal of information can be stored in a single DNA-molecule. A human baby develops from a single cell. The nucleus of the cell has DNA. This is the means of initial storage of all inborn information. There are no neural networks there, no memory traces.

A baby bird is born with knowledge of how to call for food, and knowledge of many other things. The bird developed from a single cell. The bird is born with knowledge of how to fly. From the nest, it watches its parents, but this certainly could not inform the baby of the complex muscular control signals that its brain must provide to enable flight. The baby bird will fly when its body and wings develop sufficiently.

The situation is similar with a human baby. A human baby is born with knowledge of how to walk. After about one year, the body and legs have developed sufficiently to enable walking. The first steps are wobbly and crash prone. Very soon thereafter, the baby becomes a toddler. The baby could not have learned the complex muscle controls within a few days. The control system must have been embedded in the baby's memory system before birth. The control system needs to adapt, change, and evolve as the baby grows. In order to allow change, it seems that the control system must be stored in the memory's "software," and not hard-wired into the brain. The initial control system was probably stored in the DNA of the original single cell, inherited from the parents. The initial control system is generic, not tailored to the individual baby. That is probably why the baby's initial steps are wobbly and remain so until the system adapts to the baby. The initial control system is stored in DNA that exists all over the body and remains intact through life, except in the brain where it evolves and undergoes change. Adult DNA still carries the baby's initial walking system.

DNA is remarkable. Simple computers have been built of DNA, capable of performing logic and memory. Playing the game tic-tac-toe (noughts and crosses) is a typical application of a DNA-computer.

21.2 The Design of a Cognitive Memory

A cognitive memory can be designed that would have certain behavioral characteristics like that of human memory. The design would be good engineering, and it may give insight to the working of human memory.

A machine version could be built electronically [1]. Its construction is something that nature might actually be able to implement with brain parts.

The design has a means for recording pattern data and a means for associative pattern retrieval. Figure 21.1 shows a portion of the design of a pattern storage system. Figure 21.2 shows a portion of the design of an associative retrieval system. Figure 21.3 shows the overall design of a cognitive memory capable of storage and retrieval, a key element is the clustering network.

In Fig. 21.1, patterns are sensed and transmitted throughout the memory by an input pattern bus. A bus is a communication channel that carries signals over distances. The illustrated sensors are visual and auditory. For machine memory, the input data and patterns are stored in conventional digital memory folders. In the living system, input patterns are probably stored in DNA, digital long-term memory. Input patterns, visual and auditory, are stored together like video recording.

The recording system goes out in the real world, sees and hears, and records video in folders. Pieces of the video are stored chronologically in the folders. When the folders are full, the recording continues and stores video in the next segment.

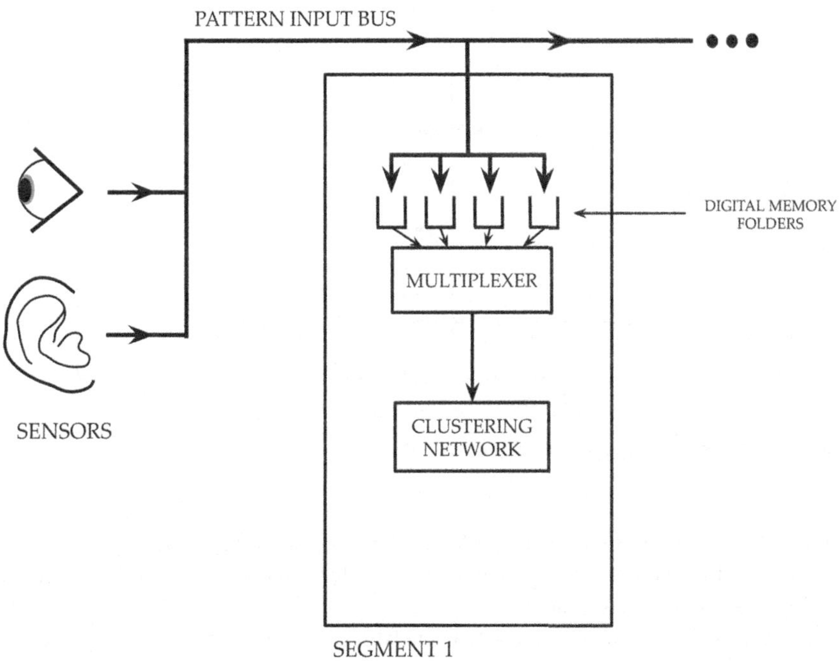

Fig. 21.1 Recording and clustering training. One segment shown

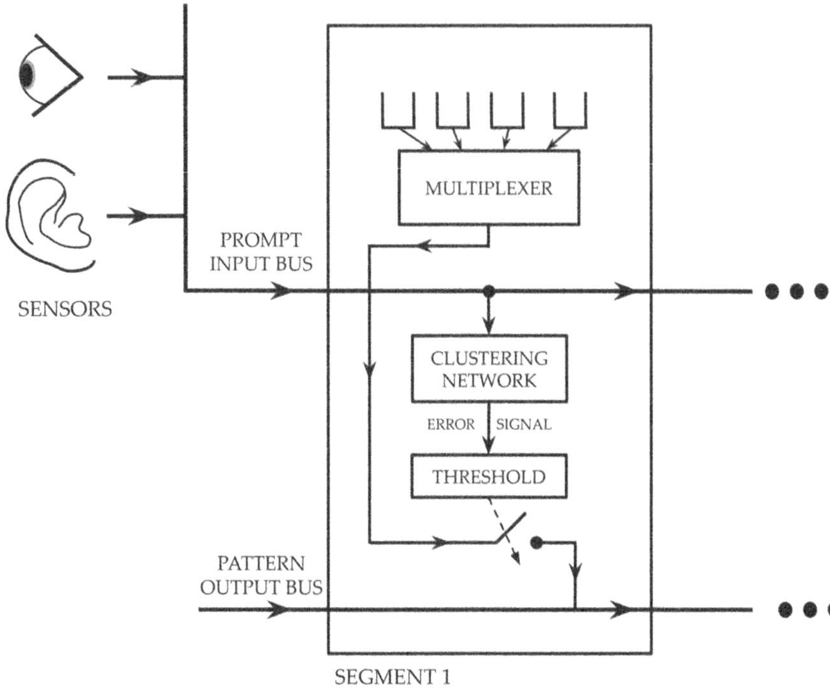

Fig. 21.2 Pattern retrieval in response to prompt input patterns. One segment shown

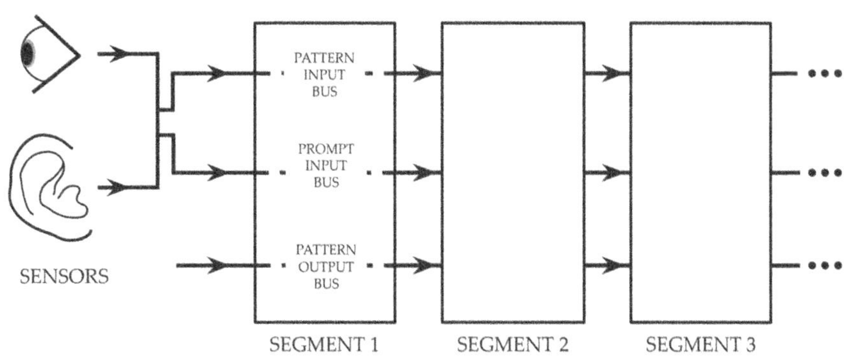

Fig. 21.3 A design for a cognitive memory

Figure 21.1 shows a single segment of the memory system. More identical segments will be used to increase storage capacity. Very large capacity can be acquired by adding more and more segments.

Each segment has its own clustering network. A multiplexer, connected to the folders, scans over and over again obtaining the video and audio patterns that would be inputted to the clustering network. The multiplexer scanning and clustering network is trained on the patterns stored in the segment's folders. There are two separate networks, one for visual and the other for auditory patterns. Simultaneously sensed visual and auditory patterns are stored separately but in the same folder. With experience in the real world, the segments begin to fill with data and patterns and the clustering networks will begin training.

Retrieval of stored patterns when required is accomplished by the system shown in Fig. 21.2. Prompting by visual and auditory input patterns causes the memory to seek patterns stored somewhere in the folders of a segment or possibly two, stored patterns that are associated with current input patterns. Prompting could be done by visual patterns, by auditory patterns, or both.

The question is, deja vu? Are there patterns stored in the memory associated with current input prompt patterns.

The prompt patterns travel through the entire memory on the prompt input bus to provide inputs to all the trained clustering networks. A network's useful output signal in response is its error signal. If any error signal corresponding to a prompt input is below threshold, a switch closes and allows patterns stored in the segments' folders, scanned by the multiplexer, to be outputted on the pattern output bus. The output of the entire memory system is the set of patterns delivered by the memory output bus.

When a prompt input is presented, a parallel search is made of the entire memory to find a folder having an associated pattern and if deja vu is positive, the contents of the entire folder are outputted by the cognitive memory system. The cognitive memory is both a memory and a pattern recognition system. A diagram of a cognitive memory system is illustrated in Fig. 21.3.

The cognitive memory represented in Figs. 21.1, 21.2 and 21.3 is a construct that is biologically feasible. With neurons, synapses, and DNA, nature could build such a system. The cognitive memory is proposed as a highly simplified model of long term human memory. Think of how this memory functions, then think about experiences with one's own memory. Deja vu is the key. Memory and pattern recognition are intertwined.

Learning goes on continuously day and night in the memory segments. Some researchers associate learning with growth of new synapses. It seems that evidence for this is scant. Slow changes in neural network circuitry due to cell death and due to addition of new synapses are disturbing but easily accommodated by the constant adaptation and learning that normally takes place.

21.3 Summary

A cognitive memory, whether human or machine, is content addressable. Stored data or patterns are retrieved in response to input prompt patterns. When there is an association between a prompt pattern and one or more patterns previously stored in memory folders, there is a hit. The contents of the folder or folders containing hit patterns are delivered to the memory output bus. A hit is a positive deja vu. Searching for a hit is done simultaneously, in parallel, among all the segments of the memory. Data and patterns are retrieved without any knowledge of where they were stored.

Reference

1. Widrow, B., Aragon, J.C.: Cognitive memory. Neural Netw. **41**, 3–14 (2013)

Part VI
Growth of Plants

Introduction to Part VI

At my desk at home, I can look out a large window and see shrubs and some redwood trees. The trees are almost 60 years old and are very big in diameter and tall, about 100 feet high. By comparison, the shrubs are very small.

The redwood trees are of the same species as the giant redwoods located in the Mariposa grove at the Yosemite National Park in Northern California. These trees are almost 300 feet tall and some of them are 3,000 years old. In 3,000 years, these trees grew to be much taller than my redwoods, but for all that time they are only a factor of 3 taller. Something slowed and stopped their growth. It seems that there is a certain range of heights that these trees like. Homeostasis regulates the size of these trees, it would seem.

My shrubs have grown to a given height over 60 years and reaching that height, stopped growing. Homeostasis regulates the size of these shrubs. Being a different species, they grow to their favorite heights different from that of the redwoods. They all have their favorite sizes regulated by homeostasis.

Thinking of homeostasis, one might wonder if % Hebbian-LMS has something to do with growth regulation. This question is addressed in Part VI with some ideas of how this might work. The ideas are highly speculative with nothing proven or even similar to anything in the field of plant biology. If the ideas turn out to be correct, they could have a significant impact on the world's agriculture.

Chapter 22
Trees

Abstract In this chapter, an analogy is drawn between a synapse in the brain and a tree in the sun. The synapse has neurotransmitter input signals and learning involves upregulation and downregulation of neuroreceptors. The input signals to plants and trees is sunlight and the receptors are the leaves. An adaptive algorithm controls the population of leaves and size of the tree or plant. An ohmic current results from the static charges from friction of air moving over the leaves. A hormone current, analogous to the synapse's ionic current, is powered by sunlight. It is argued that an adaptive algorithm perhaps like % Hebbian-LMS, is employed by nature to regulate the size of plants and trees. The ideas expressed in this chapter are highly speculative and should be regarded as a "thought experiment." If these ideas turn out to be correct, strong enhancements to agriculture will be possible.

Keywords Tree hormones, hormone currents · Tree cell to cell communications · Tree growth · Tree roots growth · Effect of wind on trees · Tree homeostasis · Seasonal effects

22.1 Introduction

Plants, shrubs, and trees share a common ancestor with animals, an ancestor that lived perhaps a half billion years ago. There are certain commonalities between trees and animals. Both are made of cells that have DNA in their nuclei. Both use hormones to communicate with and control their various parts. There are many other similarities that lend credibility to the common ancestor theory. Over time, plants and animals evolved separately and developed many differences. For example, trees have no means of locomotion (except for Stanford University's mascot "the tree"). Every cell of a tree can generate and release hormones, whereas in animals only cells in specialized organs generate and release specific hormones. There are many examples of similarities and differences.

Figure 22.1 is a diagram of a tree with its roots, trunk, branches, twigs and leaves. At great risk and with a healthy uncertainty, an analogy can be made between a tree and a synapse. A synapse has receptors. The tree also has receptors, its leaves. The

B. Widrow, *Cybernetics 2.0*, Springer Series on Bio- and Neurosystems 14,
https://doi.org/10.1007/978-3-030-98140-2_22

Fig. 22.1 Tree

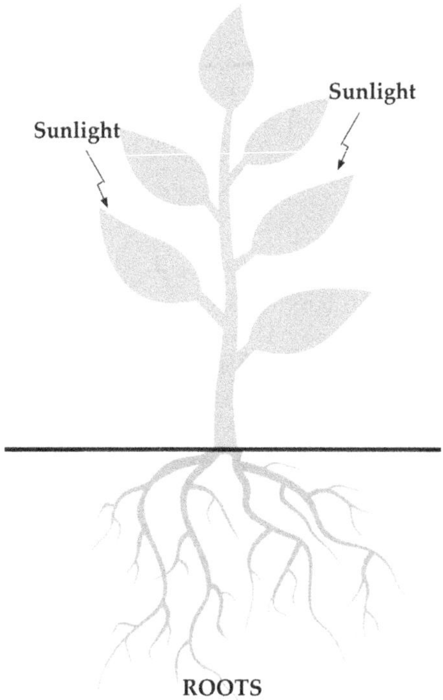

leaves are receptors of sunlight. The brightness of sunlight is the input signal, the X-vector. Unlike the inputs to a neuron's synapses, excitatory and inhibitory, the input to all the leaves is the same sunlight, and this is excitatory. Sunlight is not only the input signal but it is also the source of energy for the tree, its power supply.

Carrying the analogy further, Fig. 22.2 shows a hypothetical drawing of a growth regulating system for the tree. Growing taller, growing more branches, and growing more leaves increases the number of receptors. This increases the "weight" by which the sunlight interacts with the tree.

In Fig. 22.2, just below ground level and within the tree trunk is a mechanism that implements a growth algorithm. The mechanism's inputs are an ohmic current, analogous to the synaptic ohmic current, and a hormonal current driven by the energy of sunlight, that is analogous to the synapse's ionic current. The sources of the tree's ohmic current will be described below.

The tree and its leaves are analogous to the neuron's synapse. This analogy is apparent when viewing Fig. 22.2 in light of Figure 5.1 and 5.2 of Chapter 5.

In animals, the hormones and their receptors open gates in the membrane, allowing the flow of ions and creating the ionic current. In trees, the hormones and their receptors do not open gates and allow ion flow. They perform other functions that are essential for cell operations. Sun powered hormones are available to the mechanism. These hormones are generated in the leaves and branches and travel down to the base

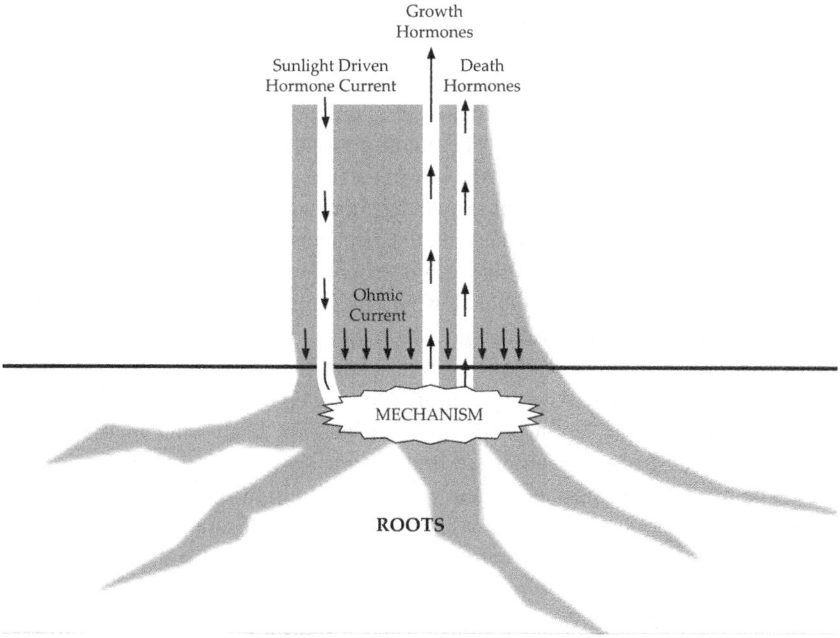

Fig. 22.2 A tree's growth regulating system

of the tree. Trees have elaborate hydraulic systems that allow fluids to travel up and down the trunk, into and out of the branches, twigs and leaves. The sunlight driven hormones provide the power to operate the decision-making and functioning of the mechanism.

The outputs of the mechanism are hormones that flow upward into the tree; one causes growth and the other causes death. Increasing growth results in an increase in the ohmic current. Regulating the ohmic current regulates the size of the tree, the number of leaves, the number of receptors.

Every species of tree has a set point for the ohmic current. If the ohmic current is greater than the set point, some leaves will die and fall off. If the ohmic current is less than the set point, the tree will grow branches and grow more leaves in order to bring the ohmic current up to the set point. If branches were cut, the tree will grow more branches to re-establish homeostasis. The error is the difference between the ohmic current and the set point. The rate of growth is proportional to the product of the error and the amplitude of the sunlight driven hormone current.

The ohmic current is a flow of electrons down the trunk and out into the roots and the Earth. It originates at the leaves and is proportional to the number of leaves. The ohmic current is very small and difficult to measure. The ohmic current is created by static charges that develop on the surfaces of the leaves due to contact and fluidic friction of particles of water vapor in wind flowing over the leaves. Resulting movement of the electrons charges the leaf surfaces, but static charges do not accumulate on the leaves because they are discharged to ground by traveling from leaf stems to twigs to branches to the tree trunk, and down to earth through the root system, having passed through the mechanism.

The amount of current from a single leaf is minute, but this is magnified by the large number of leaves on the tree. The ohmic current is still very small.

Regulating this current regulates the size of the tree. To increase the ohmic current, the mechanism releases and pumps growth hormone up into the tree and out into the root system. To reduce the ohmic current, the mechanism pumps death hormone into the tree system. The rate of growth/death is proportional to the product of the error and the rate of sun-powered hormone flow. Strong sunlight makes things happen fast. At night, the rate of growth or death goes to zero.

High winds cause stronger ohmic currents and tend to energize them over the set point on a full grown tree. A young sapling growing up in a high wind area will not grow as big as the same sapling within a low wind area. High-wind areas are along beaches and waterfronts in California. In these areas one sees scrub plants and few trees. The trees are of special species, probably with high set points, and adapted to high wind. Every species of tree will have its individual set point. In California, there are giant redwoods that grow in many places not far from the sea, in foggy climates. They always grow in forests and groves. They protect each other from the wind. Trees in forests do very well as they shield each other and keep their ohmic currents low.

Deciduous trees lose their leaves in the autumn and grow new ones in the spring. A possible explanation for this is that their set points vary with ground temperature near the earth's surface. If the set points in their mechanisms lower as the ground temperature is lower, that would explain the phenomenon of leaf loss in the Autumn. They would find themselves with an ohmic current higher than the set point. They have too much tree. Their mechanisms would send out death hormones and the leaves would fall off.

In the springtime as the earth temperature rises, their mechanisms send out growth hormones and their twigs develop buds and subsequent leaves. Flowering fruit trees will develop buds and flowers in the springtime and during the summer, fruit. Growth of fruit is important for propagation of the species. Growth of fruit would not immediately affect the ohmic current however, but is simply a longer-term manifestation of growth hormone.

In a commercial fruit or nut orchard, the trees are densely populated. This helps shield them from the wind. These trees are maintained at low height. If they were taller, their absorption of sun light would not change but their wind exposure would increase, reducing production.

Evergreen trees such as redwoods and conifers do not lose leaves in the autumn. They do drop small amounts of needles and debris all year around. They apparently have set points that are not substantially affected by ground temperature. They are unique that way.

To conclude this chapter, a comparison between a synapse and a tree can be reiterated. With a neuron and its synapses, control of the numbers of receptors in the synapse is done by a mechanism residing in the cell and its membrane. The inner workings of the mechanism are not completely understood. The inputs of the mechanism are an ionic current and an ohmic current. The ohmic current is proportional to the membrane potential, and the ohmic current is the controlled variable. The difference between the ohmic current and the set point, its equilibrium point, is the error which is to be reduced to establish homeostasis. The error is controlled, up or down, by increasing or decreasing the number of receptors. The inputs of the neuron are neurotransmitters whose effect on the membrane potential are weighted by the number of receptors and the neurotransmitter concentration. The numbers of receptors are adjusted up or down by the mechanism such that the error is reduced. The ionic current is the power supply for the mechanism. Control of the number of receptors, the weights, is well modeled by the % Hebbian-LMS algorithm.

In the tree, the "neurotransmitter," the "input signal," is sunlight. The receptors are the leaves. The "ionic current" is the sun-driven hormone flow which is proportional to the number of leaves and the sunlight intensity. The "ohmic current" is the ohmic current caused by wind blowing. The "error" is the difference between the ohmic current and its set point. Error reduction is accomplished by increasing or decreasing the number of leaves by a mechanism. The mechanism is located just below ground, incorporated in the tree trunk. It is the keeper of the set point. The inner workings of the mechanism are not understood. Its inputs are only the ohmic current and the sundriven hormone current. This hormone current is the power supply for the mechanism. The mechanism's outputs are hormones that cause growth or decay in the number of leaves. Error correction establishes homeostasis, controlling the number of leaves and the size of the tree. The error correction process is well modeled by the % Hebbian-LMS algorithm.

The mechanism at the base of the tree has a fundamental purpose, to control growth or decline, primarily to regulate the overall size of the tree, invoking homeostasis. If the mechanism sends a growth signal up through the trunk of the tree, where would the growth take place? If a new branch erupts from the trunk, what determines its location? On the new branch, where would the twigs sprout? Where would the leaves form? The mechanism does not control all this. Local control systems all over the tree decide answers to these questions. There is a lot of "intelligence" all over the tree. It seems that similar intelligence is present in animals, utilized during development and wound healing.

The mechanism does not "care" about the location of new branches. All it "knows" about is its hormonal current and its ohmic current that result from the branches. To maintain homeostasis, the mechanism controls the ionic current by sending growth and death signals up the trunk. These are not instantaneous currents but currents averaged over weeks, months, and years.

% Hebbian-LMS theory applied to neurons and synapses is an implementation of Hebbian learning by means of the LMS algorithm. Hebbian learning is accepted in the field of neurobiology. In an extended form, it is supported by experiments such as those by Bi and Poo. The LMS learning algorithm is accepted in the field of digital signal processing. It is used by billions of people everyday, when logging into the internet or using a cellphone. % Hebbian-LMS theory is on solid ground. It accounts for homeostasis in living neural systems and offers an explanation for many phenomena relating to addiction and mood disorders. Extended to hormonal systems in the body, it provides new insight into homeostasis and feedback control of physical variables such as blood salinity, blood pressure, and body temperature. It introduces new concepts in control theory.

Nature seems to use the same tricks over and over again. Accordingly, this chapter applied % Hebbian-LMS theory to growth in plants and trees. It introduces concepts such as ohmic current and sun-driven hormone flow as inputs to a mechanism that establishes homeostasis to the growth of trees and their leaves. These concepts were introduced by analogy to phenomena in animal systems, and they have not yet been observed experimentally in plants. They can be regarded as a prediction.

The ideas and concepts put forth in this chapter are intuitive and do not exist in the literature. They explain certain phenomena related to tree growth. If it turns out that the ideas have merit, they suggest ways of increasing agricultural productivity.

22.2 Roots

The root system of a plant or tree is often more complex and convoluted than the branching of the above ground plant or tree itself. One can observe this if a plant is uprooted or if a tree is uprooted or pushed over by a hurricane, typhoon, or cyclone. Plant biologists study root systems and develop theories of how and where roots grow, and how they compete with neighboring plants.

Studying pepper plants growing singly or growing in neighboring pairs, Cabal et al. have discovered that roots near the base of the plant grow outward and extensively near the base of the plant when solitary, but refrain from growing outward toward each other when in pairs. The plant's objective seems to be avoidance of competition. The Cabal et al. paper was published in the December 4, 2020 issue of Science magazine and is entitled "The exploitative segregation of plant roots." [1].

For a solitary plant, the root growth "decisions" are simple. For a neighboring pair of plants, the growth decisions are more complex. Analytical models based on game theory are proposed. The model assumes complex decision making by the plants. The model explains observed phenomena and has real credibility.

A much simpler model is proposed here that will equally account for the phenomena observed by Cabal et al. This model requires much less intelligence on the part of the plant or tree.

The ohmic current hypothesized as flowing down the trunk of a tree must flow into the earth via the root system. A cross-section of a root shows the core to be soft and damp, highly conductive, and an outside sheath that covers the root and is probably not very electrically conductive. The hypothetical conclusion is that most of a root's ohmic current flows from the tip of the root. The higher the current, the greater the growth at the tip. The solitary tree or plant will have roots growing in every direction so that well separated adjacent roots will not interfere with the earth's voltage gradient such that current of one root does not diminish the current of the other root. The roots spread and grow new roots until the total current, split among the roots, discharges into the earth making the currents small at the root tips, and growth then slows or stops. That will limit the spread, but done such that they avoid competing electrically.

If there were several years of drought and a tree needed to be watered, and if the water were applied over one sector at the tree's base, not all around, the roots where the water is will grow and the roots in the dry earth may atrophy. How does this work? Do the roots have sensors that detect water and decide to grow where the water is? That seems unlikely. More likely maybe that the wet earth becomes highly conductive and the ohmic current readily flows into the roots in the wet sector and not into the roots in the dry sector.

More recently, household sewer lines are being made of PVC plastic. These lines do not leak. Older sewer lines made of sections of cast iron pipe soldered together develop leaks at the joints because of corrosion and earth subsidence. Roots of adjacent trees will find the defective joints, grow right into them, and plug the sewer line. Raw sewage is good plant food. Are the roots really smart in seeking out food? Probably not. They are probably growing where there is low electrical impedance to ground. The leaking fluid makes the earth highly conductive, and the metal pipe and the fluid therein is a solid connection to ground. (When the pipe plugs, the home owner calls "roto-rooter").

The why and wherefor of root growth proposed here is a simple algorithm. It does not depend on tree "intelligence." It simply says that root growth takes place in response to the flow of electric currents.

No one knows for sure what causes roots to grow. Explanations and models exist. The model described here is highly speculative. So are all the other models.

22.3 Summary

In this chapter, an analogy is drawn between the synapse of a neuron and an entire tree. Ohmic currents in the tree trunk, believed to be a result of static charges from air blowing on the leaves, will cause a growth signal to the branches and leaves if the ohmic current is less than homeostatic and will cause a retreat signal if the ohmic current is greater than homeostatic. Homeostatic regulation on the size of the tree comes from an adaptive algorithm, possibly % Hebbian-LMS.

For the tree, sunlight is both the power source and the input signal. The leaves receive the power and at the same time are the receptors for the input signal. This signal is proportional to the intensity of the sunlight multiplied by the number of leaves. The rate of growth is proportional to the amount of power absorbed from the sunlight on the leaves. At night, growth stops. There are many more elements to the analogy and they seem to fit both trees and synapses.

Counterfactual is a word with many definitions. One such definition is extrapolation beyond facts. This chapter extrapolates, it imposes hypotheses about upregulation and downregulation of neuroreceptors in the synapse upon plants and trees. Their leaves are the receptors. This unproven extrapolation predicts certain aspects of plant growth that seem to fit natural observations. All this is yet to be proven.

22.4 Questions and Experiments

The following experiments challenge the hypotheses of this chapter and if successful offer practical agricultural enhancements.

1. Figure 22.3 shows an apparatus for speeding the growth of a tree and increasing its size and number of its leaves. A metal ring is placed on the tree trunk. Current injected into this ring will add to the ohmic current. Depending on amplitude and polarity, this current can be made to cancel the ohmic current or to partially cancel it. Reducing the sensed ohmic current will cause the tree to grow faster. Reversing the polarity of the injected current, the sensed ohmic current can be increased. Increasing the ohmic current will cause the tree to slow or stop growing or to possibly die back.
   The apparatus of Fig. 22.3 has a battery, a polarity reversing switch, and ground conductors. A series resistor R will limit the current. The current should be quite small. With R being a large resistance, perhaps 0.1 megaohms, the placement of the ring contact may not be critical. With a battery voltage of a hundred volts, a current of 1 milliamp will result, and that might be a good place to start. The placement of the ring and the placements of the ground conductors might need to be varied, as well as the current level and its polarity.
   Be careful with the battery. One hundred volts dc can be dangerous. A smaller voltage could be used with a smaller valued resistor, but then the placement of the ring contact and its coupling to the tree would be more critical.
   If rate of growth is enhanced with the various parameters optimized, that would

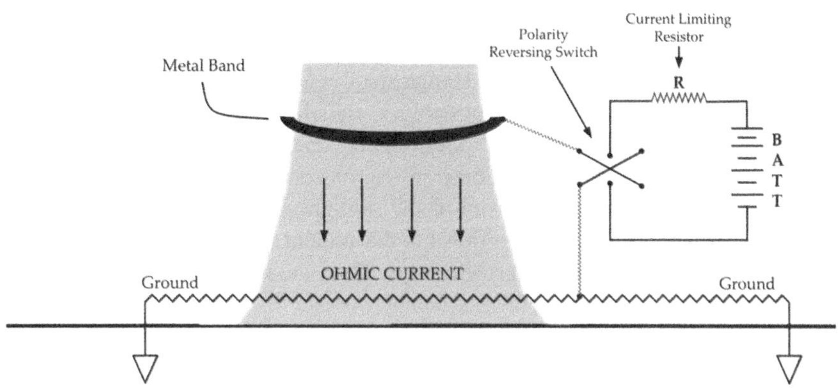

Fig. 22.3 Electrical stimulation of a tree trunk

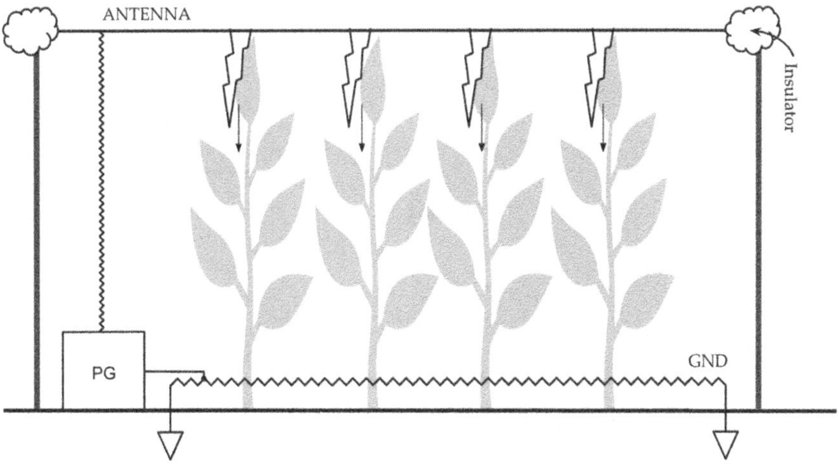

Fig. 22.4 Electrical stimulation of a corn field, capacitive coupling

be amazing. In an orchard, production of tree nuts or production of fruit could be increased. Application of current would need to begin in early spring, before buds appear. Then it could have influence over the development of branches, twigs, buds, flowers and leaves.

2. Another experiment is pictured in Fig. 22.4. Growth in a cornfield is enhanced by inducing currents into the corn stalks. An overhead wire, an "antenna" is connected to a pulse generator that is capacitively coupled to the corn stalks. Variables are pulse shape, frequency, amplitude, and polarity. With capacitive coupling, no average current can be induced in the corn stalks, but sharp asymmetrical pulses may cause the cornstalks' mechanisms to think that the ohmic current has been reduced and this will produce accelerated growth. Making this work is a distant possibility, but worth trying.
3. This experiment is designed to test the hypothesis that wind blowing on the leaves generates static charges that discharge into the tree and cause an ohmic current. Refer to Fig. 22.5. An ohmic current in the tree trunk would create a voltage gradient. Drive two spikes (large nails) into the trunk and connect to a microvoltmeter. Check to see if there is a voltage gradient and check it with the wind blowing and with no wind. Also check to see if sunlight has an effect on the voltage. The measured voltage might be millivolts, microvolts, or hundredths of a microvolt. This it not known presently.
4. The purpose of this experiment is to determine if it were possible to "fool" a deciduous tree regarding the season of the year. This is a test of the idea that the set point of a deciduous tree is temperature sensitive, varying with the earth temperature at the base of the tree.

   Encircle the base of the tree with a long spiral of tubing buried about 1 foot or about 30 cm below the surface of the earth. In the middle of the summer, pump

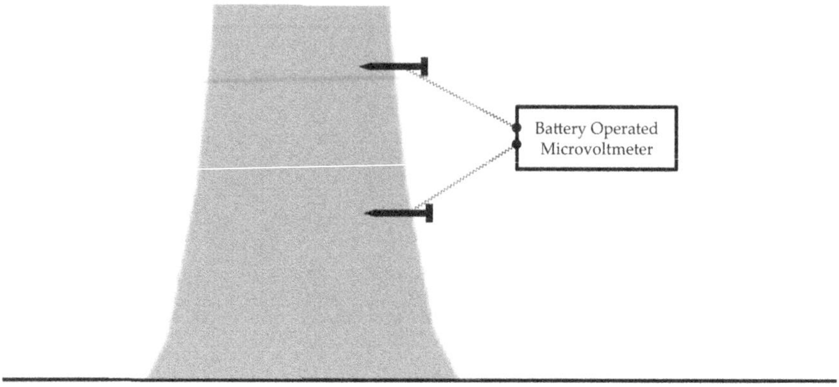

Fig. 22.5 Measurement of voltage proportional to ohmic current

ice water through the tubing for several weeks and observe if the leaves turn and fall off. At the end of winter and beginning of spring, pump hot water through the tubing for several weeks and observe if buds and leaves form prematurely.

These experiments, each with many variables will require time. Doing many experiments in parallel, each with different experimental parameters, will speed the testing. If any of these experiments were successful, that would lend great credibility to this chapter's hypotheses.

Reference

1. Cabal, C., Martínez-García, R., de Castro Aguilar, A., Valladares, F., Pacala, S.W.: The exploitative segregation of plant roots. Science **370**(6521), 1197–1199 (2020)

Part VII
Norbert Wiener

Introduction to Part VII

Part VII focuses on Norbert Wiener, a great mathematician and the inventor of the word "cybernetics." His contributions to mathematics are many fold, as are his contributions to science, technology, and engineering. Of special interest here is his work on statistically optimal signal processing filters, known as Wiener filters. He introduced least squares methods to the design of filters for noise reduction and signal prediction. Least squares methods probably started with Gauss, but Wiener formulated this in a way that enabled filter design. Analytical formulas were devised for this purpose that were in accord with Wiener filter theory.

In Wiener's day, all filters were analog. Digital versions came later. Adaptive filters came even later, and they are basically digital Wiener filters that have the capability for self design making use of learning algorithms.

Wiener was MIT's most famous professor. He was a colorful character. He was seen by everyone wandering the halls of the interconnected buildings. He was on the campus all the time, puffing on his cigar, wandering the hallways. This must have been his method for creating the right conditions for thinking.

There were many stories about him from innumerable witnesses. He was funny to watch sort of waddling like a duck, walking very slowly down the hall. Many stories about him were well known on the campus. Some were real classics.

A selected set of Wiener classics are presented here. In addition, there are stories of personal interactions and observations by Prof. Widrow. Students in Widrow's Stanford classes loved to hear these stories. Telling these stories was prompted by mention of Wiener in connection with a subject of discussion in the class. The stories are presented here to give the reader an unusual view of the inventor of cybernetics.

These stories came from direct experience and as such are presented from a first person point of view.

Chapter 23
Wiener Filters and Adaptive Filters

Abstract Wiener was the first to develop a means of statistical filter design. Previous methods were deterministic. Wiener filters were optimal in the least squares sense. Applications were to filtering noise, to prediction, and to prediction of signals in the presence of noise. Wiener design required knowledge of autocorrelation and crosscorrelation functions of signal and noise. Adaptive filters are an offshoot of Wiener filters. They do not require knowledge of input signal statistics. These filters self design, based on input signals, to create optimal least squares solutions. They learn from the input data. They are used for filtering noisy signals, prediction, noise cancellation, adaptive antennas, and adaptive control systems.

23.1 Introduction

This book is called Cybernetics 2.0. The central theme of this book is homeostasis in living systems. Homeostasis is a critical element of the various regulatory systems in the brain and in the body. Norbert Wiener recognized this in his book "Cybernetics." Homeostasis was a central theme of his book. It has to do with feedback, control, and communication of physical variables all over the body.

Wiener's Cybernetics was published in 1948. This book had a subtitle "Control and communication in the animal and the machine." Wiener coined the word cybernetics, taken from a Greek word for steersman, governor, pilot, rudder, or governance. The word cybernetics has currently been shortened to cyber and used as a generic term for computer related things and actions. We now have cybersecurity, cyber warfare, cybercrime, etc. Wiener would probably be horrified by some of the uses of his cyber [1].

Wiener's 1948 Cybernetics was based on knowledge at that time of physiology and homeostasis, and mechanical and electronic feedback control systems. He was a visionary and many of the subjects that he wrote about have since become practical reality. His work was philosophical, not getting into any detail of how the mechanisms actually worked. His book was a major contribution to science.

B. Widrow, *Cybernetics 2.0*, Springer Series on Bio- and Neurosystems 14,
https://doi.org/10.1007/978-3-030-98140-2_23

Wiener made other major contributions to mathematics, statistics, and engineering. He worked on Fourier theory, harmonic analysis, statistical signal processing, and nonlinear systems analysis. These were all highly original works and had great influence in their fields, influence that exists to this day.

23.2 The Wiener Filter

In 1949, Wiener published a book entitled "Extrapolation, Interpolation, and Smoothing of Stationary Time Series." This book described a signal processing method that today is known as the Wiener filter. The development of the subject began during World War 2 when Wiener was working on radar signals, making them less noisy. The book was revolutionary and brilliant.

Up to the time of its publication, filters, which were analog, were designed to be low-pass, high-pass, band-pass etc. Realization of these filters often involved synthesis techniques called Butterworth filtering or Chebyshev filtering. The mathematics was highly developed and well understood by design engineers. The input signals were deterministic.

Wiener's approach was quite different. The input signals were stochastic. His theory allowed the design of filters that were statistically optimal. No one else had ever come up with design techniques for random inputs. Wiener's filters were linear and optimal in the least squares sense. He developed a general theory that gave a formula for the impulse response of the optimal linear least squares analog filter. A simple formulation realized filters that could be non-causal, having two-sided impulse responses. A more practical but more complicated formulation gave the impulse response of a one-sided filter, constrained to be causal. The optimization theory was based on calculus of variations.

Typical Wiener filtering applications are to noise filtering and prediction. Noise filtering involves separation of signal and noise components from a noisy input signal. Prediction can be applied to an input signal that is correlated over time, predicting the future of this signal.

Given a noisy input consisting of signal plus noise, the Wiener filter will produce an output that is a best least squares estimate of the true signal. To do the Wiener filter design, the autocorrelation function of the noise and of the signal needed. The crosscorrelation function of the signal and the noise is also needed. Given these correlation functions, they can be "plugged into" the Wiener filter formula and the result will be an impulse response of the optimal linear filter. This filter will do the best possible job of delivering an output that is an estimate of the true signal, even if the spectra of the signal and noise overlap and even if there is correlation between the input signal and noise components. A noisy input signal will still give one a noisy output signal, but the output is improved over the input in the best way. Nothing like this had ever been done before.

Prediction is another application. An input signal that is correlated over time can be predicted with minimum mean square error by a Wiener filter. One needs to know the input autocorrelation function and how far into the future the prediction should be. Plugging into the Wiener formula, the result will be the impulse response of the best linear predictor.

Another problem is prediction of the signal component of a noisy input. Given an input consisting of signal plus noise, knowing how far into the future one wishes to predict the signal, knowing the autocorrelation function of the input signals component, knowing the autocorrelation function of the input noise component, knowing the crosscorrelation function of the input signal and the noise, one can plug all this into the Wiener filter formula and get the impulse response of the best linear filter for predicting the input signal component with the minimum amount of output noise. The filter output will be a best least squares estimate of a predicted version of the input signal component. Noise filtering and prediction are addressed simultaneously.

My first introduction to Wiener filtering took place in 1953. I was a graduate student at MIT and took a course from Professor Y. W. Lee on the subject "Statistical theory of communication." Professor Lee was a disciple of Professor Wiener. Wiener was in the Mathematics department. Lee was in the Electrical Engineering department. Professor Lee was translating Wiener's theories into the language of electrical engineering.

Lee taught Wiener filter theory for analog filters with continuous inputs. In 1957, a set of notes were written by my colleague Robert W. Sittler presenting Wiener filter theory in discrete form for digital Wiener filters. I taught Wiener filter theory for many years with Sittler's notes as a text. The theory is much simpler in discrete form.

When doing homework problems for Lee's class, autocorrelation functions and crosscorrelation functions of inputs were given. I was able to do the homework problems, but I was troubled. In the real world, who is going to give you the correlation functions? How could you use Wiener filter theory in practice? I found answers to these questions later when I began to work on adaptive filters.

23.3 Adaptive Filters

I received the Sc.D degree in electrical engineering in June, 1956, at MIT. I joined the MIT EE faculty as an assistant professor. I was researching in the area of my doctoral thesis on quantization noise and teaching a course on what today would be called a combination of digital signal processing and digital control systems.

In the summer of 1956, there was an all summer long seminar on the subject of artificial intelligence at Dartmouth College. One of my colleagues, Ken Shoulders, told me about this. He said it is a conference on artificial intelligence. I asked him, "what's that?". He said, "I don't know but it sounds interesting." "Would you like to go with me?". So we drove from Cambridge, MA to Hanover, NH.

All the pioneers of the field were there. You could speak if you wished, otherwise listen. They were talking about making machines that think. I was captivated. I am still captivated to this day.

Back at MIT, I spent six months thinking about thinking. I came up with ideas that I thought were good. I realized that at least 25 more years would be needed before electronic components would be capable enough to begin building a thinking machine. (I was optimistic. Sixty three years have gone by and there is no thinking machine yet in sight).

I realized that a 25 years time horizon was too long for a successful career. But I had a "monkey on my back." I couldn't get the idea of a thinking machine out of my mind.

Teaching with Sitter's notes and knowing about digital Wiener filters, an idea came to me. Instead of trying to build a thinking machine, why not build a machine that can learn simple things. Why not a digital filter with variable parameters that could adjust itself to minimize mean square error. This would be a digital Wiener filter that learns and constantly varies its parameters to improve its least squares performance. The result was an adaptive Wiener filter.

My MIT graduate students and I were simulating adaptive filters on an IBM 701 mainframe computer. To my surprise, the mean square error turned out to be a quadratic function of the filter parameters. We were using stochastic gradients to optimize, to find the bottom of the quadratic bowl. We called the optimal solution the Wiener solution.

Figure 23.1 is a diagram of an adaptive digital filter. It is an FIR (finite impulse response) filter. The input signal is x_k, a sequence of input samples. The time index is k. The output signal is y_k. The weights are w_{1k}, w_{2k}, etc. The inputs to the weights are the present input sample x_k, the previous input sample x_{k-1}, the previous sample before that x_{k-2}, etc. The set of input signal comprise the input vector X_k. The set of weights comprise the weight vector W_k. The output signal is $X_k^T W_k$. The desired response is d_k. The error ϵ_k is the difference between the output and the desired response. The mean square of the error is a quadratic function of the weights. The gradient is a linear function of the weights. An iterative gradient algorithm, based on the method of steepest descent, can be used to find the Wiener solution, the weights that minimize mean square error, such as the LMS algorithm:

$$W_{k+1} = W_k + 2\mu\epsilon_k X_k \tag{23.1}$$

$$\epsilon_k = d_k - X_k^T W_k. \tag{23.2}$$

Other least squares algorithms could also be used for this purpose, but LMS seems to be the world's favorite.

A symbolic representation of the adaptive digital filter is shown in Fig. 23.2. This compact representation is convenient when drawing circuits that incorporate adaptive filtering.

Statistical prediction is one of the Wiener filtering problems. This can be done with an adaptive filter. When the input autocorrelation function is not given and all that is available is input data, the adaptive filter that learns the statistical characteristics of the input data can be used for prediction. A diagram of an adaptive predictor is shown in Fig. 23.3.

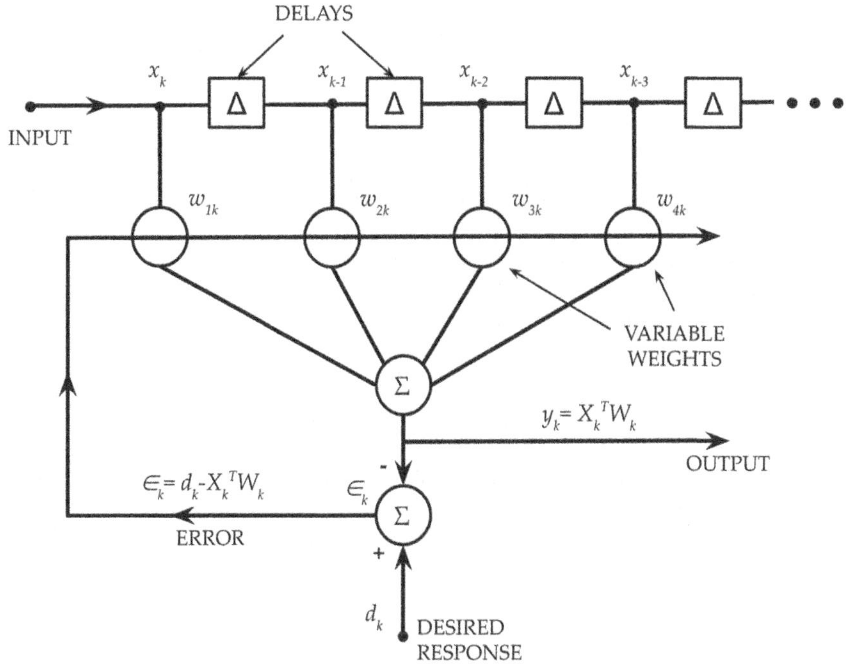

Fig. 23.1 An adaptive digital filter

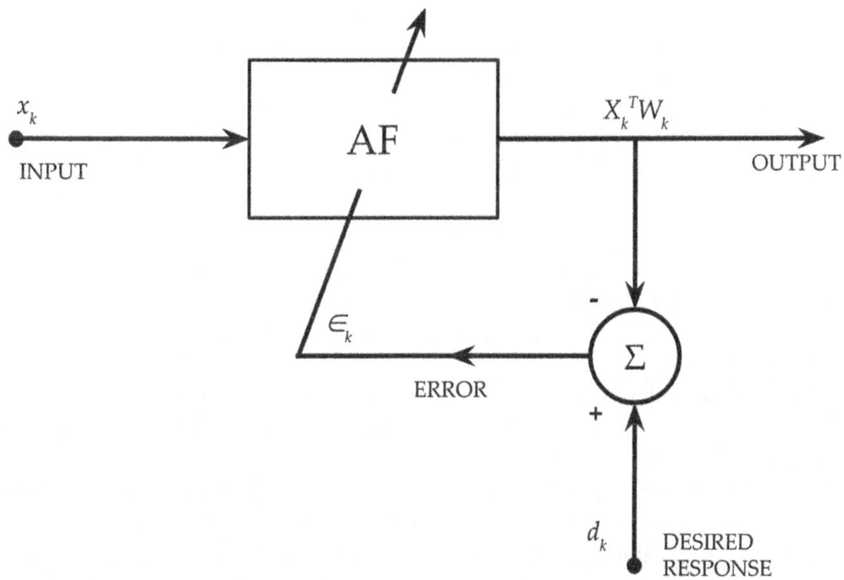

Fig. 23.2 Symbolic representation of the adaptive filter

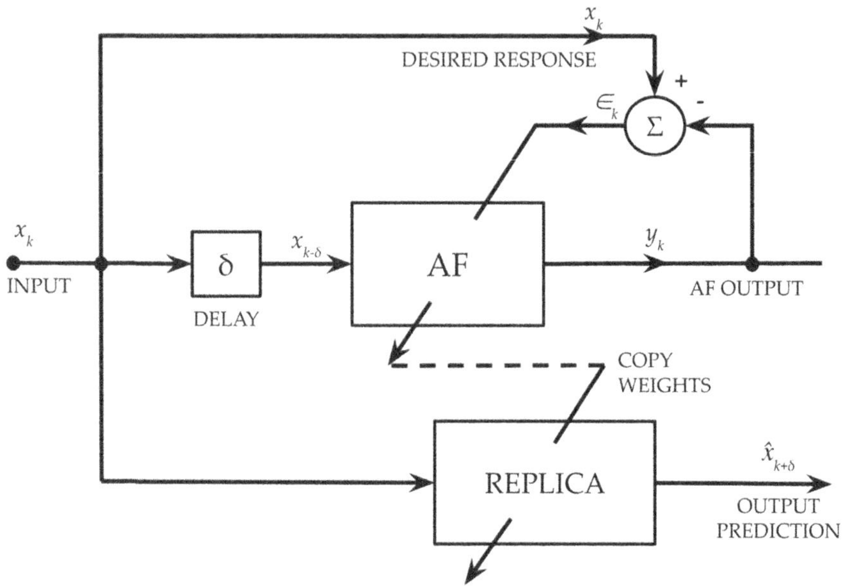

Fig. 23.3 An adaptive predictor

The input signal to be predicted is x_k. This signal is delayed to form $x_{k-\delta}$, which is presented as an input to an adaptive filter AF. The desired response for this filter is the original input signal x_k without a delay. The filter AF will adapt to become a best least squares predictor of the input signal. It learns to overcome the effects of the delay δ and becomes a Wiener predictor of the input, δ units of time into the future. The weights of the adaptive filter are continually copied into a "replica" filter whose input is x_k without delay. The output of the replica filter is the useful output, a best least squares prediction of the input x_k, δ units of time into the future.

A common example of the application of the adaptive predictor would be the prediction of the Dow Jones average. Each day at closing bell, the Dow Jones value could be taken. Today's Dow Jones sample is x_k. Streaming the samples into the predictor, adapting with each new input sample, the predictor will gradually converge and start predicting.

A predictor of this type was not anticipated by Wiener. During his time, there was no such thing as an adaptive filter, nor did digital filters exist. He could not have devised the predictor shown in Fig. 23.3.

Wiener filters have input signals, and they deliver output signals. The idea of the desired response was Wiener's. The desired response was hypothetical, it did not exist per se. According to Wiener theory, the desired response as a signal was not needed. What was needed was the crosscorrelation function between the filter input signal and the hypothetical desired response [2].

The situation is quite different with adaptive filters. The correlation functions are not known. What is needed is the input signals, input data, and the desired response as an additional input signal. A question naturally arises, if one has the desired response signal, why would one need the adaptive filter? The answer is that although the desired response exists, one still needs the adaptive filter. An example is the adaptive predictor. A big issue with the practical application of adaptive filters is, how does one get a desired response? It is hard to answer that question in general, but one can learn how to do this by seeing samples of application. There are many.

One application is the adaptive predictor, diagrammed in Fig. 23.3. In this case, the desired response for the adaptive filter is the original input signal, without delay. The output of the adaptive filter is not used except for comparison with the desired response to create an error signal. The error signal is necessary for the purpose of adaptation and learning. The output of the adaptive filter is not a useful output, but the filter itself is useful. Its weights are continually copied into the replica filter that in turn produces the useful output, the prediction of the input signal. So here is a case where one has a desired response but the adaptive filter is still needed.

The book "adaptive signal processing" by Widrow and Stearns [3] has many applications. This book was published in 1985, is still in print. It was the first book on this subject and is arguably the simplest and easiest to understand. Some of the applications are adaptive equalization of communication channels, adaptive noise cancelling, adaptive antenna arrays, and adaptive inverse control systems. Many books have been published in this field and they describe many more applications. One way or another, a desired response must be found for each application. In some cases, a good deal of invention is required to find a desired response signal [4].

23.4 Summary

Wiener filter theory addresses certain signal processing problems such as prediction, separation of signal from noise, and combinations of these. His design method was based on linear analog filters whose outputs were optimal in the least squares sense. He used the calculus of variations to develop formulas that gave the impulse response of the optimal filter. His formulas require knowledge of the autocorrelation functions of the input signal and of the noise, and of the crosscorrelation function between the signal and the noise. Statistical filter design had never been done before.

A discrete form of Wiener's theory was created for design of optimal linear digital filters. The mathematics is based on ordinary calculus, setting derivatives to zero for optimization. Wiener's theory is much simpler in discrete form and is implementable with software.

Adaptive filters are self-designing Wiener filters. They were first developed by B. Widrow and students, first at MIT in 1956 and further, incorporating the LMS learning algorithm, in 1959 at Stanford [5]. These filters seek the Wiener impulse response by means of an iterative process using input data, not correlation functions. Wiener theory is in the background, not directly used in the design process. The filter leans to adjust its own parameters with the goal of minimizing a measured mean square of its error signal.

23.5 Questions and Experiments

Make a log of the Dow Jones average closing value each day for the past 3 months. Code an adaptive predictor with 10 weights. Train the predictor with the oldest 2 months of data, test predictor with the most recent month's data. Do you think the stock market is predictable?

References

1. Wiener, N.: Cybernetics, Second Edition: Or the Control and Communication in the Animal and the Machine. The MIT Press (1965)
2. Wiener, N.: Extrapolation, Interpolation, and Smoothing of Stationary Time Series. The MIT Press (1964)
3. Widrow, B.: Adaptive inverse control. In: Adaptive Systems in Control and Signal Processing 1986, pp. 1–5. Elsevier (1987)
4. Widrow publications. https://isl.stanford.edu/~widrow/publications.html. (Accessed on 06/18/2021)
5. Widrow, B., Stearns, S.D.: Adaptive Signal Processing. Prentice Hall, Englewood Cliffs, NJ (1985)

Chapter 24
Norbert Wiener Stories

Abstract Norbert Wiener was the most famous professor at MIT when I was a student there and an assistant Professor there. He was an eccentric. Brilliant and kind, he was smoking a cigar, deriving equations, in a cloud of smoke. Some of these stories about him were based on my personal experiences observing him and once helping him. Seven stories describe a man totally absorbed in his work, at the peak of his renown.

24.1 Introduction

When I was a student at MIT between 1947–1956, and a faculty member between 1956–1959, Norbert Wiener was our most famous professor. He was a lovable eccentric. He seemed completely oblivious and disconnected from the world around him. He loved being famous. He was very happy when people recognised him and knew who he was. He was an egotist, but not disliked, because of his innocence. On the contrary, he was a most likable gentle person.

He was a pacifist despite his important contributions to fire control of artillery and anti-aircraft guns during the second world war. This experience was fundamental to his development of Wiener filters.

He was a child prodigy. He received his undergraduate degree from Tufts University in 1909 at the age of 14. He received his Ph.D. in philosophy from Harvard University at the age of 18. He was a genius. He was fluent in a dozen languages. He made fundamental contributions to the statistical theory of brownian motion and harmonic analysis. He was on the MIT faculty for many years and lived to the age of 69.

Everyone at MIT knew some Norbert Wiener stories. The stories told here are some of the classics that everyone knew, and stories that I can relate from direct personal experience.

B. Widrow, *Cybernetics 2.0*, Springer Series on Bio- and Neurosystems 14,
https://doi.org/10.1007/978-3-030-98140-2_24

24.2 Professor Wiener at MIT

Norbert Wiener was one you couldn't miss. He was a rather enormous man—I would estimate his weight at around 300 pounds. He had a goatee beard. How tall was he? I would say probably about 5 foot 9 inches. We'd see him there every day, and he always had a cigar. He'd be walking down the hallway, puffing on the cigar, and the cigar was at angle theta—45 degrees above the ground. And he never looked where he was walking. Because of his size, I don't think he could see his feet. So he would walk along with his head cocked back and his cigar at angle theta, down the long, long hallways at MIT—all the buildings at that time were interconnected with hallways. It seemed that the only way he could guide himself and not bump into the walls was by looking up to see where he was in the corridor. I think he was navigating based on the line between the ceiling and the wall. But he'd be puffing away, his head encompassed in a cloud of smoke, and he was just in oblivion. Of course, he was deriving equations. So he'd come down to the end of the hallway, at the end of which were steps going down, and the steps were concrete, and the edges were steel. It was an unforgiving hallway, and if he ever took a fall, it would not be very nice. And here would be Prof. Wiener coming down to the end of the corridor, and he'd be puffing away on his cigar, and looking up and not looking down—not knowing where the heck he is—and his head busy with mathematics. And now what do you do? You can see he's going to kill himself—he's going to fall down those steps—but if you disturb him, you might break his train of thought and set science back like ten years! There was always that problem.

24.3 The Absent-Minded Professor

This story is a classic. Everyone knew this one. Wiener was the absolute absent-minded professor. I heard one story about him—whether it's true or not, I don't know—that showed this very well. It was around lunch hour, and he was going on the sidewalk across a green on the campus, heading away from the communal dining hall cafeteria, called Walker Memorial, where one had lunch. As usual, he doesn't know where he is, or where he's going, or what he's doing. So he's walking along, and he stops a student and says to him, "Am I going to Walker Memorial, or am I coming from Walker Memorial?" And the student said, "You're coming from Walker Memorial." And Prof. Wiener said, "Oh, good, then I must have eaten."

24.4 The Quiz

This is another Wiener classic. Teaching a math class, Professor Wiener gave a quiz. When he returned the papers, one of the students in the class raised his hand and asked to see how one of the quiz problems is solved. Wiener said, OK. He erased the blackboard and stood in front of it for several minutes and proceeded to write the answer to the problem on the blackboard. The student raised his hand again and said that he didn't understand how the problem was solved. Wiener was puzzled, but he said that he will solve the problem another way. He erased the blackboard and stood in front of it for several minutes, and scratched his head. Then he wrote down the answer. He turned to the class with a big smile and said, "there, I solved it another way."

24.5 The IBM Typewriter

This is a personal story, not a classic. My wife told me this story about him from our days at MIT. It was around 1955, and at that time, she was my girlfriend. I was in a laboratory group, and she was the secretary of our group. We used to call her "The Chief" because she ran the place. We were located on the third floor of Building 10, which is the building under the famous big dome at MIT. In those days almost the whole electrical engineering department was in Building 10. Now, of course, it's spread out all over the place. Well, she always had her door open, and it was right on the main corridor, and she'd often see Prof. Wiener come wandering by, puffing on his cigar, in a cloud of smoke. One day he just came wandering into her office. She had a typewriter called an IBM Executive, which at that time was state of the art in IBM typewriters—IBM was primo in those days in typewriters. She had removable keys so that she could pull the keys out and replace them with keys that had Greek letters to type mathematics. Norbert Wiener was fascinated with her typewriter—the fact that you could replace the keys and type mathematics with a typewriter, without having to handwrite the Greek symbols in the manuscript. So he just had a big talk with her about that and wanted her to demonstrate, and she did. Then he just wandered out and disappeared down the hallway.

24.6 Wiener at the Smith House

I had two close friends at MIT—one was Mark Beran and the other was Victor Mizel—and the three of us were in the same dormitory. On Sundays, they didn't serve food in our dormitory, so we had to go out and get food. On one particular Sunday, we didn't get up too early, so we went out to get brunch. A typical place where we went to get brunch was a place called The Smith House, which was located on Memorial Drive, facing out over the Charles River. My understanding is that The Smith House is no longer there—it's in The Smith House heaven. In any event, we all got into my car, which was a Plymouth, and drove over to The Smith House to get brunch. As you came in through the main entry from the parking lot, right there was this L-shaped bar—actually more like a counter—with stools. Now you could eat at the lunch counter or you could go straight back and get a booth, which the three of us did. But when we went by the lunch counter, who did we see perched up on a stool, eating his brunch, was Norbert Wiener. Well, I looked at that and I said to myself, "What an unlikely looking scene that was! To see the distinguished Prof. Wiener sitting on a stool having brunch at The Smith House." It looked like he could fall off easily. So we went back and had brunch, and when we finished, we paid the check and took off for the parking lot. I had the keys to my car out, and I was dangling them, talking with the other two boys, and we noticed out in the parking lot, sitting in his battered old Chevrolet, was none other than Prof. Wiener. As we were going right by him to get my car, he rolled down the window and said to me, "I say, I can't get my car started. I wonder if you could give me a push?" I said, "Prof. Wiener, absolutely! I'd be happy to give you a push." As soon as I called him "Prof. Wiener," you could see a big smile on his face. He was happy because people knew him and he was with family. I think he assumed that would be the case, I guess because we looked like students. So I got my little Plymouth behind the battered old Chevrolet. In those days, practically all cars were stick shifts. So I got behind his car, bumper to bumper. I was in first gear because I needed the force to push him. I started to let the clutch up, and I found my engine slowing down and practically ready to stop—and he and I are not moving. So I told the boys in my car, "I know exactly what's happening—he's got that car in first gear and he's got the clutch up." And that's exactly what you don't do to get a car rolling and started. You've got to get it rolling first and then gradually let the clutch up. So I said to the boys, "What should I do? Should I go try and explain to him what to do—how you start a car by pushing? He'll never understand anyhow. So I'll just gun it and practically burn my clutch out, but we'll get him going and it'll be easy." So I just gunned it, and let the clutch out gradually, and slowly I started to creep forward, and he started to creep forward. And all of a sudden, after just a few feet of pushing, his engine caught. You see, when you're in first gear, and the clutch is up, even if the car is moving slowly, the engine is turning fairly rapidly. And because he was in first gear and had his foot on the gas, he took off like a rocket. Meanwhile, he's blowing his horn and waving his hand to thank me, and because of his great weight, his car lurched forward under him. Now we just deduced this . . . we didn't talk with him about it . . . as he was in his car and we were in ours. So as his car lurched forward, the speed of the car under him caused his body to lurch backward, and that probably took his foot off the gas pedal. And when you're in first gear, the moment you take your foot off the gas pedal, the engine braking practically stops you. So the car practically stops and his body lurches forward, and his foot hits the gas pedal again, and he goes lurching backward again. We can hear his engine going "Vroom . . . vroom . . . vroom." Meanwhile, he's still busy waving his hands—"Thank you, thank you!"—and blowing his horn, and he goes shooting out of the parking lot and right onto Memorial Drive, with two-way traffic in each direction and no island in the middle. And we can hear the brakes screeching and the horns blowing, and we can see he's still waving to us, and I said to the boys, "My God, we've killed him!" You know, Prof. Wiener wrote a book on cybernetics, which has to do with how feedback in man and machine is similar, and how all this works. And here you've got man and machine with feedback, and the thing was practically unstable and into oscillations, and he had no clue. Somehow he kept going forward and managed to turn left onto Memorial Drive, crossing three lanes of traffic, and the last we saw he was headed back to MIT.

24.7 Norbert Wiener on Harry Truman

It was around 1955 that Professor Wiener held a seminar that I attended. He was rambling along when he made a sudden strong statement like this: "The most dastardly act ever committed by man was done by President Harry Truman ordering the atomic bombing of Japan." I was taken aback. Wiener was a Jew. Hitler murdered six million Jews. That was about as bad as it gets. Of course, Wiener knew this.

I discussed the atomic bombing with my father. He said that this ended the war and saved millions of Japanese lives and an enormous number of lives of American soldiers. My father said that what Truman did was an act of great courage.

I thought further about this. If the United States had the bomb and didn't use it, we would be a "paper tiger." No one would be afraid to used it against us. There would be no nuclear deterrent. But since we of all people used it and the horror of it was revealed, no country has used a nuclear weapon since. The trick is to keep these weapons out of the hands of terrorists.

24.8 Wiener (and I) at the 1960 IFAC World Congress in Moscow

The International Federation of Automatic Control (IFAC) held an enormous conference in Moscow in 1960. This was the first time that the Soviets allowed their scientists to mingle with scientists from other countries on a significant scale. Allowing Americans into Russia was most unusual. There were 2000 foreign visitors and 2000 Russians at the conference. Norbert Wiener was one of the Americans. So was I.

In 2001, I was interviewed by Ms. Barbara Field on the subject of the IFAC meeting, which was historic. The interview was published in the IEEE Control Systems Magazine, June 2011, pages 65–70. This was almost 20 years ago. My recollection of events from 1960 was better in that interview of 2001 than today, but I remember that the Russians were all excited about cybernetics. That was all they wanted to talk about. Norbert Wiener was the star of the show.

I probably could have talked with Professor Wiener about my work on adaptive filters. But he seemed to be off in dream land with his head shrouded in cigar smoke, glowing about cybernetics. That would have been an interesting connection between least squares and adaptive least squares, but it never happened.

I presented a paper on adaptive filtering. No one seemed to have any idea of what I was talking about. I described gradients of the mean square error function, derived crudely before the invention of the LMS algorithm, which at that time was only one year old. I didn't want to give the Russians the LMS algorithm just yet.

My recollections of the trip to Moscow, the IFAC conference, and observations of Wiener were best recorded by the Barbara Field interview. The following story is based on that interview.

A number of years later—I had already gone to Stanford and Wiener was still at MIT, still doing all the usual stuff, as far as I knew—I had gone to the first IFAC Conference in Moscow. This was my first time going to Europe, in fact, the first time I'd ever been outside the United States. In those days, if you were traveling on contract funds, you had to fly a government-run airline called MATS, an acronym that stood for Military Air Transport Service, or something like that. So we flew to New York, and from there to New Jersey, to McGuire Air Force Base, where you picked up your airplane. We flew in a DC-6, a four-engine propeller plane, to London, with a stop in Labrador and another stop in Scotland for refueling. Those planes didn't have the range of a modern airplane. So we landed in London and spent a few days there. It was 1960, and there was still damage in London—neighborhoods that had been bombed out in the Second World War. From there, we flew to Paris, where we got on Air France to fly to Moscow, with a stop in Warsaw. The airplane was a French Caravelle, a twin jet, which was up and flying well before the Boeing 707. It was the first time I'd ever been in a jet airplane, and it was quite an experience. We landed in Warsaw and went into the airport terminal, a one-room terminal. Out the window we could see the Polish commercial air fleet—a whole bunch of DC-3's—and were trying to figure out where the Polish had gotten them. We realized they must have been leftover Lend-Lease equipment that we gave to Russia during World War II, and they ended up in Warsaw as the Polish commercial air fleet. There was sort of a gift shop where you could buy little Polish things, and to our surprise, the currency in the little gift shop was dollar bills. I was amazed, as this was 1960, during the height of the Cold War, and was feeling sorry for those dollar bills so far from home. What happened next is that the French pilot who flew us to Warsaw took the copilot's seat, and this Russian pilot—a short, squat guy with medals all over his chest—came strutting out of the Warsaw airport with his black briefcase to fly that Caravelle. I guess it was not possible for anyone to fly over Russian territory except a Russian pilot. Now the time frame was June, as I recall, and two weeks before that flight, a U-2 aircraft piloted by Francis Gary Powers was shot down over Russia, and I was saying to myself, "I've got to be crazy." But everything seemed up and on, so we went into Moscow. It was extremely hot when the Caravelle landed. Things were not like they are now, where they plug in an air conditioning unit and you immediately get cold air on the airplane. It took quite a while from the time we landed before we could get on the bus to take us to the gate, and we were baking in that airplane. I mean, absolutely roasting! There was no air movement, nothing, and we're going nuts. The Caravelle is smaller than a DC-9 or the smallest MD-80. I think the DC-9 was modeled after the Caravelle, although the designers would probably deny it, but that was sort of the layout. As I was getting off the plane, I saw that up ahead of me was Prof. C. Stark Draper. Today at MIT there's a lab called Draper Lab, named after him. He was a heavyset man with rimless glasses and a very pleasant person. It was so hot, and we were all drenched and carrying our bags. I saw that he had a lot of hand carries, so I introduced myself to him and said, "Prof. Draper, can I help you with your bags?" "Oh, that would be wonderful!" he said, and I could see by the look on his face that he was grateful there was someone who knew him and could help him. I got his bag into the terminal, where it was much cooler. We got checked in through Customs, and the bus finally took us to our hotel.

All the foreigners were staying at a hotel called the Hotel Ukrania. The architecture of the Hotel Ukrania was exactly the same as that of the University of Moscow and of City Hall. In fact, there were a whole bunch of major buildings that all had a seemingly identical design—you know, "the system." I thought the hotel was quite fine, but then I was not much of a world traveler at the time, so I didn't have much to compare it to. This hotel had 2000 rooms, and the lobby was enormous. I was there chatting with some friends when in the distance I saw Prof. Wiener, wandering along, puffing on his cigar. He seemed to be talking to himself, muttering away. Now the Russians absolutely lionized Norbert Wiener. He wrote a book called Cybernetics, and in those days the Russians called the whole field of control theory "cybernetics," and the esteem they had for Norbert Wiener was far greater than that accorded him anywhere else in the world. In the United States, we didn't use the word cybernetics very much. At MIT, Prof. Wiener was lionized by engineers, but I think the fact that engineers were interested in his work was not a plus among his fellow mathematicians. In other words, if the mathematics threatens to be useful, it's probably not too good. Prof. Wiener was just tremendously overwhelmed, I suspect, by the reception he got from the Russians. I could see he was talking to himself while puffing on his cigar, so I told my friends, "I'm going to go and see what this is all about." Now, he's not going very fast, just waddling away like a duck, so I had no trouble catching up. As I got behind him, I could hear him muttering to himself, "I spawned all this." Now, he was a person who was not particularly modest. He really understood his accomplishments. But he was such a kind person, a lovable person—and he was such a fuddy-duddy— that everybody loved him anyway. Everybody knew that he was immodest, but he was immodest in the nicest way, so it didn't matter. I think there were at least 2000 attendees at the first IFAC conference, about half of them Russians and half foreigners. At that time, there were tremendous political problems between China and Russia. They had already had a total falling out. We would have all three meals in the Hotel Ukrania, and one day at lunch a friend asked me a rhetorical question. "When the Chinese commissars sit down to have lunch with the Russian commissars," he said, "what language are they speaking?" And I said, "Gee, I don't know. Probably Russian." He said, "No, the language they're speaking is English." Today, I think that's easier to accept, because English is so ubiquitous, but not back then. We found that our Russian colleagues spoke English with heavy accents, but they all spoke English. Our Chinese colleagues all spoke English as well. The politics was that the Chinese would refuse to speak Russian and the Russians would refuse to speak Chinese, so they all spoke "politically correct" English. We were their common enemy. With regard to Francis Gary Powers, in Gorky Square, the Russians had organized an exhibit of the U-2 airplane. This was a top-secret U.S. airplane. Eisenhower was president at the time, and before the U-2 got shot down, the Russians were publicly complaining about American aircraft flying spy missions over Russia. Eisenhower absolutely denied it and denied it, but they kept complaining. What happened was that the U-2 flew so high that the Russians couldn't shoot it down. Then they developed missiles that went up higher, so along came the U-2, and, bingo, they shot it down. Poor President Eisenhower had a big red face. Powers survived it, and they captured him, and he was in prison there for years. What they had on display in Gorky Square was pieces of his airplane. They also had his flight jacket, his log book, part of his flying outfit—all sorts of stuff. It was a major exhibit, and the people moved four abreast for ten blocks, waiting to see it. I didn't see it, but some of my American colleagues told me that the Russian cops would grab the Americans or other foreigners who came near the line and take them to the front. There they would push aside the Russians who were a few people back from the exhibit—to make sure the foreigners got to see it and didn't have to stand in line like the Russians. Can you imagine in this country if a cop tried to break into a line and put somebody ahead? Back in the hotel, we had noticed the elevator never stopped at the 13th floor. One day I was in the elevator alone, and by accident, it stopped at the 13th floor and the door opened. I could look out and see down the hallway—racks and racks of gear with glowing tubes. The whole floor was electronics. So this is why they put foreigners in this hotel. Need I say more? One day, I was in my room and got a phone call from a young Russian guy who spoke with a very heavy accent, but spoke very good English. He wanted to come up to my room and talk with me, so I said, "Fine, sure. Come on up." He was a young Communist who was trying to talk to me about Communism. Now, I wasn't famous, just a young brand-new assistant professor at Stanford, but I think they wanted to see if they could make a convert. After two hours of this, he gave up, as he couldn't make any progress. I remember discussing elections with him and saying to him, "You don't have elections." "Yes, we have elections," he said. "We have one candidate, and the people vote for him. When you have an election, you have only two candidates, so you have only twice as many as we have." I said, "We don't have twice as many as you have, we have infinite times as many as you have." Well, I wasn't getting anywhere, and he wasn't getting anywhere, but it was all probably being caught on tape by those electronics. I didn't see any tape going anywhere, but I saw the electronics. The main benefit of attending that first IFAC meeting was the personal contacts that you made. I made a personal contact with a colleague who became a friend for life. That was Prof. Tsypkin—Yakov Zalmanovich Tsypkin—who died a few years ago. The way he wrote his name was Ya. Z. Tsypkin. The Russians who knew him well would call him Yakov Zalmanovich. "Yakov Zalmanovich did this, and Yakov Zalmanovich did that." When I came to the meeting, he was looking for me. He knew all about my doctoral thesis, which was on the theory of quantization noise, and he had a small group of doctoral students working on that theory. I was astounded that he would know about the work of an insignificant young guy in the United States. How would they pick that up so far way in Russia? Although at the time Russia was a superpower, these scientists were in the backwater. I realized that they were much more systematic about studying our literature than we were about studying theirs. For one thing, they knew English, but we didn't know Russian. We tend to be like that. We're very provincial and think the sun rises and sets in America.

Wherever we went, the people of Moscow welcomed us with open arms, despite the fact that just two weeks earlier an American spy plane had been shot down. They couldn't have been happier to see us. This was the first really big scientific meeting of the Soviet era, where they actually opened up the country and let some foreigners in. We were very carefully controlled. We were programmed where to go and where not to go. One of our colleagues had taken some pictures out of bounds, and they took him down to the jailhouse, took the film out of his camera, chewed him out, and then let him go. You could take pictures, provided you took them in the right places. I took pictures all over Red Square, but you were not supposed to take pictures at the airport, for instance. When we landed at the airport, we could see, far out on the horizon, a whole fleet of heavy aircraft, probably Russian bombers. Their military and civilian bases were kind of intermingled, whereas we keep them totally separate. It was an interesting period of history, and the reaction of the Russian people to us was just tremendously welcoming. Regardless of the Cold War, I think the Russian people remembered us as allies in the Second World War. We had helped them and given them so much stuff through the Lend-Lease Program, and they had just never forgotten what we did. Even though you have conflict from government-to-government, it seems that, regardless of the propaganda, the people see right through it and are friendly and hospitable. They wanted to shake hands, to touch us, as if to be sure that we were real.

24.9 Summary

These Norbert Wiener stories were favorite of my students. They would be prompted from my memory by some subject related to what was being taught in the classroom. The favorite of the favorites was the story about giving him a push to start his car. So much for feedback in animal and machine.

Part VIII
Conclusion

The conclusion of this book is not a summary but contains additional material about influences that gave me a sense of direction in my life and in my research. The famous people I met were inspirational. The students that I worked with were very inspirational.

In my 50 years as an EE faculty member at Stanford, I was principal advisor for 89 Ph.D. students. Many of them subsequently had brilliant careers. Some became professors, some started successful companies, some became medical doctors. Some assumed leading roles in government departments. Two became admirals in the US Navy. My first Ph.D. student was Ted (Marcian) Hoff. He is credited with being the inventor of the microprocessor. For this, he was awarded the Kyoto Prize from the Emperor of Japan and the National Medal of Technology and Innovation from President Obama.

The final chapter has archival photos of my lab at Stanford, 1960-1965. YouTube has my collection of videos, some of them date back to this period.

Chapter 25
Famous People Stories

Abstract Here are thirteen stories about famous people who touched my life and influenced me. These people were leaders in their fields. Many of them founded new fields. I told many of these stories to my Stanford classes, stories that were prompted by the subject matter of the class. The material of this book would prompt these stories. They are not only people stories but they reflect history of many developments in science and technology. When I would ask my students in class if they would like to hear a story, a loud roar of "yes" would come from the assemblage. Enjoy!

25.1 Introduction

I am now 91 years old. I have known and in some cases worked with some truly remarkable people. My long-term memory is still working, and I have some stories to tell. Here are a few selected examples.

25.2 Ken Olsen

In June, 1951, I received the S. B. degree, a bachelor's degree in Electrical Engineering at MIT. I was fortunate then to have an assistantship with the Digital Computer Lab so I could continue and study toward the master's degree. The deal was that you work in the lab half-time while taking two graduate courses. My assistantship assigned me to this Lab that was located in the Barta building, about two blocks up on Massachusetts Avenue away from building 7, in the direction toward Harvard. The Barta building housed the famous Whirlwind computer. I was assigned to the magnetic core memory group led by Bill Papian. I was the junior kid of the group. The intellectual leaders were Ken Olsen and Dudley Buck.

I had never before met men like Olsen and Buck. Both were charismatic, extremely creative, brilliant inventors who came up with new ideas for magnetic core memories at a very high rate. I thought everybody was like that. I didn't know. I looked up to

B. Widrow, *Cybernetics 2.0*, Springer Series on Bio- and Neurosystems 14,
https://doi.org/10.1007/978-3-030-98140-2_25

both of them. They were a pair. I was gasping to keep up with them. Olsen was only three years older than me and Buck was only two years older, but they seemed far older, wiser and mature.

Olsen was the chief inventor and leader. Whenever Olsen came up with an idea, everyone in the group wanted to work with him on it and help bring it to realisation. He was a charismatic.

The magnetic cores are little toroids. They were made of a ceramic ferromagnetic material. They had hysteresis loops that were almost square. Magnetization at the bottom of the loop was a "zero", at the top of the loop a "one". The objective was square loop, fast switching from one to zero etc., and small size. Several material scientists were in the group working together with outside industry (General Ceramics) to find the right materials. We needed and industry needed electronic equipment to test cores. Olsen developed this.

In order to prove the magnetic core memory concept, Olsen designed a special computer called MTC, Memory Test Computer. The core memory and its associated electronics was mounted in a special rack called the "shower stall." It was approximately 1 m × 1 m with a height of over 2 m. The electronic current drivers for the core memory were high power vacuum tubes. Everything else in the computer was transistorized. The innovation in this machine was extraordinary. This was probably the world's first transistor computer. In contrast, Whirlwind had about 10,000 vacuum tubes.

A log book was kept for the MTC. This computer ran for months without a single memory alarm, a real breakthrough in RAM technology. There was nothing in the world like it.

Meanwhile, Whirlwind was having troubles. It's RAM was implemented with electrostatic storage tubes. They were highly unreliable. Whirlwind was run 24 h per day on national air defense problems. The only difficulty was reliability. It would be a lucky day if Whirlwind was able to run as long as two hours before getting a memory alarm. When that happened, computation needed to be repeated hoping for no memory alarms. Meanwhile, MTC was solid. No memory alarms.

A decision was made to install the core memory from MTC into Whirlwind. Preparation for the change took weeks, with Whirlwind running all the while. Whirlwind had to keep running, 24 h per day.

The core memory was moved and placed between the memory racks of Whirlwind. All the cables were ready for the changeover. I was picked together with one other member of our group to do the installation. Whirlwind was shut down, power off. Gasp. It took the two of us 36 h nonstop to do the installation. We only took time out to go to the bathroom or to go to the drugstore across the street to get tuna fish sandwiches.

In Whirlwind's control room there was a loud speaker connected to one of the flip flops of the CPU. Programmers could listen to their programs while running. If something went wrong, they would know it from the sound.

We successfully installed the core memory and I was in the control room when Whirlwind was powered up and a first program ran. I can never forget the expression on the faces of the programmers when they heard their program running five times faster. Their programs would run even faster if the rest of the computer circuits were faster.

The magnetic core memory was fast and reliable. No other form of RAM could compare.

Olsen finished his master's degree and was ready to go out into the real world. Together with several of the guys in the group, he founded Digital Equipment Corporation (DEC). Their product was magnetic core test equipment and other pieces of computer hardware, flip-flops, gates, etc. Eventually they went into the computer business.

Their first computer was the PDP-1. It was a great success. It was a mini computer that fit on a desktop. It was a handsome piece of apparatus. Ken always had a keen sense of style. DEC always made good looking computers. The DEC computer was always the showpiece of a lab. The machines were also highly reliable and high-performing. Other traits of Ken Olsen. He predated by many years another charismatic with a keen sense of style, Steve Jobs.

Olsen's computers really ate into IBM's market, taking over the mini computer business. The IBM 1620, a fine machine in its day, could see the handwriting on the wall.

Eventually Steve Jobs' micro computers ate into DEC mini-computer business. After years of successful leadership, Olsen was pushed out of his company. DEC lost its soul and never recovered. The same thing happened to Steve Jobs, and Apple lost its Soul. However a few years later Jobs came back, Apple found its soul, and took off like a rocket. Olsen never returned to DEC.

25.3 Dudley Buck and the Cryotron

It is really hard to invent and develop a new electronic device. Dudley Buck came up with a new device that he thought would revolutionize computer devices and circuits. Dudley was a colleague of mine at MIT. He was a fellow graduate student and subsequently a fellow Assistant Professor. He invented a superconducting device that he called the Cryotron. It operated in a bath of liquid helium at cryogenic temperature, close to absolute zero. The device was able to switch between superconducting with zero resistance and normal not-superconducting with finite resistance, giving an infinite on-off ratio. It was ultimately to be a thin-film device that could be made into an integrated circuit. This was long before the invention of silicon chips.

This was sensational. IBM became very interested and spent years on the Cryotron. Labs all over the world that had access to liquid helium were working on Cryotrons. US industry and the US government spent hundreds of millions of dollars, perhaps billions, on the Cryotron.

I was with Dudley when an audio oscillator that he made with a Cryotron was first tested. He had it mounted on a long probe and had it connected to an audio amplifier and loudspeaker. When he put the probe into the liquid helium, a moment went by as the Cryotron cooled and then we suddenly heard a pure tone from the loudspeaker. This proved that the Cryotron was an amplifier with a power gain greater than 1. This was a major demonstration. We looked at each other with big smiles.

In the summer of 1956, both Dudley and I were being recruited to join the Stanford faculty in Electrical Engineering. We both had a lunch date with John Linvill who was doing the recruiting. Dudley didn't make it. He was in the hospital with pneumonia. He died the next day. His lab mates and I at MIT were devastated. He was 32 years old. He left a wife with three little children.

Without Dudley, the Cryotron did not do well. No one was ever able to make an integrated circuit with Cryotrons that worked. Meanwhile, a steamroller came along, the silicon integrated circuit pioneered by Bob Noyce, who was soon to be the founder of Intel.

25.4 Marvin Minsky

In 1956 ,when I just became a faculty member at MIT, I lived in an apartment in Cambridge not far from Harvard Square. The building was located on a quiet tree-lined street, off the beaten path. Minsky lived across the street in a single-family home. One day I went outside and saw Minsky walking his dog. I looked at the dog and said "Marvin, that dog is incredibly pregnant." He said no, she is not pregnant. She has pseudocyesis, false pregnancy. He explained that the dog has all the outward signs of pregnancy and on the "due date" will go through labor and water will come out but no puppies. He explained that this is very rare and that it could even happen with humans as well. This was all new to me. This conversation was the longest I ever had with Marvin.

Marvin was at Harvard. With his colleague Seymour Papert, a book entitled "Perceptrons" came out several years later. I had a copy and gave it a quick look. It was a piece of scholarship. But the way it was written and the conclusions drawn seemed to me that the book had a political purpose. It was a "hatchet job" against the field of neural networks. I did not pay much attention but the book was evidently influential. It received credit for federal defunding of the field. Minsky and Papert were trying to kill neural networks. The big source of money was DARPA (maybe ARPA at that time). Neural networks got nothing but AI (coined by Minsky and John McCarthy) got millions. After about 50 years of funding the AI Community, DARPA got no AI. Their money wasn't totally wasted however. Their funding provided a lot of support for the computer industry. Every major AI lab had to have a DEC PDP-15 computer which was a pretty big machine and cost a lot. The AI community favored DEC computers and they had the money to buy them. Those of us in the neural network community survived on nickels and dimes, wherever we could find them.

The field of AI hit a brick wall. The AI Community failed to produce Artificial Intelligence. About 25 years ago, some members of the AI Community discovered neural networks. These members were not the original founders and apparently didn't adopt the neural network taboo. The field of neural networks has flourished and it is now ironically called AI.

Neural networks really work. Minsky and Papert created a fine piece of scholarship but they were on the wrong track. They were able to prove all sorts of things that you couldn't do with a single neuron. They did not study networks of neurons. Networks easily circumvented the objections of Minsky and Papert, but the damage was done.

About 30 years ago, Minsky was invited to present a seminar at Stanford. He was at MIT at that time and was secretly being recruited by Stanford. John McCarthy had already been at Stanford for a number of years. If Minsky were here, Stanford would have the two founders of AI on its faculty.

The famous Professor Minsky gave a seminar. I can only recall that he presented no new ideas or no new insights. He talked about the human brain. He called the brain a "meat machine." My takeaway from his talk was nil. The recruitment rumors seemed to stop.

The last time I saw Minsky was on 26th of April, 2001. On that day he received the Benjamin Franklin medal for Computer Science and I received the Benjamin Franklin medal for Electrical Engineering. These awards were made by the Franklin Institute of Philadelphia. A group of eight received Franklin medals for physics, chemistry, medicine, etc. Among the recipients were Irwin Jacobs, founder of Qualcomm, and Paul Baran, inventor of packet switching. In the past, some of these awards have been precursors to the Nobel. Our Physics laureate subsequently received a Nobel prize.

The award ceremony was a black tie affair in a great hall at the Franklin Institute. The social elite of Philadelphia filled the hall. The laureates were up on a stage with the Mistress of ceremony Cokie Roberts. Looking out from the stage, you see a huge seated statue of Benjamin Franklin, as big or bigger than the statue of Abraham Lincoln at the Lincoln Memorial in Washington. Looking at that beautiful statue and thinking about Benjamin Franklin, I was humbled. I was thinking, my God, what am I doing here?

Minsky was an eccentric. No doubt about it. He was wearing a tuxedo with a piece of rope tied around his waist. I talked to my wife about this after the ceremony. We were both aghast. We were guessing that maybe he lost his cummerbund.

It was a magical ceremony. My wife, my daughter and their husbands, and my grandsons were there. My grandson Jeffrey Sclarin, 10 years old and grandson Adam Sclarin, 8 years old, looked like little gentlemen dressed in their tuxedos. We have pictures. I remember little Adam mixing in with the crowd telling everybody about his grandfather.

Minsky played a pivotal role in the founding of the field of AI. He was charismatic and brilliant, not humble. He became famous. His fame allowed him to be the big booster of AI. He thought deeply about thinking and brain structure. With his lecturing and writing, he laid down his version of what thinking is and this became the basis of what he thought artificial intelligence should strive to emulate. He attracted a large following, and substantial support. Marvin was a philosopher, not an engineer. The promise of thinking machines was never realized, yet his ideas permeate AI today. His death was a great loss to science.

25.5 Claude Shannon

I had just completed my doctoral studies and started to think about thinking after attending the first AI conference at Dartmouth in the summer of 1956. I wrote a few papers about quantization noise, the subject of my Sc.D. thesis. I had developed a statistical theory of round off error.

After a distinguished career at Bell Laboratories Shannon "retired" to a faculty position in Electrical Engineering at MIT. He was teaching a very advanced course in information theory in a field he had founded. He was the father of Information Theory. Years earlier, he had worked on quantization noise. I was eager to see what he thought of my work.

I saw him in his MIT office. I tried to discuss quantization noise and how I approached it. His mind seemed to be on other things. I noticed a box, a cube on his desk. It was about 6 in × 6 in × 6 in, not very big. I asked him what it was. He pointed to a toggle switch on one side of the box and said turn it on. I snapped up the switch and then heard a mechanical sound from within the box. It sounded like gears turning. Suddenly the top lid of the box began to rise, a mechanical hand came out, reached over to the toggle switch, pressed down on the switch, shut it off, then reached up, then withdrew itself into the box, then the lid closed and the mechanical sound went silent. Evidently, Shannon liked gadgets. This was a great one.

He told me that he was working on two projects. I was expecting Information Theory, but no. He had a Volkswagen Microbus and he was devising a shower for it. He must have liked camping and needed a shower. This was years before there was such a thing as a motor home. He was ahead of his time, of course. The second project had to do with swimming. He explained that his home was up on a hill. At the bottom of the hill was a lake, nice for summer swimming. He mounted a telephone pole out into the lake. He attached a pulley to the telephone pole and another pulley to his house. He placed a long rope over the two pulleys. The idea was to grasp the rope at the top of the hill and glide down fast to the lake and let go of the rope before hitting the telephone pole. Splash. So as he described this to me, he was working on some of the details. Nothing to do with Information Theory.

When I knew that I was going to Stanford, I saw him again to tell him. He had spent the previous year at Stanford at the Center for Advanced Study in the Behavioral Sciences. He said to me, "Bernie, you are going to God's country. You need to get a chef's hat, a great apron, and a barbecue, then you will be all set." He was expressing an East Coaster's view of California. Laid back, everybody on permanent vacation. He must have spent his time here at the Center which is on the Stanford campus but far away from the main quad and the principal academic buildings. If he could come back now and see Stanford and Silicon Valley, he would be surprised.

25.6 Frank Rosenblatt and the Perceptron

Frank Rosenblatt was the father of the Perceptron. This was a trainable classifier that had gained considerable notice in the press, in the New York Times and among other publications. The Perceptron was severely criticized in the book by Minsky and Papert, without rational justification. Government funding agencies were persuaded by the Minsky and Papert book and funding for neural network research dried up. This probably set the field back by 10 years.

Rosenblatt and I were contemporaries. We attended the same conferences and presented competing approaches. He lectured on the Perceptron and I lectured on ADALINE and MADALINE. They were all different forms of trainable classifiers.

The Perceptron had two layers. The first layer consisted of a large number of neurons with small numbers of fixed random weights that were randomly connected to the "retina", the pattern input vector. The outputs of the first layer were the inputs to the second layer, the single neuron that delivered the Perceptron binary output signal, i.e. the class of the input pattern. The first layer was not trainable. The second layer was trained in accord with the desired binary output of the input pattern. Rosenblatt had developed a training algorithm for the second layer that became known as the Perceptron Learning Rule.

MADALINE was a two layer network, but unlike the Perceptron, MADALINE's first layer neurons were trainable and the second layer was fixed. The LMS algorithm was used to train the neurons of the first layer. The trick was to take the known desired response of the second layer and from it, derive desired responses for the first layer. I worked out a method for doing this, and MADALINE could be trained. I was never satisfied with this however because I wanted to train both layers. Rosenblatt would also have liked to be able to train both layers. Neither one of us succeeded with this.

I didn't like the idea of the Perceptron. I argued with Rosenblatt about the effect of his first layer neurons randomly sparsely connected to the retinal inputs. Patterns on the retina that were similar in shape and were in the same class would be scrambled by the random first layer making training of the second layer more difficult. He argued that the Perceptron must have a random first layer because the human eye has layers of neurons that are randomly connected. He was a neuroscientist trying to emulate biology. I was an engineer trying to make something that worked. Now, it is known that the neurons in the eye are not randomly sparsely connected but comprise a highly structured network.

Rosenblatt did his original work at the Cornell Aeronautical Laboratory. He went to Cornell University in 1959, the same year that I joined the Stanford faculty. In 1962 he published a book entitled "Principles of Neurodynamics: Perceptrons and the Theory of Brain Mechanisms." This became a sensation in the press and in the scientific world. The Perceptron became famous.

A few years later, I had occasion to visit him at Cornell. I wanted to see the Perceptron in action and compare it to ADALINE. In my suitcase, I brought the memistor ADALINE shown in the photo of Fig. 26.15 of the next chapter. In his lab, he had the Mark-1 Perceptron. The adaptive weights were motor-driven potentiometers. I had portable ADALINE next to the Perceptron which was mounted in a full-size 19-inch rack.

Frank chose a pattern and presented it to the Perceptron, and trained it to classify the pattern as a 1 or a 0, as he desired. You could hear the gears going around as a Perceptron was adapting. I put the same pattern into the memistor ADALINE (it will be described in Chap. 26) and it trained. Another pattern was presented to the Perceptron and it just barely trained. The same pattern into ADALINE trained with no problem. A third pattern into the Perceptron didn't train at all. Into ADALINE, it trained easily. ADALINE was saying, give me more. Perceptron was saying enough.

Theoretically, the Perceptron should have had far more capacity than demonstrated. I think the problem was with the many mechanical parts of the Mark 1. I was never sure. A single layer, a single neuron, little ADALINE easily beat the rack mounted multi-layer Perceptron.

From all the work on the Perceptron and ADALINE and MADALINE, the modern-day survivor is the LMS algorithm. It is at the foundation of the backpropagation algorithm that powers the neural network AI Revolution. The Perceptron played a major original role in stimulating research on artificial neural networks.

Frank Rosenblatt had a difficult time of it. He alternated between being celebrated and being slammed. The price of fame. I supported him whenever I could. His life ended tragically. He went off in his sailboat on the Chesapeake Bay and never returned. He was 43 years old.

My impression of Frank Rosenblatt was that he was brilliant, a gentle soul who became caught up in the hype surrounding the Perceptron in the news media and popular press. He was a true believer in the Perceptron and perhaps overplayed its significance as a brain model. This was sensational to the laymen of the world and at the same time, the opposite to his colleagues. When he was being considered for promotion, his department at Cornell asked me for a letter describing his contribution to science and his position in his field. I wrote a strong letter describing the significance of his work. This letter certainly contributed to a successful promotion. Years later, I spoke with one of his students who was a close collaborator, asking what happened on the Chesapeake Bay and he had no idea. Rosenblatt's death was tragic, a great loss.

25.7 Bill Shockley

Bill Shockley retired from Bell Laboratories after a long and distinguished career. With a shared Nobel Prize in physics for the invention of the transistor under his belt, he arrived in Palo Alto, CA, and started Shockley Semiconductor. He was a notoriously bad manager of people. His key employees that he subsequently called "the traitorous eight" could not tolerate his supervision and left to form Fairchild semiconductor. Their sponsor was Fairchild Camera. Ultimately they split and two of them, Bob Noyce and Gordon Moore, started Intel. Shockley never recovered from this and his company folded. He became Professor of Electrical Engineering at Stanford.

Shockley bought a house on the Stanford campus, on Esplanda way. I also owned a house on Esplanada way. We were next-door neighbors. In the fall of the year, both of us would be outside raking leaves. I chatted with him many times. I had built a sandbox in my backyard for my two little girls, Leslie and Debbie. Checking the sandbox one day, I saw some things I didn't like. Shockley's cats were getting into the sandbox. I went next door to tell Bill about it and he said "herding cats is like trying to confine the molecules of a gas." I changed the sand and built a screen cover for the sandbox.

I had many discussions with Bill on technical subjects. He was clearly one of the few absolutely brilliant people that I have ever met. When Ted Hoff and I were developing the memistor device (to be described to the next chapter), I had many questions for him. He was always interested and helpful. He seemed to know everything, Electrical Engineering, physics, chemistry, etc. He was a very smart guy.

Because of his anti-black racial theories he fell into disrespect as a scientist. The social scientists at Stanford and around the world insisted that he was out of his element and that he didn't know what he was talking about. I knew that he was much smarter than many of them and he did a lot of reading and study in their field. He knew how to understand data and technical arguments. No doubt about that. My colleague Jim Angell knew Shockley well. To me he said simply that "Shockley is a bigot." Knowing what you are looking for could have a strong effect on how you interpret data. But I thought that Shockley was too smart to be fooled, but he was fooled.

One day while raking leaves, I asked Shockley about his racial ideas. A hot-button issue. He said that the capability of black people depended on how much white they had in their ancestry. His favorite people were Jews and Asians. He himself was neither. So there, you have Bill Shockley.

25.8 Arthur Samuel

During a six-month sabbatical in 1967 in Belgium, my house on the Stanford campus was rented to Mr. and Mrs. Arthur Samuel. Dr. Samuel and Dr. Shockley became next door neighbors. I left the screen over the kid's sandbox to keep out Shockley's cats while my wife and I were away.

Arthur Samuel had just retired from IBM where he had been doing amazing work on IBM computers that he coded to learn to play the game of checkers. The Samuels were renting to learn the area before buying their own home. I knew about his work from an article he wrote for MIT Technology Review. I was a great admirer of Dr. Samuel. In retirement, he became a staff member in the Computer Science Department at Stanford, continuing his research on machine learning. He was the first person to use the words "machine learning" in a scientific paper.

In my opinion, Dr. Samuel's work on machine learning was the best piece of work in the field of artificial intelligence that was ever produced, perhaps to this day. It is the only AI work that actually performs with human-like reasoning. The AI Pioneers never seemed to adopt Samuel's thinking. They came up with expert systems and they of course invented agents. None of this gave us anything like machine intelligence. The present work on AI is focused on artificial neural networks and although highly useful, does not give us machine intelligence. Samuel's work does and it should be pursued.

Samuel's Checker Player looks at the pieces on the checkerboard as the present state and gives it a numerical value. The objective is to maximize this value. Playing against a human opponent, when it is the Checker Player's turn to make a move, the Checker Player considers all possible moves and derives a new numerical value for each. For each of these moves, it contemplates possible opponent moves and chooses the best move for the opponent, the move that minimizes the numerical value. For all best opening moves, it contemplates its own possible move and computes the corresponding numerical value. It stops here with a "ply" of three. It chooses the state with the highest numerical value and then works its way back through the decision tree to find the optimal first move that has the highest probability of optimizing the numerical score after a ply at three. Better results were obtained with a ply of six, contemplating moves further into the future.

The higher numerical value was a value advantageous to the Checker Player program. It was based on a count of the number of pieces on the board, their values and their positions. The pieces on the board comprised a pattern. Samuel selected the features of the pattern. These features were linearly weighted, creating a sum that was the numerical value. He devised a learning algorithm that adjusted the weight values from playing games, winning and losing. Over thousands of games, the weights were adapted toward winning. Samuel said that the Checker Player could beat him but not the best human checker players.

The clearest presentation of Samuel's work is the paper "Machine Learning," MIT Technology Review, November 1959, pp. 42–45. I highly recommend this paper for anyone interested in artificial intelligence.

25.9 Sukhanova Conference, John McCarthy, Karl Steinbuch

I cannot remember the exact date but it was in February 1965 or 1966. I was invited to attend a small Conference of about 50 people, half Russians, half foreigners, to be held at Sukhanova, about 50 mi from Moscow. The subject was AI and learning systems.

I landed in Moscow at about 9 pm. A young Russian man came up to me and asked, "Professor Widrow?" I said yes, and there were four of them helping me with my luggage. I was surprised to be met. I knew of no such arrangement. Since they new my name, I wasn't worried. Without them I wouldn't know how I would get to Sukhanova on a pitch black night in the dead of winter in Russia.

They had a tiny little car, a Moskovitch. We piled in. I was in the backseat with Russians on both sides of me. In the front was the driver and the navigator. The leader of the group was Semyon Meerkov who succeeded in getting out of Russia years later and finally joined the Electrical Engineering faculty at the University of Michigan. We became lifelong friends.

The road was covered with snow. It must have been rutted. It was a bumpy ride to Sukhanova. My Russian hosts were all graduate students at, I think, Moscow State University and the KGB. They were firing questions at me about ADALINE and MADALINE all the way to Sukhanova. It must have been 11 p.m. or midnight when we arrived. I was greeted by a famous Russian scientist, Professor Mark Aizerman, with a gigantic bear hug. He said, you are hungry, you must eat. (I already had dinner on the airplane). He turned on all the lights in a big kitchen and roused a team of Babushkas, and they started producing food and vodka. So I had another dinner. It was good.

They guided me to my room and hauled the luggage. In this room I had a roommate, John McCarthy. We were room-sharing for about a week during the conference. He was very eccentric and we hardly said a word to each other. I don't think he had any idea that we both were Stanford Professors. He didn't know me from Adam.

Sukhanova is a beautiful estate. It was called "Rest home for Architects." It was the place where all the architects from the Moscow area had their vacations, their holidays. In the Soviet Union, everything was controlled. The estate belonged to Duke Andrei Nikolayevich Bolkonsky, the hero of Tolstoy's novel War and Peace. The Duke was a real person. The estate must have been seized during the Russian Revolution. Now it belongs to the Russian government.

There were many fine buildings on the estate. The meeting room and the dormitory rooms occupied one building. Some distance away, another building had the kitchen and dining facility. We had all our meals there. It was a fine conference, good people, and a lot of camaraderie. I met a number of people there who became good friends for life.

I met Karl Steinbuch there. He was Mr. Computer Science of Germany. We had many nice conversations. I remember one day walking with him after lunch toward the main building to rejoin the conference. It was snowing. Snow in Russia in February is something to behold. It was so thick that you could hardly see a meter ahead. The snowflakes were bigger and more juicy than any I had previously experienced. I commented on this to Professor Steinbuch. He said "Bernie, I'm going to tell you something not to be discussed with our Russian friends. During the second World War, I was drafted into the German Army and was sent to the Russian front. I saw snow like this. It was awful. We were not prepared for the cold. We were freezing. You couldn't move a cannon because the axle grease was frozen. You couldn't start a vehicle. I don't ask for sympathy. We inflicted horrible suffering on the Russian people. There was nothing I could do. You fight in the army or they will kill you."

In 1967, I spent a six-month sabbatical at the University of Louvain in Belgium. With my wife and two little girls, we traveled in Europe when my academic duties were finished. We visited with Professor and Mrs. Steinbuch at their home in Karlsruhe. His and our children had a great time playing together. We had a nice dinner. Earlier that day, he took me to see the University of Karlsruhe. We visited the building of the Electrical Engineering Department. This is where Henrich Hertz discovered Hertzian waves, electromagnetic waves, radio waves, wireless waves. He is memorialized with the calling of the unit of frequency Hertz or Hz. Outside the building, there is a statue of Heinrich Hertz. To me it was very moving to see all this.

A few years later, Steinbuch visited Stanford for a month or so. He and I wrote a paper together. He developed the Learnmatrix, a storage array that was a content-addressable memory. In our paper, we compared ADALINE with the Learnmatrix. They are quite different concepts as we explained in the paper. The Learnmatrix was a popular concept in Europe at the time when the Perceptron was a popular concept in the US.

25.10 First Snowbird Neural Network Workshop, John Hopfield, David Rumelhart

Sometime in the late 1980s, I think, a neural network conference was held for the first time at Snowbird, in the mountains near Salt Lake City, UT. This was an invitational workshop organised by a group at Bell Laboratories and a group at California Institute of Technology. A Stanford student told me about this and suggested that I contact Bell labs and ask for an invitation. I called a man at Bell labs and spoke with his secretary. She said she will check with her boss and that I should call again tomorrow. I did and she was very enthusiastic and said "you bet you can come."

The workshop had about 50 attendees. It was held in the middle of winter, ski season. Sessions were held each morning. We all had breakfast and lunch together. After lunch was skiing. I took long walks. We reassembled for dinner and after dinner the sessions went on almost to midnight. The conference lasted about four or five days. Everyone had a chance to present his/her work. It was a lot of camaraderie. We as neural network researchers were almost like a persecuted minority.

The first paper of the first session was presented and during the question period after the talk, several people were saying didn't Widrow do something like that. They went back and forth. When they were finished, I stood up and announced that "I am Bernard Widrow, greetings to you all." After the second paper, there was coffee break time. They crowded around me, wanting to shake my hand. You are real! You are still alive! This was a whole new crop of neural network researchers and I didn't know them until this conference.

I met for the first time John Hopfield, Professor of Chemistry at Caltech and David Rumelhart, Professor of Psychology at Stanford. They were both distinguished researchers in the field of artificial neural networks. Hopfield had invented the Hopfield Network, a recurrent form of neural network. Rumelhart and his colleague James McClelland in 1986 had published two "PDP" books. They were "Parallel Distributed Processing: Explorations in the microstructure of cognition" and "Parallel Distributed Processing: Psychological and Biological models," MIT Press. These books popularized the backpropagation learning algorithm. This algorithm is used in all modern neural networks.

The inventor of backpropagation was Paul Werbos. His contribution was not recognized at the outset. It was his Harvard PhD thesis presented to the Faculty of the Department of Economics. Years later, Werbos told me the story about his thesis. It almost didn't get him a PhD since the Econonmics faculty couldn't make sense of it.

The first evening of the Snowbird conference was chaired by John Hopfield. I wasn't scheduled to speak, but he put me on the program. I had a memistor ADALINE in my suitcase, the ADALINE shown in the photo of Fig. 26.15, to be described in the next chapter. I demonstrated ADALINE by training in a set of training patterns. This was quite interesting since the people had never seen anything like it. ADALINE was making binary classifications having a sharp quantizer as an activation function. Professor Hopfield convinced me that a sigmoid is a better activation function since it is more biological and it is used in the back propagation algorithm and used in the Hopfield Network. This was new to me and I took his advice very seriously.

At this meeting I first heard about the backpropagation algorithm. With this algorithm, it was possible to train all the neurons of all the layers of a multi-layer neural network. That is something that for years I tried to do and could not think of how to do it. To me this was quite a revelation.

At the time of this conference, I was not putting serious effort into neural network research. I was troubled by the thought that student research on the subject would not be accepted by my colleagues in Electrical Engineering. Neural networks, networks of non-linear devices, nonlinear because of the activation functions, could not be treated analytically. A thesis on the subject would present the results experimentally, not mathematically. Electrical engineers do things analytically. I was focused at that time on adaptive filters that have no activation functions. Adaptive filters can be treated analytically and were easily accepted by the Electrical Engineering faculty.

25.11 Lotfi Zadeh

Lotfi Zadeh was the father of fuzzy set theory and fuzzy logic (Professor Zadeh himself was not very fuzzy since he was bald and shiny as long as I knew him).

I first met Lotfi around 1955. I was introduced to him by my MIT Mentor Bill Linvill. We were on a sidewalk in New York City about to attend an IEEE meeting there. I was 25 years old. He was 10 years older and a senior Professor of Electrical

Engineering at Columbia. I was an MIT graduate student attending his first scientific conference. Over subsequent years I met Professor Zadeh many many times. I think he always remembered me as a graduate student. He would introduce me to the other people as "....". I cannot remember the exact expression but it was something like "precocious youngster."

His mathematical work on fuzzy set theory led to applications of fuzzy logic. Prof. Zadeh was an engineer, but deep in his heart he was a mathematician. He was a good friend for many years. 1959, the year that I came to Stanford was the same year that he left Columbia and came to the University of California, Berkeley. There has been a long-term rivalry between Stanford and Berkeley, on the football field, but not at the professional level although maybe there is a little bit remaining.

The first time I heard him give a lecture on fuzzy logic was at a small IEEE conference sometime in the 1970s. There were about 100 people, at the University of Florida, Gainesville. The conference went on for several days. On one of the evenings, a group of our Florida colleagues arranged a car caravan for a group of us "foreigners" to go to a Florida nightclub. Fine. But it turned out to be a "strip club." We were sitting around a large table having drinks when suddenly appeared a very pretty young lady wearing not much clothing and she popped up on our table and started dancing to the music. I was enjoying the scenery but then I became concerned about Professor Zadeh, that he would be embarrassed. He was fine and enjoying the scenery like the rest of us. Bad Boys! The next day, the technical sessions went on as usual.

The last day of the conference broke up after lunch. Lotfi and I shared a taxi to the airport in order to catch a flight to San Francisco. We both were booked on the same flight. In the taxi, I had a chance for a one-on-one discussion with Lotfi about fuzzy logic. I made the mistake of saying that fuzzy set theory was just like Probability Theory, and what you can do with one you can do with the other. Lotfi was a gentleman, otherwise he would have blown a fuse. It's fuzzy set theory. An object could belong to two sets, with a certain degree of membership, a fraction, and in one set and another degree of membership, a fraction, in the other set. The fractions would add to one. To me, it seemed like the same idea could be represented by probabilities. No, no, no, no!

Lotfi became a world traveler, evangelizing fuzzy logic. He would fly to attend a conference, give his paper, then fly home all in one day. He had tremendous energy and never seemed to slow down. This went on for years.

We both were invited to a conference in Kyoto, sometime I think in the 1990s. Japan was really excited about fuzzy logic, but not as excited about neural networks. The conference was on both of these subjects. Fuzzy logic in Japan became a household item. A Japanese housewife would show with pride her rice cooker, controlled by fuzzy logic. Her dishwasher was controlled by fuzzy logic. Her Canon camera was controlled by fuzzy logic. The subway trains in the city of Sindai when coming to a stop did this without jerk, since the braking system was controlled by fuzzy logic. Lotfi had become a national icon. His portrait was seen in every post office in Japan. I was impressed with this and amused at the same time. If his portrait was shown in every post office in the US, he wouldn't be an icon, he would be a criminal.

At the conference, the commercial exhibits were really amazing. Everything was run with fuzzy logic. Most of the exhibitors were famous companies like Panasonic or Sony. There was an air conditioner that had a motor and compressor that could run efficiently at variable speed rather than like the usual air conditioner that turned on and off giving temperature that was either too cold or too warm. The fuzzy one would give a constant temperature, not too cold or not too warm. This air conditioner also had outside sensors in addition to the indoor sensor. When outside temperatures start rising, the air conditioner began cooling early so that it would not need to catch up later. This was a high-tech device, better than any other one I have seen since.

The washing machine was very clever. A clear plastic U-shaped tube was attached to the bottom of the machine filled with wash water. Light of different wavelengths shined through to be received by photo detectors. The purpose was to measure the dirtiness and the greasiness of the wash water. This data was fed to a fuzzy controller whose objective was to make the clothes perfectly clean while using a minimum amount of soap, water, and electric power. I have never seen anything like this since.

These are impressive products. I was impressed by the fuzzy logic. I bought a Canon camera with fuzzy logic for autofocus and image stabilization. I was however mainly impressed with the Japanese desire to make a better product in addition to clever improvements in the sensors and, mechanical and electronic parts that make all this possible. It seemed to me that there would be many ways to do the controls other than fuzzy logic but fuzzy was doing a good job and this was impressive.

My friend Professor Bart Kasko, who was a fuzzy logic researcher told me that fuzzy was very successful in Japan because fuzziness fit Japanese culture. Answers are not black or white but always some shade of gray. Although fuzzy logic was sensational in Japan and Asia, it hardly caught on in the US. This was a great source of frustration to Lotfi. That is probably why he did so much travel and evangelizing.

Lotfi and I had quite an experience at the banquet on the final night of the Kyoto conference. He was Mr. Fuzz, I was Mr. Neural. On a stage, there were two huge wooden barrels of sake. Our hosts put aprons on Lotfi and me. They paid us a great honor. They gave Lotfi a big wooden mallet, and another one for me. The woodentops of the barrels were pre-sewn to make breakage reasonably easy. They gave the signal and Lotfi and I raised our mallets and crashed them down hard and broke the barrel tops. Sake flew everywhere and the crowd yelled “Banzai.” It gave me the creeps. It was just like a World War II war movie. We were highly honored and proudly did the job. Everyone had sake.

Over the years attending conferences, I would present results of new neural network experiments and projects from my lab. A year after presenting something, the next year Lotfu would show the same results from his lab with fuzzy logic. This happened a number of times and I had the feeling that he was tracking me. Finally, we developed a real mechanical truck backer. This was a toy truck that was an exact replica of a huge trailer truck. The toy truck was controlled by radio. This truck had two trailers and was able to back up without jackknifing while under human control. A neural network was trained offline to act as an interface between the human and the truck’s steering mechanism. An onboard computer, an Intel 386, implemented a neural network whose weights were obtained by offline simulation. My student who built all this was Marcelo Lamego, a very bright Brazilian. He made a video of me walking alongside the truck steering it with remote radio control to travel backward and do loop-de-loops without jackknifing. It would be impossible to do this without the neural interface. I spoke with truck drivers who drove trucks with double trailers and they only drove them forward. They all told me that the backing up a double trailer is impossible. The Ford 150 pickup truck is equipped to back up a single trailer. I think I know where they got this idea.

The neural network of the toy truck did no learning. The weights were fixed. Offline learning found the weight values. If this control system were to be implemented with fuzzy logic, it could probably be done but with great difficulty. A fuzzy controller is a rule-based approach. First, one needs to find the rules, then to implement the rules with fuzzy logic. Finding the rules for a human to machine interface system for a truck with trailers backing up would require a great effort, almost impossible. To implement the rules with fuzzy logic would be very difficult. Two difficult procedures can be compared to the neural network that was easily trained with the back propagation algorithm. There were no evident rules. Comparing neural network implementation with fuzzy logic implementation, the neural network does the system design by a learning algorithm. The rule-based approach of fuzzy logic that does not learn, requires a great deal of human effort. The Berkeley folks were never able to make a real mechanical truck that could back up a double trailer. The video of our truck is available as part of the collection of Widrow videos on YouTube [1].

In 2009, Lotfi Zadeh was awarded the Benjamin Franklin medal for Electrical Engineering from the Franklin Institute. He received this for the invention of fuzzy logic. The award was presented at a ceremony at the Institute. Recipients are called laureates. Former laureates in all fields of science are invited each year to attend the award ceremony. Mrs. Widrow and I attended in 2009 so that we could congratulate Lotfi. We had dinner with Dr. Zadeh, a new Laureate, and Fay Zadeh, his wife of many years. A good time was had by all.

The last time I saw Lotfi was in 2016 at a conference at Stanford. He was an invited speaker and he talked about fuzzy logic. He was about 95 years old at the time. He spoke softly and slowly. It was hard to hear him and understand what he was saying, but he was still Lotfi. Everyone was very happy to see him. About a year and a half later he was gone. I am pleased that worldwide research and fuzzy logic is continuing strongly. There is an IEEE Transactions on Fuzzy Systems. It is a tribute to Lotfi.

25.12 Richard Feynman

In 1987, DARPA initiated and funded a study to evaluate the status of neural network research and applications in the US primarily, and in the world. They had not invested in this work but wanted to know about it. They contracted with MIT Lincoln laboratory to do the study. Lincoln lab asked me to chair the study. They assured me that my required effort would be minimal. Lincoln lab staffers would be scouring the world, interviewing neural network researchers all over the US and a selected few overseas. They would bring their interview notes back to Lincoln lab for consolidation. The ultimate deliverable would be a report to DARPA.

They wanted me to conduct a review meeting at Lincoln lab every other month for a study for six months. They wanted me to be available to attend a few of the interviews. The review meetings were to reflect progress with the interviews. The review meetings were attended by a very distinguished panel that included two previous directors of DARPA, two previous directors of Lincoln laboratory, the four staff of Lincoln lab who are doing the interviews, Dr. David Hubel, a Nobel Laureate from Harvard Medical School, and others. The panel consisted of scientists and technical people, but not people who were doing neural network research.

The real hard work was done by Lincoln lab staff people who conducted the interviews. My job was to meet with them the day before each review meeting. I distilled what they had learned into a coherent presentation with slides. The only person whose interview was conducted by me was Richard Feynman.

One member of the panel was an engineer who worked for Hughes aircraft. He arranged the meeting with Dr. Feynman. Someone met Dr. Feynman at his home and drove him to the meeting place, a conference room at one of the Hughes plants. The panel was assembled and anxiously waiting for him. The great man entered and we all sat around the conference table.

The motivation for Lincoln labs to interview Dr. Feynman was based on a rumor that he had been interested in neural networks and that he had done some work on the subject. This made sense since there was a group of neural network researchers at Caltech. Leaders were John Hopfield in the Chemistry Department and Demetri Psaltis in the Electrical Engineering Department. Feynman was in the Physics Department.

We were sitting around the conference table and the air was electric. We were expecting to hear some amazing words about neural networks from Dr. Feynman. He was silent. Then he spoke with a question. "What's a neuron?" The panelists were petrified. Everyone was quiet, and they all started looking at me. So I went to the blackboard, picked up a piece of chalk and drew ADALINE, an artificial neuron that is a model of a living neuron. I mentioned that the weights are adjusted by an adaptive algorithm, a learning algorithm. Then I sat down and there was silence.

Dr. Feynman broke the silence with another question, "What's an adaptive algorithm?" Once again, everybody stared at me. So I went back to the blackboard and described the LMS algorithm as one of many algorithms, but that LMS was the simplest and most popular. I derived the LMS algorithm for him. It only required a few lines of algebra. Then I sat down.

Again there was silence. Then he popped another question, "What is a neural network?" I went back to the blackboard and sketched a multi-layer network. On small-scale, I drew a diagram of the Perceptron and explained how its learning worked. Then I drew a simple MADALINE and explained how it learned. And I sat down. Then it was time for lunch which we had in the conference room.

Lunch was provided by our host, Hughes aircraft. It was a simple lunch, sandwiches, cokes, and potato chips. Feynman and I pulled up two chairs and sat down to have lunch together. We discussed many subjects. It was a real privilege for me to have one-on-one time with Dr. Fynman, the world's most brilliant living scientist. Within a few minutes with him, you would realize how amazing he was.

In the course of our conversation he mentioned that he liked the way I responded to his questions. He said that, "your answers were simple and clear. I like simple." This came from the man who invented Quantum Electrodynamics. I didn't tell him what it felt like having to answer his questions. It felt like an oral PhD qualifying exam.

The meeting broke up after lunch. In the course of the day, he was true to character. There was a sparkle in his eyes. He was full of the devil. We knew that he was suffering from terminal cancer. He seemed perfectly normal. An amazing man. I wasn't sure if he knew anything at all about neural networks. He wasn't there to talk about this. He was there to give us a quiz.

Two months later, the panel met again with Dr. Feynman, this time for lunch at the Caltech Faculty Club. Nothing technical was discussed. It was a pleasure to see him again. He had the sparkle in his eyes, and was full of the devil. Lively person. He died two weeks later at the age of 70.

The DARPA study concluded and a report was prepared by the Lincoln laboratory staff. That was the end of it. No money of significance followed to neural network research. Looking back on the study, I can recall several conversations with Dr. David Hubel. He shared the Nobel Prize for the discovery of neurons in the visual system of cats that respond to certain patterns, for example a vertical line wherever it appeared in the visual field. Another set of neurons responded to horizontal lines. This work stimulated the development of feature detectors for pattern recognition. It has had a lasting effect on the field.

David Hubel pointed out to me that ADALINE is much more than a neuron. It is a neuron plus its synapses. Since then, I have been more careful in describing ADALINE.

25.13 Senator Henry Jackson, Admiral Hyman Rickover

My daughter Debbie began her freshman year at Stanford in the Autumn of 1981. On moving in day, she was back and forth hauling stuff between our home on the campus and her dormitory. She was moving into a co-ed dorm called "Flo Mo". The original donor of the building, Florence Moore, stipulated in her will that the residents have ice cream offered with lunch and dinner every day, so this was a famous dorm. While moving in, my daughter became friendly with Anna Marie Jackson who was moving in 6 doors down the hall. They chatted and Anna Marie told Debbie that her father is Senator Henry Jackson from the state of Washington. Debbie told Anna Marie that Senator Jackson is her father's favorite Senator. The two girls became friends and they have been best friends for life.

The next day, my phone rang and it was Senator Jackson calling. I was starstruck and dumbstruck. Here is the man with the weight of the country on his shoulders, and he's calling me. We had a nice conversation and he asked me if I ever get into Washington DC. I told him I have government sponsors so I do go to Washington from time to time. He said that the next time I go to Washington, let him know. We can have lunch at the Senate dining room. I told him that I would contact him before my next trip. I knew that he really wanted me to contact him. It was not at all like Mae West in a movie saying, "Come up and see me sometime."

A few months later, the need for a trip to Washington arose. I needed to see a sponsor. I called Senator Jackson and he wanted to know the arrival date, time, flight number. I landed at Dulles airport and he was there waiting for me. He insisted on carrying my luggage. He was an elderly gentleman and I was relatively a lot younger. But he insisted and you don't argue with a U.S. senator. He loaded the luggage into his car and we took off.

His car was an old Chevrolet, dented and scraped and battered. It reminded me of Norbert Wiener's car. Driving with the senator was pretty scary. He was not a very good driver.

We got along together very well, two Democrats with similar outlooks. He was very strong on national defense and very liberal on domestic issues. You could call him a conservative Democrat. He was my kind of Senator. We arrived safely at his home.

Mrs. Helen Jackson, the lady of the house, had prepared dinner for all. The Jackson's son Peter was there. Anna Marie wasn't there of course, she was at Sanford. During dinner and after, we talked about a lot of things. I knew of the senator's military interest so I told him about an experience I had spending a week under the sea in a nuclear submarine. He said Admiral Rickover would like to hear this. I will invite him to join us for lunch at the Senate tomorrow. I said that he's probably way too busy for that, but the senator said no, he has plenty of time, he is retired. So he called the Admiral and invited him for lunch.

Rickover was the father of the US Nuclear Navy. He became a very famous Admiral and was frequently written about in newspapers. He had a very crusty personality and this became very well known. These are good reasons for the Navy to fire him. The only trouble was that he was more competent than all the other Admirals put together and most important, he had a special relationship with Congress. Nobody knew what this was, but the rumor said this was true. So I asked Senator Jackson what Rickover's special relation with Congress was. Jackson said "the special relation was me." The two men had been great friends for years. Year after year when the Navy brass tried to fire him, Jackson was able to save him. Jackson was the chair of the Senate Armed Services Committee. Jackson many times saved Rickover's nuclear submarine project.

Finally it seems that the Navy got rid of him. I asked the Senator how did that happen? Simple. A few weeks after Reagan became president, he summoned the famous Admiral Rickover to the White House. He just wanted to meet him. Rickover insulted the president, and Jackson couldn't save him.

That evening, Jackson told me another Rickover Navy story. There was a meeting of the Joint Chiefs, the top military leaders of the country. The subject of nuclear propulsion for submarines came up. The navy's top Admiral said that this would be a long way off, 25 years at least. But what the Joint Chiefs didn't know was that the world's first nuclear submarine, SSN Nautilus, on that very day was out at sea having its first sea trials under nuclear power. It was amazing that Rickover could pull this off and keep the secret from the navy's top brass.

The next day, I arrived at Jackson's office at noon time and his secretary informed the Senator that I had arrived. He popped out of his inner office and said Bernie, it will be a few minutes before I could join you. I have a couple of things to clear up before lunch. Okay. A few minutes later in came this little man that I easily recognized as Admiral Rickover. He was seated near the doorway to the hall of the building and the constant in and out traffic was annoying him. He got up and sat down next to me, away from the noise. I said to him, Admiral Rickover, I am Professor Bernard Widrow from Stanford University. You and I are supposed to have lunch with Senator Jackson. He replied with "Humpf!".

We sat in silence for a few minutes until the senator came out and invited us into his inner office. He was the most senior Senator so he had a gigantic office. One whole wall was bookshelves filled with law books. Rickover took off like a rocket and went to the bookshelves. He pulled a book and brought it over to me, ran his finger over the top of the book and showed me with triumph, "See the dust on my finger. He never read these books." That was my introduction to Admiral Rickover.

I knew some things about Rickover since stories about him were often in the press. I also knew about him from talking with the Captain and Officers on the submarine, SSN Trepang. Every officer on a nuclear-powered submarine or nuclear surface ship had to be approved by Admiral Rickover. They all interviewed with him and he asked tough questions. He was a tough guy.

We were walking from the Senate Office Building to the Capitol, the location of the Senate and the House of Representatives. Jackson said to Rickover, Admiral you are an electrical engineer and Bernie is a Professor of Electrical Engineering. Bernie might want to ask you some questions about Electrical Engineering. Rickover started to cough and choke. I didn't ask him any questions, but I thought of the naval officers who had to face the Rickover quiz.

Coming into the Senate dining room, Jackson was introducing us to his Senate friends and colleagues. I recognize a lot of famous Senators. It was a nice experience. Jackson explained that the Senate is a club of colleagues, Democrats and Republicans. His Senate friends were a mixture of both. They liked each other, respecting each other, socialized together, and often disagreed on policy. That made no difference. If Jackson were alive today he would be heartbroken. The Republicans considered the Democrats the hated enemy. The Democrats don't hate but they look at Republicans as cretins, kowtowed by a big fat bully.

We had a nice lunch and for me it was a memorable day. I tried to tell Admiral Rickover a little bit about my successful trip in the submarine and about the experiments that we performed and he was hardly listening. It's too bad. He would have been really pleased with what we accomplished. Every submarine and surface ship of the U.S. fleet is carrying now the LMS algorithm.

Another recollection. When we were out to sea on the submarine, the word spread to the sailors that a Stanford Professor was on board. Before you know it, I was holding office hours in the torpedo room. The most popular question was, how do you get into college?

25.14 Edward Teller

Edward Teller was widely known as the father of the hydrogen bomb. He played a major role in the development of the first atomic bomb at Los Alamos, NM. He pushed hard for the development of the thermonuclear weapon, the hydrogen bomb. When the development was finally approved he led the project. Years later, during the Reagan presidency, he persuaded the president to pursue a program for defense against intercontinental ballistic missiles. The program was known as "Star Wars". Billions was spent with no missile defense resulting.

Toward the end of his life he joined the highly-regarded Hoover Institute at Stanford, a "think tank" with evident far-right leanings, a seeming anomaly on the liberal Stanford campus. Being on the Hoover staff, he was allowed to own a home on the Stanford campus. His home was walking distance from my home on the Stanford campus. He lived in this house until his death in 2003 at the age 95.

I first met him when he was 94 years old. His wife had died a number of years previously, so he was living in the house alone except that he had three full-time ladies who provided care 24 h per day. He could hardly see and hardly hear, but his mind was sharp. He was a nice old gentleman.

I was invited to his home to try to help him with his hearing. He had been to an Audiology clinic and had an audiogram. This is a plot of hearing loss measured at about a dozen frequencies within the audio spectrum. He wore behind the ear hearing aids (BTE) equipped with telecoils. These are coils of wire inside the hearing aids that are capable of receiving signals by magnetic induction.

Telephone receivers generate magnetic fields modulated by the telephone signals. By placing the telephone receiver close to the hearing aid, the telephone signal could be heard through the telecoil and the hearing aid giving sound matched in spectrum to the hearing loss shown by the audiogram. The telecoil signals are loud and clear and enable communication by telephone.

I was the inventor of a hearing device that enabled person-to-person communication even if the wearer had a hearing loss in the profound range, about 80 decibels or so below normal. The device was an array of six microphones that were configured to form an audio beam about 60° wide. The device was worn like a necklace. The supporting loop was conductive, and the array's output drove the loop which created a magnetic field carrying the audio signal. The magnetic field was a wireless link between the microphone array and the telecoil. The array's signal could be heard through the hearing aid.

I went to Dr. Teller's house and brought an array with me. I had already programmed it to match his hearing loss. I placed the array on him, turned it on, switched his hearing aids to telecoil position and I was speaking with him through the array and his hearing aids. He was able to understand every word.

He had a lot of technical questions about the array, how does it work? Why is it so clear? How does the internal signal processing work? He was a physicist and he wanted to know. I explained to him that the array only receives signals in the beam and therefore most of the surrounding noise was eliminated. I also explained that the beam almost totally eliminated the effects of reverberation. Also, the array eliminated feedback that could cause a hearing aid to howl. All this made speech clearer and louder, enhancing intelligibility.

He always sat in a big stuffed chair. I explained to the ladies where to sit so their voice would be in the beam. Several visits were necessary in order to optimize the frequency response of the array for him. I had to coach him on the use of the controls of the array. He was trying to second-guess how to control the array. I straightened him out and it was a success. He could easily understand what the ladies were saying.

In all the visits, I never discussed politics or his political history. I would have liked to ask questions about his role in development of atomic and hydrogen Bombs. I was very curious about his interactions with President Reagan, how he persuaded the president to make Star Wars the signature piece of his administration. I never asked him about this. I did not think that appropriate. I was there to help him.

He used the array to hear the women when they read to him. The reading was physics texts in Hungarian, his native language. He never lost his thirst for knowledge. I was very sad when I heard about his death. My "patient" died.

25.15 Women

The famous people of this chapter are all men, brilliant leaders and many were the fathers of this or that technology. There were no mothers. I only reported direct experience, and there were no mothers.

Women are now taking their place as innovators of science, technology, and engineering. I hope soon to become acquainted with a mother of this or that technology. It's coming.

25.16 Summary

Here are more stories that were favorites of my students. A very favorite story was about my meeting Richard Feynman. He was the most brilliant person that I had ever met during my lifetime. I met with a group from MIT Lincoln Laboratory while doing a study for DARPA on the state of the art of artificial neural networks in the USA and in the rest of the world. We met in a conference room. There were about 15 of us. We were expecting words of wisdom from Professor Richard Feynman. But this was not to be. He started by asking a question, "what is a neuron?". I was the chairman of the study so everybody looked at me to answer the question. There were more questions after that and I was "on the spot." I found myself taking the Ph.D qualifying exam all over again.

Reference

1. (32) truckbacker - youtube. https://www.youtube.com/watch?v=Ym9Ts4GBt7Y. (Accessed on 10/04/2020)

Chapter 26
Ancient History

Abstract Original photos and drawings from my laboratory in the early to mid-1960s, at the dawn of the field of artificial neural networks, are shown in this chapter. The photos depict apparatus that we built and experiments we had done in developing what today would be called machine learning. Shown are the original ADALINE (adaptive linear neuron), and MADALINE (multiple adaline), a small two-layer network of ADALINEs. Adaptation was done with the LMS (least mean squares) algorithm. An electronically controlled synapse called the memistor (resistor with memory), based on electroplating, was developed. With it an electronic neuron was possible and was demonstrated. Applications were to trainable systems. One of these was the now famous "broom balancer". In addition, an IBM1620 computer was used to simulate small networks.

26.1 Introduction

I would like to end this book with the beginning, with ancient history. I have a collection of original photos and drawings from my lab in the early 1960s. From this early work, many subsequent developments in artificial neural networks, adaptive signal processing, and adaptive control systems came about. Also this was the period when first experiments with unsupervised learning were done with ADALINE. Years later, this led to the % Hebbian-LMS algorithm.

26.2 Learning Experiments with ADALINE

Figure 26.1 shows a block diagram of ADALINE (adaptive linear neuron). The input signal vector is X_k. The weight vector has components w_0, w_{1k},...w_{nk}. The bias weight w_{0k} has a fixed input a $+1$. The linear output that we have been calling (SUM) is the product of the input vector and the weight vector. The LMS algorithm

B. Widrow, *Cybernetics 2.0*, Springer Series on Bio- and Neurosystems 14,
https://doi.org/10.1007/978-3-030-98140-2_26

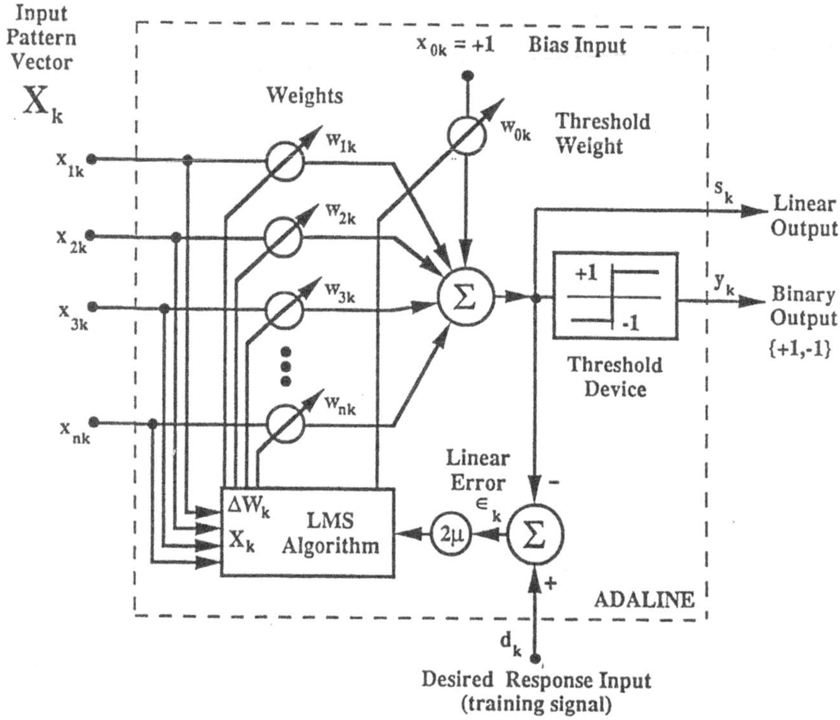

Fig. 26.1 A block diagram of ADALINE

is used to adapt the weights. The error is the difference between the desired response and a linear output, the (SUM). The LMS algorithm adapts the weights to reduce error. The LMS algorithm is represented by a functional box whose inputs are the components of the X_k vector and the error ϵ_k multiplied by 2μ. The outputs of the LMS box control the adjustments of the weights. This ADALINE is a trainable binary classifier. A binary output, either $+1$ or -1, is obtained with a quantizer, a threshold device, whose input is the linear output or (SUM).

The LMS control signals are analogous to hands reaching up to turn the weights of knobby ADALINE, pictured in Fig. 26.2. The symbolism was inspired by Claude Shannon's box with the mechanical hand.

The ADALINE of Fig. 26.1 resembles a neuron and its synaptic inputs. The weights are like synapses. The summer is analogous to the neuron's summation which is done electrochemically in the neuron. The threshold device is like the non-linear activation function of a neuron.

ADALINE is configured for supervised learning. For every input vector or input pattern, there needs to be a desired response input. For every input pattern, you need to tell ADALINE that the input pattern should be in either the $+1$ class or the -1 class. Nature's neuron learns without supervision. Nature's input signal vectors or input

Fig. 26.2 Knobby ADALINE

patterns have all positive components, not positive or negative. Nature's neuronal outputs are zero or positive, not positive or negative. So ADALINE is similar to nature's neuron and synapses, but not quite the same. Nevertheless, ADALINE was a working trainable binary classifier.

A photo of knobby ADALINE is shown in Fig. 26.2. This ADALINE was built in 1960 at Stanford by an electronic technician of uncommon skill. The circuits were designed by Ted Hoff (Marcian E. Hoff Jr), my first Stanford Ph.D. student. The two of us worked in close collaboration in 1959 and thereafter. We both invented the LMS algorithm and were granted a patent.

For any given set of training patterns and their associated desired responses, the mean square error or MSE is a quadratic function of the weights. The negative gradient is followed step by step in order to find the bottom of the quadratic bowl where the MSE is minimum. The weight vector that minimizes the MSE is the Wiener weight vector. Figure 26.3 is a vintage drawing of the quadratic bowl with steepest descent.

For a given weight vector, the components of the gradient were originally determined one component at a time. Each weight was varied forward and back and the difference in their respective values of MSE was observed. The MSE difference divided by the amount of weight change gives one a gradient component. This is elementary differentiation. The process had to be repeated for each gradient component. To determine the gradient, a considerable amount of data would be required.

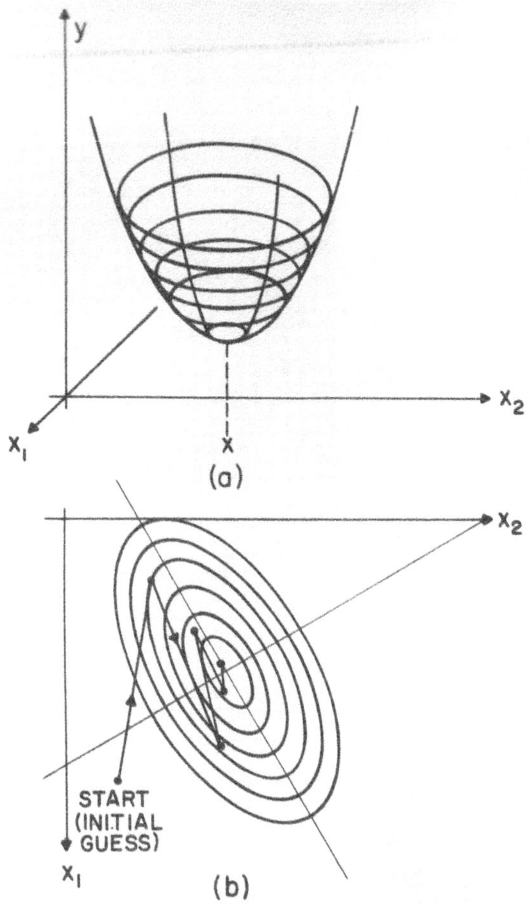

Fig. 26.3 Quadratic MSE surface

The LMS algorithm is also based on the method of steepest descent. It derives the gradient in a very different manner however. The basic idea is to take a single sample of error, square it, and call that the mean square error. This is of course not true. The square of a single sample of error is an extremely crude representation of the true MSE. Nevertheless, finding the derivative of the square of the single error sample, doing this algebraically, a crude estimate of the true gradient is found. All components of gradient are found at once. This led to a learning algorithm that is so simple and easy to implement in code or in hardware that it is used worldwide.

Years later, I was able to prove that the expected value of the LMS gradient is equal to the true gradient, that the LMS gradient is unbiased. When used in adaptive filters, the LMS algorithm uses input data in the most efficient way possible if the eigenvalues of the R-Matrix are all equal. It is also most efficient with non-stationary statistical inputs. Hoff and I didn't know all this in 1959. We couldn't have imagined that our simple learning algorithm would become the world's learning algorithm.

But when I first wrote that algorithm on the blackboard during my first meeting with Tedd Hoff, I knew that we had discovered a learning algorithm that was simple but quite profound. I thought that we had discovered the secret to life.

Referring to the photo of knobby ADALINE in Fig. 26.2, one can see a 4×4 array of switches with which input patterns were inserted. A 4×4 array of neon lights above the switches gave an indication of the inserted pattern. The weights were implemented as a 4×4 array of potentiometers. The extra potentiometer was the bias weight whose input was fixed at plus one.

When an input pattern was presented with the array of switches, the (SUM) was formed as the inner product of the input pattern and the weights. The (SUM) was displayed by the microammeter. The middle of the right half of the scale was $+1$, the middle of the left half was -1. The desired response for each input pattern was given as either $+1$ or -1. The weights were changed to bring the meter reading closer to the desired response. The input pattern was added to the weights to make the meter reading more positive, if that were needed. Otherwise the input pattern was subtracted from the weights to make the meter reading more negative, if that were needed. This is correcting the error, making it smaller with each adapt cycle.

Over a set of training patterns, reducing the error for each pattern, repeating the training set over and over again, this iterative process converges with the weight vector brought close to the Wiener solution.

The results of a training experiment are shown in Fig. 26.4. The experiment was done in 1960. The training patterns had 4×4 pixels, each pixel having the value of $+1$ or -1. The two T-patterns were trained for $+1$, i.e. the (SUM) was meant to be $+1$. The two F-patterns were trained for -1. The two G-patterns were trained for 0. In this experiment, ADALINE was trained for three linear output (SUM) levels. The outputs were not binary but ternary. LMS algorithm obliges.

A learning curve is plotted in Fig. 26.4. The horizontal axis is calibrated in terms of number of training pattern presentations. The vertical axis is calibrated in terms of sum of the squares of the errors over all the training patterns. These patterns were selected for presentation in a random sequence. With each presentation, the weights were adapted slightly in a direction to reduce error for that pattern. After each adaptation, without changing the weights, the training patterns were inputted one at a time and the errors were noted. The sum of the squares of the errors were plotted. The resulting learning curve is theoretically a sum of noisy exponentials. The number of exponentials corresponds to the number of eigenvalues of the R-Matrix, the input autocorrelation matrix. If all the eigenvalues are equal, the learning curve would be a single noisy exponential. The noise comes from the noisy LMS gradient. The instantaneous gradient is noisy, and adaptation needs to be iterative with small steps. On average, the small steps progress along the negative gradient.

The training patterns are binary vectors. They happen to be linearly independent vectors. As such the theoretical asymptotic error should be zero. The learning curve shows that the error did not go to zero. This is because of the quantization of the weights. ADALINE's weights were not smooth potentiometers. They were implemented with 19-position rotary selector switches. The quantization was introduced deliberately to ease the mental calculation when manually adapting the weights. For

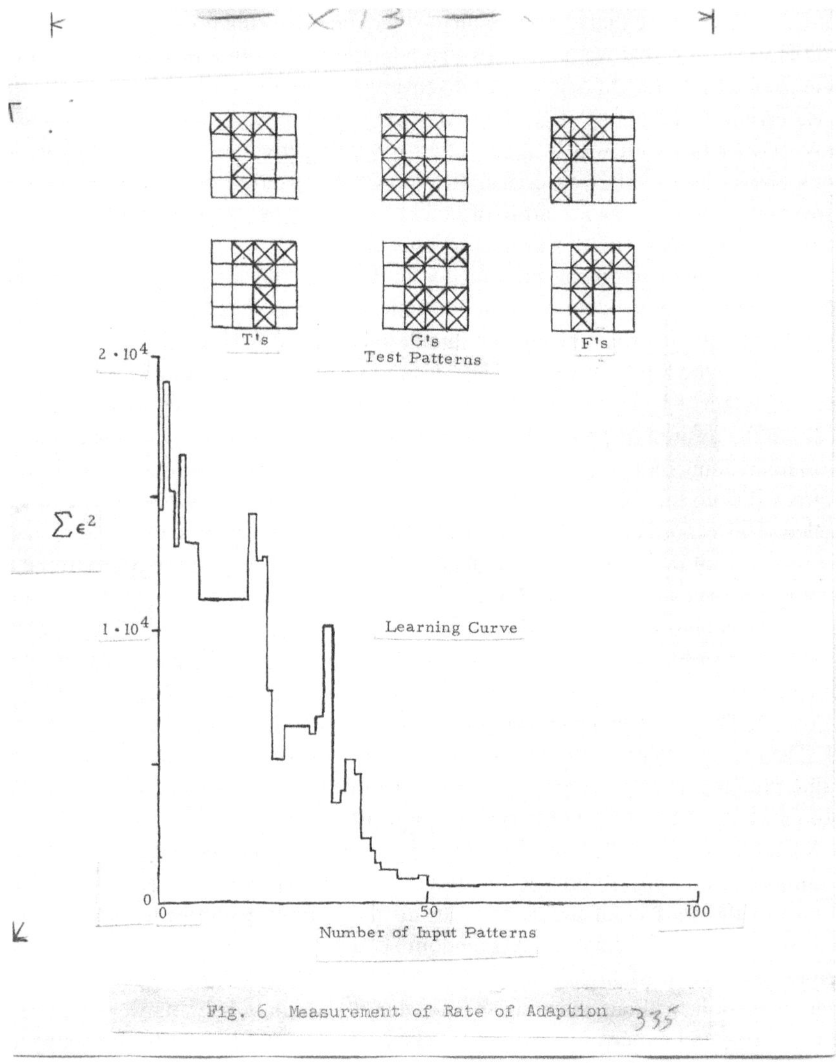

Fig. 26.4 A learning experiment

a given input pattern, the size of the error could be seen from the meter. A small error called for one click's worth of change. A bigger error would get two clicks, bigger yet would get three clicks. The weights were clicked up or down. The input pattern was added to the weights to make the (SUM) more positive if that were required to reduce error. Otherwise the input pattern was subtracted from the weights. When the errors of all the training patterns become small so that further change would not be

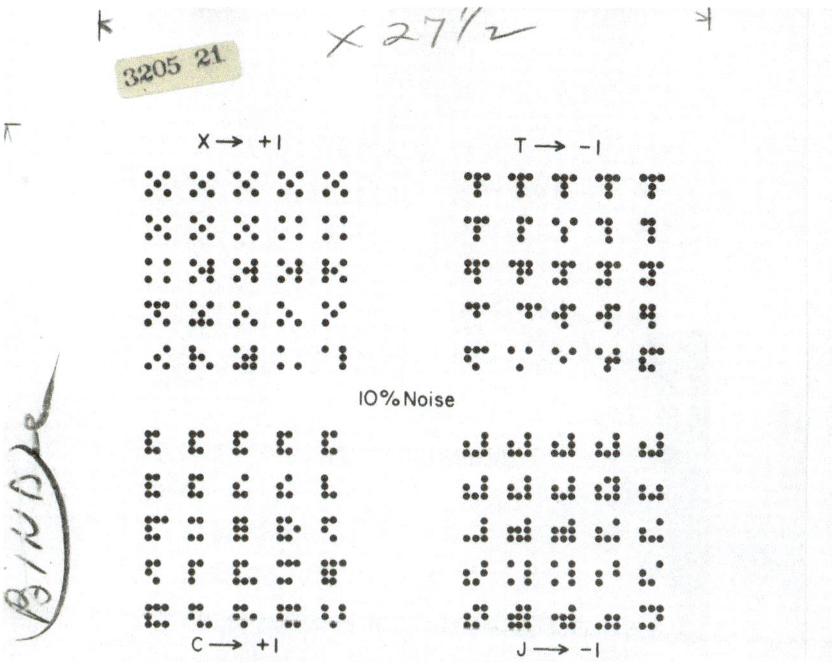

EXPERIMENT #	PATTERNS ADAPTED ON	NUMBER OF ERRORS	MISADJUSTMENT
1	95, 79, 07, 60 73, 61, 08, 02, 72, 26	25	$M = \frac{25-12}{12} = 108\%$
2	70, 69, 52, 55, 32 97, 30, 38, 87, 01	19	$M = \frac{19-12}{12} = 58\%$
3	65, 12, 84, 83, 34 38, 71, 66, 13, 80	20	$M = \frac{20-12}{12} = 67\%$
4	07, 42, 85, 88, 63 35, 37, 92, 79, 22	28	$M = \frac{28-12}{12} = 133\%$

Fig. 26.5 Training with noisy patterns

fruitful, the process stalls and the weights are no longer adapted. One can see this from the learning curve.

Figure 26.5 shows adaptation with a large number of noisy training patterns. The X-patterns were trained to be +1s. The T-patterns for −1s. The C-patterns were +1s, and the J patterns were −1s. After training, 12% of the training patterns were misclassified. Not bad for a single neuron.

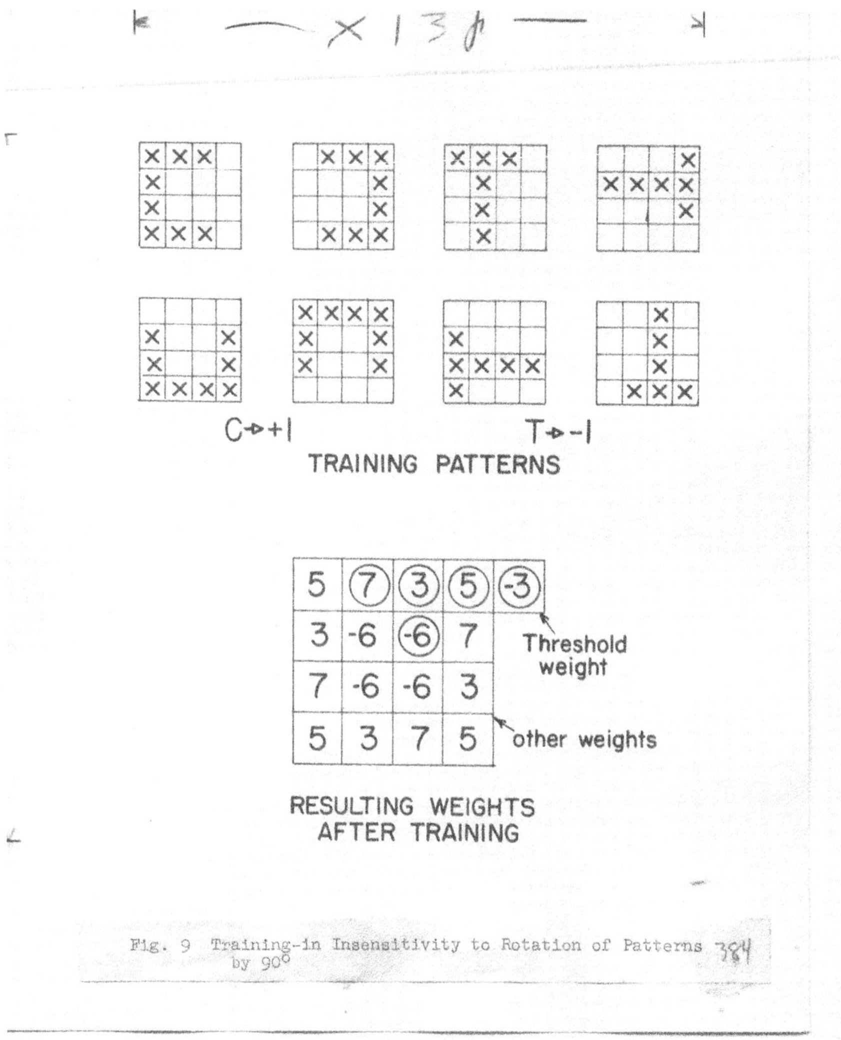

Fig. 26.6 Training-in insensitivity to rotation of patterns by 90°

Figure 26.6 shows results of an experiment intended to demonstrate a simple form of generalization, insensitivity to 90° pattern rotation. A C-pattern was presented with all four rotations and ADALINE was trained to produce a +1 output for all cases. A T-pattern was presented likewise and ADALINE was trained to produce −1 outputs for all four rotations. The resulting weights are shown in the figure. The weights automatically developed a 90° rotational symmetry. Any input pattern would produce a (SUM) output that was consistent with 90° rotations.

26.3 The Memistor

Knobby ADALINE pictured in Fig. 26.2 was trained manually. The potentiometer knobs were adjusted one at a time. Training was slow and tedious, but this ADALINE was a good machine for illustrating the working of the LMS algorithm. You could see the weights and how they change with adaptation. It was possible to do repeatable experiments with it.

I liked knobby ADALINE (I still use it to teach my students) but I really wanted a faster, more automatic ADALINE. What I needed were electronically controlled weights instead of the mechanical ones. I was thinking of a beaker containing an electrolyte solution. Two electrodes inserted into the solution would have a resistance from one to the other that would depend on electrolyte chemistry. Putting a third electrode into the solution, applying a current to it would change the resistivity of the solution and this would make weight change possible.

I described this idea to Ted Hoff and he said no, what you want to use is electroplating for weight change. I was at Ted's apartment and he had a nice electronic shop and some chemicals. He took a piece of paper and drew a pencil line on it, then contacting the two ends of the line with clip leads that were connected to an ohmmeter. The resistance was measured, about 50 MΩ. With an eyedropper, he covered the pencil line with a solution of copper sulfate and sulfuric acid. This is a solution that is used for electroplating copper. The solution is conductive and the ohmmeter was now reading about 1 MΩ. A third electrode, a piece of copper wire was stuck into the fluid and current was applied making the pencil lead negative with respect to the copper electrode thus attracting copper ions from the solution. One could see metallic copper deposited on the pencil line, and the ohmmeter reading was now about 1 kΩ. This big change in impedance was all we needed to see. The experiment had to be done quickly as the acid ate the paper and destroyed the experiment.

Instead of pencil lines, we decided to use pencil leads, graphite inserts for mechanical pencils. The next day, I went to the Stanford bookstore with an ohmmeter. I went to the counter where they had pencil leads and put the ohmmeter on the counter. The clerk asked me if I could be helped. I asked for some pencil leads that have the highest electrical resistance. With a puzzled look, the clerk produced all the different types of pencil leads that they had. I tested them and the winner was Fineline Type-H. It measured 9 Ω.

Figure 26.7 is a photo of a pencil-lead "memistor". Memistor is a resistor with memory. When the electroplating current is applied, the resistance changes. When this current is stopped, the resistance is unchanging. The value of resistance is analog memory. The resistance depends on the thickness of the copper plating on the pencil lead. The memistor is an electrochemical weight, an electronic synapse. During learning, the weight changes. When not learning, the weight value persists.

We electroplated copper tabs on the ends of the pencil lead seen in Fig. 26.8. THe ends of two flexible insulated copper wires were soldered to the tabs. Fingernail polish obtained from Mrs. Widrow was used to paint the solder joints to keep away the corrosive copper sulfate and sulfuric acid. The flexible wires were tied to a support

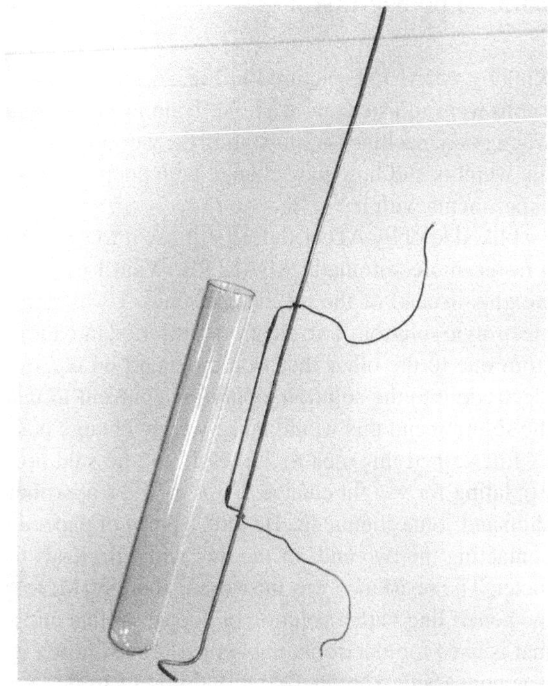

Fig. 26.7 A pencil-lead memistor

rod, a piece of No. 14 copper wire. The whole assembly was inserted into a test tube which was then filled with the copper plating solution, copper sulfate and sulfuric acid. Three wires emerged from the test tube.

An ADALINE was constructed using pencil-lead memistors. A photo of it can be seen in Fig. 26.8. Input patterns were presented with a 3×3 array of toggle switches. The number of memistors was 3×3, plus one serving as the bias weight. A glass Pyrex tray from Mrs. Widrow's kitchen was used to support the corrosive memistors that were inserted into a drilled block of wood.

The pencil lead ADALINE worked very well for about 2 weeks, when the graphite disintegrated due to the corrosive action of the copper plating solutions in the test tubes. In Fig. 26.8, a battery can be seen. Its purpose was to supply DC current for plating. The adaptation of the weights involves turning on and off (dc) current. The input signal to the weights was (ac). It was convenient to use 60 Hz from the electrical supply. The output (SUM) was of course 60 Hz and this was displayed with a Hewlett-Packard oscilloscope. The horizontal sweep was synchronized with the power line. In sensing the weight values, (ac) was used as (dc) would disturb the plating.

The results of a training experiment are given in Fig. 26.9. The pixel inputs were $+1$ or -1. One phase of the 60 Hz signal was $+1$, the opposite phase was -1. The (SUM) output was of positive or negative polarity, positive corresponding to the

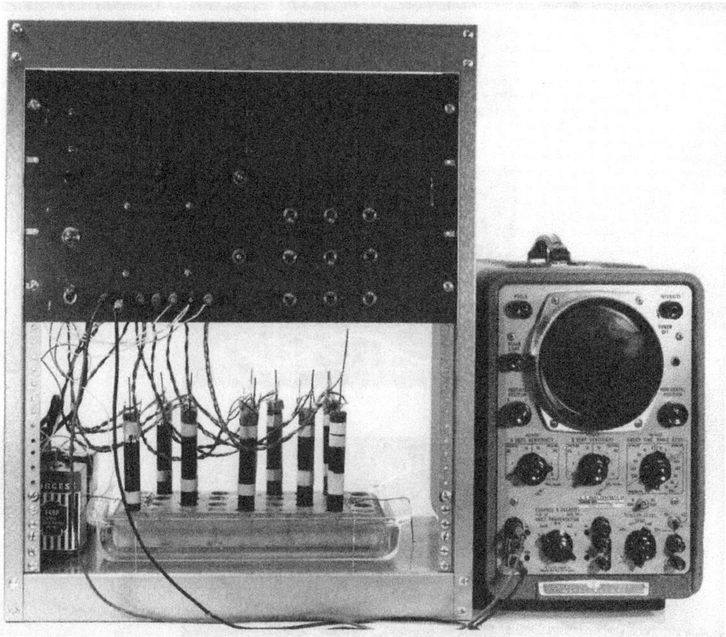

Fig. 26.8 A memistor ADALINE with pencil-lead memistors

phase of +1. The oscilloscope traces shown in Fig. 26.9 are Polaroid photos taken after training was completed.

Trained for +1 outputs were the X-pattern, the C-pattern, and the C-pattern rotated by 90 degrees. Trained for −1 outputs where the T-pattern, the J-Pattern, and the T and J-patterns rotated by 90 degrees. The polaroid photos demonstrate that the seven patterns with trained in essentially perfectly. Perfect training was possible since the weights were analog, continuously variable, and the seven training patterns happen to be linearly independent vectors.

We needed memistors that were much smaller and that lasted much longer than 2 weeks. We tried many designs with only partial success. A favorite material was glass coated with tin oxide. We could plate on to this, but sometimes the tin oxide separated from the glass substrate leading to device failure. One of the designs that we tried is pictured in Fig. 26.10. The memistors were arranged in a plastic block. Ted Hoff as seen here injecting plating solution into the memistors before being sealed with a plastic cover.

For more than a year, we tried many things without real success. I consulted with my next door neighbor Bill Shockley and he knew a lot about electroplating and many other things. He had some suggestions, but nothing really worked. Our friend Maurice Hanafin suggested that we talk to Charlie Litton who was the founder of Litton Industries and a friend of Hanafin. Charlie was retired, a billionaire living in a small town called Grass Valley, California. This town is in the Gold Country, near

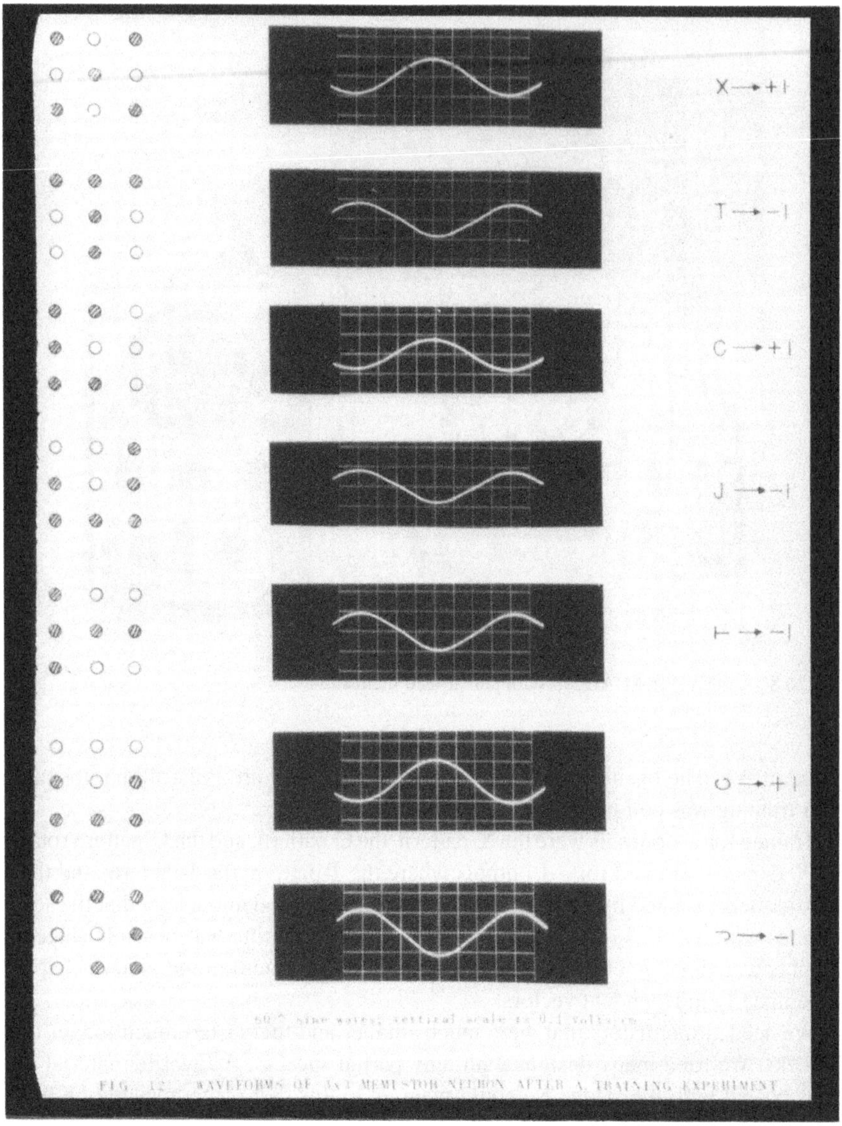

Fig. 26.9 Oscilloscope traces display (SUM) after training pencil-lead memistor ADALINE

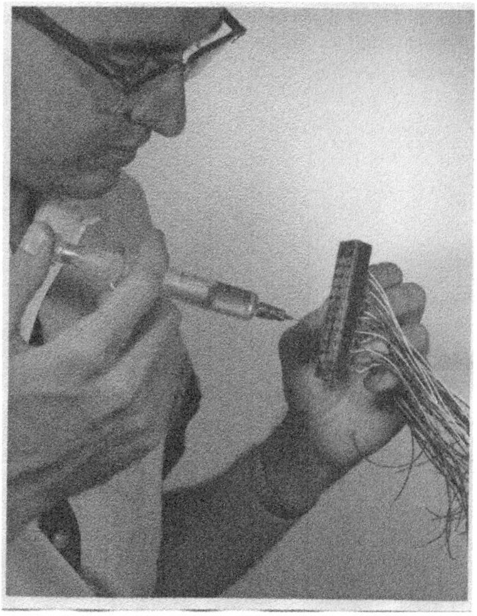

Fig. 26.10 Ted Hoff injecting plating solution into a block of memistors

where the founders of California in 1849 were mining and panning for gold. Hoff and I went with Hanifin to see Charlie Litton.

Charlie was a "hands-on" guy. He had his workshop in a huge barn. He had a glass lathe and was doing neat things with glass. We were inspired by Charlie and his powers with the glass lathe.

New idea. Make a glass bulb with a resistor inside with leads coming out of the glass, like Edison's incandescent bulb. We were able to make the resistor with a fine wire made from an alloy of noble metals. I have forgotten which, perhaps Platinum and Rhodium. The bulb was made on a glass lathe with its resistor element, and it was filled with the corrosive copper plating solution. The noble alloy was expensive but necessary to survive in the solution. A third electrode, a copper source was also placed inside the glass envelope with a wire coming out of the glass.

The glass bulb was quite small, about the size of a flashlight bulb. The three electrodes were spot welded to a T0-5 header. The T0-5 can was mostly filled with epoxy, and the bulb was quickly pressed into the molten epoxy and then turned right side up to let the epoxy draw down and cure. The result was a very sturdy device. A photograph of the finished device is shown in Fig. 26.11. It looks like a transistor. It really is equivalent to a transistor with a built-in integrator. The weight value was proportional to the conductance of the resistor which is equal to the conductance of the noble wire plus the conductance of its copper coating. The conductance of the copper is proportional to the integral of the current in the third electrode, the copper source. With no plating, the resistance was 10 Ω. With full plating, the resistance was 1 Ω. A 10 to 1 ratio was fine.

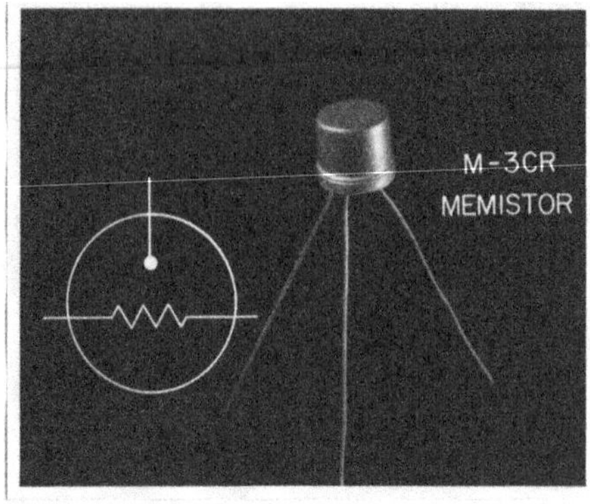

Fig. 26.11 A glass sealed memistor and its circuit symbol

Fig. 26.12 A memistor ADALINE with glass sealed memistors

Using these devices, a memistor ADALINE was constructed. It is pictured in Fig. 26.12. At the present time this ADALINE is about 55 years old. It still works very well. Input patterns can be presented with the 4×4 array of switches. Pressing the adapt control to the right, the input pattern is added to the weight vector and the meter reading increases slowly in the positive direction. Pressing this control to the left, the input pattern vector is subtracted from the weight vector and the meter reading goes in the negative direction. During training, you put in the input pattern and operate the adapt control to bring the meter reading to the desired response.

Releasing the adapt control, adaptation stops (the (dc) adapt current is shut off) and the meter reading stays. Now another pattern can be inserted and the process is repeated. All weights are adapting simultaneously, not one at a time.

The ADALINE seen in Fig. 26.12 is the very device that I carried in my suitcase to Cornell University to visit Professor Frank Rosenblatt and to see his Perceptron. I demonstrated learning with ADALINE and explained how the device and circuits worked. In the back of the room I saw attempts to make memistors by his students. They didn't know how hard it was to do this.

Some interesting observations were made from working with a memistor ADALINE. The memistors are analog devices. They are all similar to each other, but no two are alike. Theoretically, according to the LMS algorithm, all the weights should change at the same rate. The fact that this does not happen is hardly noticed when training ADALINE. You hold the adapt control until you get what you want. Repeating the training patterns over and over, the process converges. Feedback inherent in the algorithm overcomes the differences in rate of adaptation among the weights.

Some of the weights do not adapt at all. This may be due to memistor failure or to circuit failure. In spite of this, ADALINE converges and these defects are hardly noticed during training. The weights that are adapting take over the function of the defective weights. You lose a little less training capacity. You can train in a smaller number of patterns. The feedback of the LMS algorithm allows training to work even with some defective circuits. For this reason, it is hard to debug adaptive circuits. For the same reason, **THE CIRCUITS OF THE BRAIN WILL WORK WITH DEFECTIVE PARTS. THESE CIRCUITS CAN ADAPT AROUND THEIR OWN INTERNAL FLAWS.**

A circuit diagram of the memistor ADALINE as shown in Fig. 26.13. The signal flow is (ac). The adapt current is (dc). The (ac) signal flowing through the memistors are currents that are summed by a low-impedance resistor to produce the (SUM) output voltage. This is Kirchoff addition.

There is a worldwide effort underway to build neural hardware, neural chips, some digital, some analog. The analog people should find the experience here to be of direct interest.

26.4 MADALINE

I was very satisfied with the performance of ADALINE and the LMS algorithm but my interest evolved toward building networks of ADALINEs. The goal was to have greater capacity, allowing training of a large number of patterns, and to have multiple outputs so there could be a large number of output classes. ADALINE with a single binary output could afford only two classes.

With a network of multiple neurons in the first layer whose outputs provided inputs to a second layer, a question arose about desired responses for training the neurons. With multiple layers, the desired responses of the output layer neurons were given. The desired responses for all the other layers were unknown. Without desired responses, neurons cannot be trained. I could not find a way to train a multi-layer network.

Thinking of a two layer network, adapting the first layer and fixing the weights at the second layer, by knowing the weights of the second layer, I was able to transfer the desired responses of the second layer to the first layer.

Figure 26.14 shows a two layer network with a single output. For every input training pattern, the desired response at the output was given. The first layer neurons could be trained if the output layer neuron had simple weights. If all the weights of

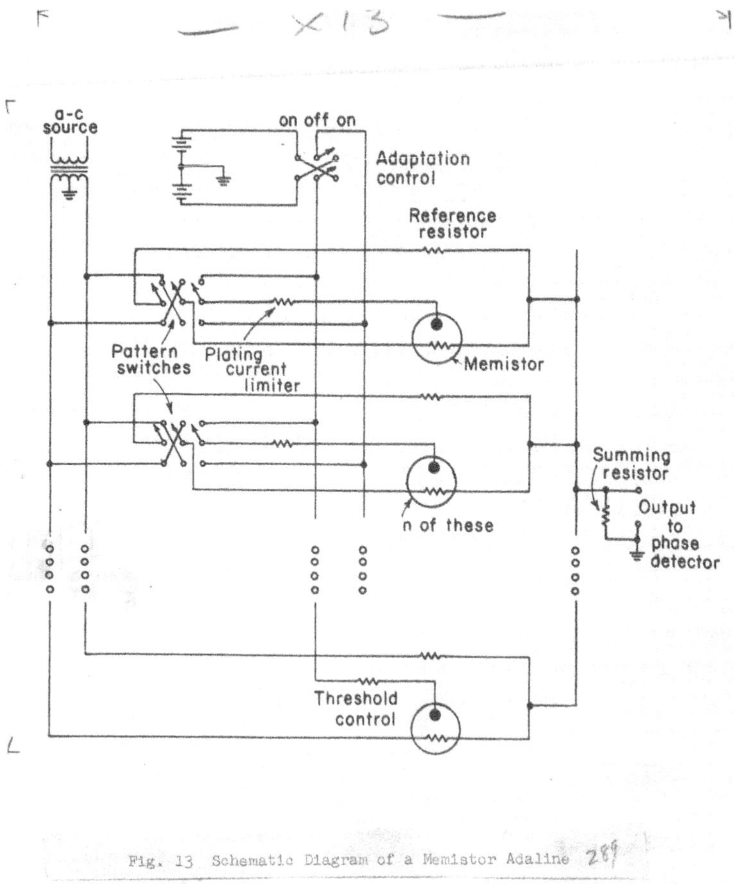

Fig. 26.13 Schematic diagram of a memistor ADALINE

the output neuron were made to be equal, the output neuron becomes a majority vote taker. With this knowledge, desired responses for the first layer could be inferred. We called this network MADALINE, for many ADALINEs.

Figure 26.15 illustrates how the desired responses for the first layer were decided. The first layer outputs were binary. Assume that the second layer output was based on a majority vote. During training, when an input pattern is presented and the desired output is given, if the binary output of the output neuron is correct, do not adapt. Adapt the first layer only to correct the response of the output neuron. If, for example, the desired output is +1 for a given input pattern and the initial output response is −1, adaptation is needed. Among the first layer neurons whose output or outputs are −1, choose the neuron whose (SUM) is closest to 0 and adapt it to give a +1 output. If that corrects the second layer output, stop there. If not, adapt the next neuron whose (SUM) is closest to 0. And so forth until the desired response is

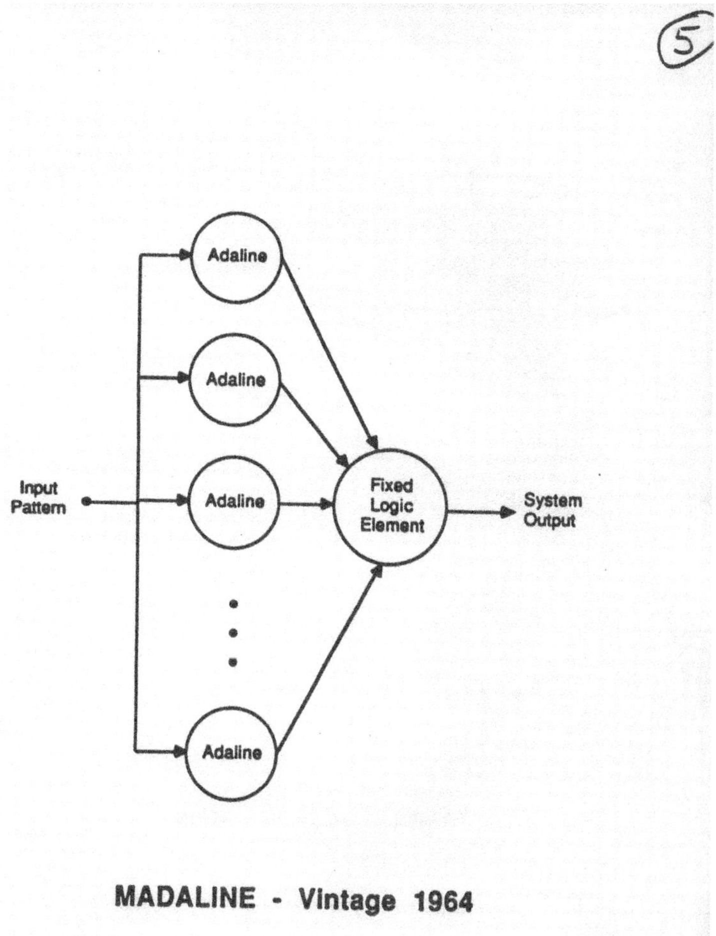

Fig. 26.14 Block diagram of MADALINE

obtained. With each training cycle, the weights are changed as little as possible, but changed to give the desired response at the output. **THIS IS THE PRINCIPLE OF MINIMAL DISTURBANCE. ALL LEARNING ALGORITHMS WITHOUT EXCEPTION OBEY THIS PRINCIPLE.**

This network trained well and greatly expanded what could be done with a single neuron, but I was not completely happy with it. I wanted to be able to adapt both layers and did not succeed. This problem was solved years later in the early 1980s by Paul Werbos, the inventor of the backpropagation algorithm.

Figure 26.16 is a photo of a rack mounted network that we called MADALINE I. It is being trained by Bill Ridgeway, one of my Ph.D. students. There were six memistor ADALINEs in the first layer, and two majority vote takers in the second

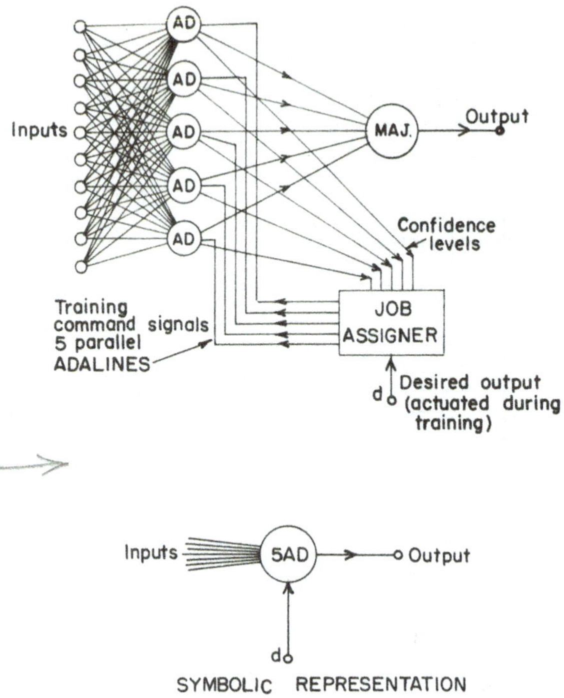

Fig. 26.15 MADALINE with majority vote-taker

layer. This MADALINE could be trained to place input patterns into one of four classes.

We were doing experiments with speech recognition and were working on several different pattern recognition problems. One of these was a control systems problem, the "broom balancer". Figure 26.17 is a diagram of the broom balancer. A wheeled cart on a track was motor-driven to balance its pendulum. Control of the motor was done by MADALINE. How MADALINE could be trained to do this is illustrated by Fig. 26.18.

Fig. 26.16 Bill Ridgeway training MADALINE I

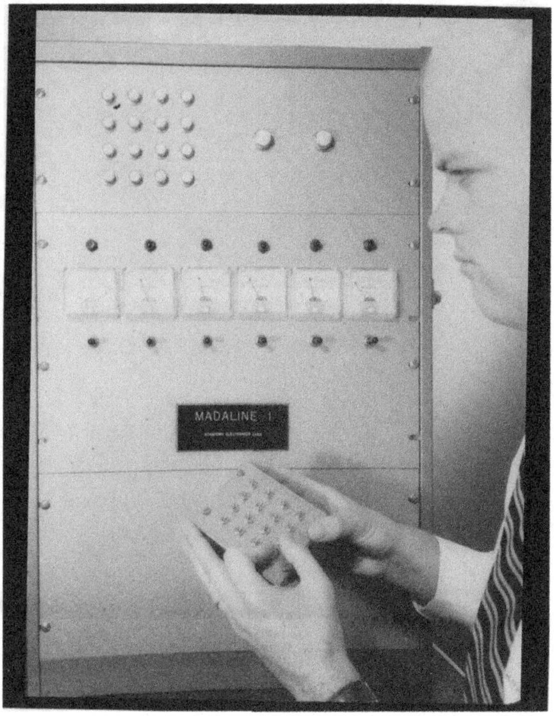

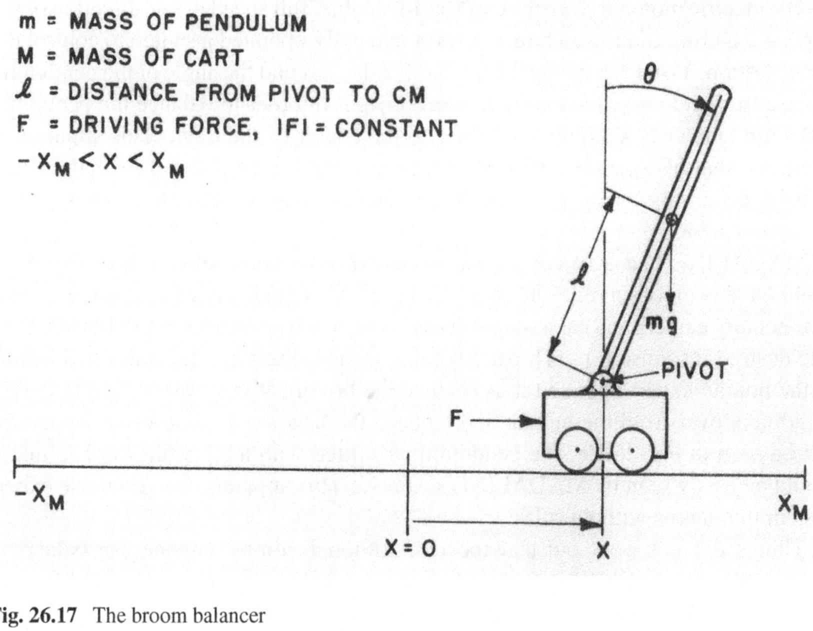

Fig. 26.17 The broom balancer

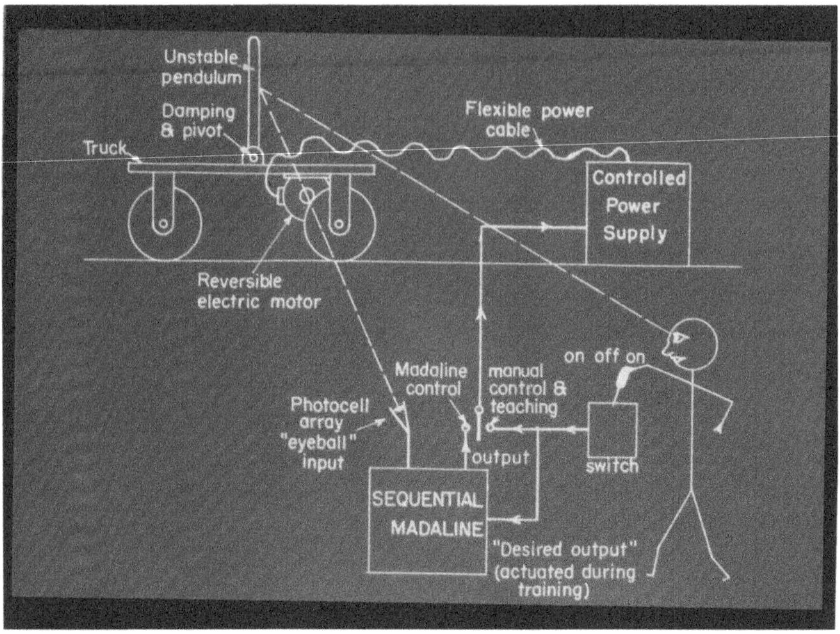

Fig. 26.18 Human "expert" training the broom balancer

The electric motor driving the cart was reversible, full speed in one direction or the opposite. During training, a human expert manually operated a switch to control the cart's motion. Visually sensing the position of the cart and the angle of the pendulum, the expert could move the cart to the left or right in order to balance the pendulum and simultaneously keep the cart close to the center of the track. This situation is similar to that of a person balancing a broom seated in the palm of the hand while visually observing the angle of the broom and moving the hand back and forth to maintain balance.

MADALINE had a camera made of an array of photocells so that the scene could have been observed. The motion of the broom and cart was captured by this rudimentary camera and each snapshot over time was a pattern fed into MADALINE. The desired response for each pattern came from monitoring the controlled switch as the human expert operated it to balance the broom. When trained, MADALINE could take over from the human and balance the broom. A photo of the apparatus can be seen in Fig. 26.19. The pendulum was fitted with a big white card so that it would be easily seen by MADALINE's camera. This apparatus is a trainable expert system that learns without rules.

Things did not work out as expected. Although almost anyone can balance a broom, no one was able to operate the switch and balance the mechanical broom. Starting with the pendulum in a vertical position one could balance it for a second or two until it crashed. Also, making the camera and feeding its signals into MADALINE

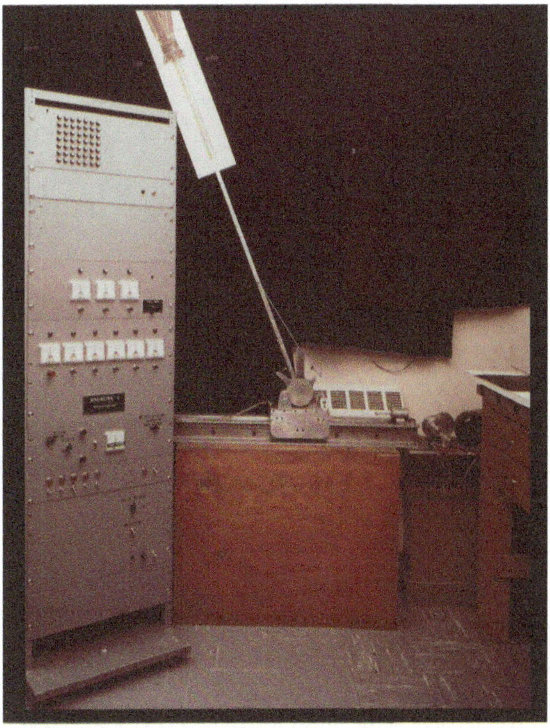

Fig. 26.19 The mechanical broom balancer controlled by MADALINE I

proved difficult. We needed to make some major changes. My students designed an automatic control system to replace the human expert. There are four state variables needed as inputs to the control system: the angle of the pendulum, angular velocity, the position of the cart on the track, and the velocity of the cart. Reluctantly, we had to put sensors on the broom and the cart measuring angle and position. Differentiating these two variables, we obtained angular velocity and the velocity of the cart.

The four state variables were crudely digitized, by "home made" quantizers, each variable quantized to a small number of quantum levels. The resulting bit patterns became inputs to MADALINE. The desired response came from the real-time decisions of the automatic control system, a "bang-bang" control system.

MADALINE was able to be trained by the automatic control system and once trained was able to take over and balance the broom. What we demonstrated was that MADALINE could be trained to do real-time control. Videos of the broom balancer in action can be seen on YouTube. A collection of my videos are there [1].

Figure 26.20 is a photo of my lab at Stanford around 1965. On the left, one can see the desk of our IBM 1620 [2] computer with a typewriter input. A digital tape drive is in the picture. Inputs to the computer came from the typewriter, the tape drive, and a punch card reader, not shown in the photo. Most of our original experiments with ADALINE and MADALINE were run on that computer. On the right of the picture is MADALINE II. With it, we were able to do analog real-time experiments.

Fig. 26.20 Photo of our lab. IBM 1620 computer at left. MADALINE II at right. Photo circa 1965

A patch panel at the lower left of MADALINE II was programmed to wire up various network configuration with the 20 memistor ADALINEs of this machine.

One disadvantage of working with analog weights is that you cannot see them or initialize them in order to do an exactly reproducible experiment. Nevertheless, this machine was able to learn, converge, and find solutions to problems. MADALINE III was a bigger machine having 1,000 memistors. It is pictured in Fig. 26.21. Input patterns were fed to it by the 1620 computer.

26.5 Post Script

Some years ago, perhaps 20–25 years ago, Intel Corp. attempted to build analog neural network chips. They did, and made them into a machine. The network architecture was programmable. The training algorithm was backpropagation. The variable weights were like semiconductor memistors. Ultimately the machine was donated to my lab at Stanford. We worked with it for some time but its performance was not reliable. Intel dropped the project. So did we.

Over the past decade, Hewlett Packard has been working on a semiconductor variable resistor with analog memory. This device is called the memristor. It is a two terminal device whereas the memistor is a three terminal device. The purpose of

Fig. 26.21 MADALINE III

these devices is similar, to have a trainable weight. HP has put a great deal of effort into the memristor, lots of publications, but no product yet.

It is really hard to develop unusual analog or digital computing devices that would be competitive with digital semiconductor computer chips. This is true of the great effort invested by IBM in the Cryotron, the effort of Intel with analog neural networks, and the work of HP on the memristor. Only mother nature has so far succeeded in mass producing analog circuits that work very well and function beyond the capabilities of digital computer chips.

Fig. 26.22 My mother, Ida Widrow, 1927

26.6 Personal Ancient History

In this section, I will present some family photos. They were scanned by my nephew, John Semel. They came from my mother's photo trove. They start with my parents. I was born on December 24, 1929. These pictures of my parents were taken around the time of my birth. There are pictures of me from babyhood to high school kid. More family photos follow. The photo captions tell the story (Figs. 26.22, 26.23, 26.24, 26.25, 26.26, 26.27, 26.28, 26.29, 26.30 and 26.31).

Fig. 26.23 My father, Moe Widrow, 1931

Fig. 26.24 My mother holding baby Bernard

Fig. 26.25 Bernard, 5 years old, on swing, 1935

Fig. 26.26 Bernard, about 10 years old

Fig. 26.27 Bernard, high school boy

Fig. 26.28 Bernard, high school boy, fixing a radio

Fig. 26.29 My mother and her brother, my uncle Charley Turner

Fig. 26.30 Uncle Charley's platoon, South Pacific, World War 2

Fig. 26.31 Our family home on Carrol Avenue, Norwich, CT

26.7 Summary

All of the figures of this chapter have been scanned from original glossy prints. There are photos of my lab in the early 1960s, photos of apparatus that we built, and photos of draftsman's drawings depicting our various experiments. There are photos of the original "knobby ADALINE". My first experiments of unsupervised learning were done with it. There are photos of several versions of MADALINE. We developed the memistor, an electronic synapse that replaced the potentiometer of knobby ADALINE, giving us an all-electronic trainable neuron. All of these devices are still working, almost 60 years later. Some of them will probably be placed in the Computer History Museum in Mountain View, CA.

A photo of the memistor ADALINE is shown that I carried in my suitcase when I visited Frank Rosenblatt and the Perceptron at Cornell University. We trained both the Perceptron and ADALINE on the same input patterns. ADALINE won!

There is a photo of Tedd Hoff loading electroplating solution into a block of memistors. Unfortunately I do not have pictures of other students in the lab. Some of them do appear however in a video demonstrating a real-time trainable speech recognition system that the students demonstrated. The adaptive neurons were implemented in software with an IBM 1620 computer. Our lab was the only one on the Stanford campus that had its own computer. All other computing was done by a single IBM "mainframe" at the Computer Center. Our 1620 had a storage access channel (SAC) that allowed direct electronic connection to the computer. This was most unusual, almost unheard of. My students constructed the speech recognition system without my knowledge. When they were ready, they asked me to come to the lab. They surprised me. About 10 of them were involved. The "ringleaders" were Paul Low and Jim Koford. Paul was an IBM employee sent to Stanford to earn a Ph.D. After graduation, he returned to IBM. Some years later, he became chief of IBM's Semiconductor Division.

References

1. (32) Trucks and Brooms - YouTube. https://www.youtube.com/watch?v=_HCOIEMwCeE. Accessed on 10 April 2020
2. (141) Speech Recognition (1962) - YouTube. https://www.youtube.com/watch?v=D9RxDukCGrU. Accessed 08 Jan 2021

The manufacturer's authorised representative in the EU is Springer Nature Customer Service Centre GmbH, Europaplatz 3, 69115 Heidelberg, Germany. If you have any concerns regarding our products, please contact ProductSafety@springernature.com

Printed and bound by CPI Group (UK) Ltd, Croydon, CR0 4YY
28/09/2026
02224071-0015